**Stripe Press**
Ideas for progress
South San Francisco, California
press.stripe.com

# The Origins
# of Efficiency

**Brian Potter**

For Mercy

Published in Belgium
by Stripe Press / Stripe Matter Inc.

Stripe Press
Ideas for progress
South San Francisco, California
press.stripe.com

ISBN 978-1-953953-52-0 (print)
ISBN 978-1-953953-53-7 (ebook)

Also available in audiobook.

Library of Congress Control Number: 2024056078
Library of Congress Cataloging-in-Publication
data is available upon request.

# Introduction

This is a book about efficiency—about figuring out ways to produce goods and services in less time, with less labor, using fewer resources. It's an attempt to understand why certain production processes become more efficient, and to identify the mechanisms responsible. In other words, it's about how things get cheaper to produce over time.

Why care about efficiency? It is not, admittedly, a sexy concept. Efficiency typically doesn't inspire passion or capture the imagination. Children don't dream of growing up to be efficiency experts. Increasing efficiency might help a company earn greater profits, but it seems like the purview of factory supervisors, middle managers, or accountants, not something of pressing concern to our day-to-day lives. In some cases, pursuing efficiency can even appear socially destructive, with cost reductions creating higher profits for big corporations at the expense of lost jobs, lower quality, and worse service. Who among us doesn't pine for the (real or imagined) days when products were made in the US and built to last, instead of being made in China out of cheap plastic?

This perspective, while perhaps not entirely mistaken, misses the larger picture. Finding ways to produce goods and services with fewer resources—improving what we'll call *production efficiency*—is the force behind some of the largest and most consequential changes in human history. Efficiency is the engine that powers human civilization.

This is a bold claim. To explain, let's start with the story of a modern miracle made possible by increased efficiency: penicillin.

### The story of penicillin

In the 1930s, Oxford University scientists Howard Florey, Ernst Chain, and Norman Heatley were searching for substances that might be used to resist infection. This was no mere academic exercise: War in Europe was looking increasingly likely, and infection had historically almost always killed more soldiers on the battlefield than opposing forces. During the American Civil War, more than two-thirds of the 660,000 deaths were attributed to infection, and infection killed between 12 and 15 percent of wounded soldiers treated in frontline hospitals in World War I.[1]

When searching through the scientific literature on the subject, the researchers came across the work of bacteriologist Alexander Fleming. In 1928, Fleming had returned from vacation to discover that some of his petri dishes had been contaminated, which had halted the growth of the staphylococcus bacteria he was working with.[2] Reasoning that there must be some agent present that was killing the bacteria, Fleming isolated the contaminant, which was later found to be a mold particle of the species *Penicillium notatum*. Fleming named the antibacterial substance penicillin, published several papers about his discovery, and preserved the mold culture. However, due to the fragile nature of the substance, he was unable to isolate it.[3] Further attempts by British scientists in the 1930s fared no better.[4]

Building on this earlier work, Florey, Chain, and Heatley were able to successfully isolate penicillin. Chain found that by carefully controlling the pH and temperature of the solution it was in, the penicillin would remain stable until it could be crystallized via freeze-drying.[5] By 1940, tests in mice showed that the substance was capable of treating what could otherwise be lethal bacterial infections.[6] During the first human trial of penicillin, it was given to an Oxford police officer with a severe staphylococcus infection, and his condition greatly improved. By 1945, penicillin was being hailed as a miracle drug, and Fleming, Florey, and Chain would be awarded the Nobel Prize in Medicine for their research.

This is how the story is often told—as a tale of scientific discovery, in which some combination of serendipity and persistence resulted in an important breakthrough that changed the world. But that story is radically incomplete.

In actuality, *Penicillium notatum*, the species of fungus the scientists used to produce penicillin, only generated the substance in miniscule amounts. In fact, over the course of an entire year, the researchers had gathered such a small amount of penicillin that they exhausted their supply during the initial trial, and the police officer they had treated relapsed and died.[7] Yes, penicillin was a miracle drug— but without a way to produce it in large quantities, it was of little use.

So the researchers' focus shifted to devising high-volume manufacturing methods for penicillin. Because such operations were difficult to carry out in Britain due to the war, Florey and Heatley traveled to the United States, where a joint effort between government (in the form of the US Department of Agriculture's research labs), industry (in the form of pharmaceutical giants such as Pfizer and Merck), and academia gradually developed mass production techniques.[8]

One of the first improvements was to the growth medium. USDA researchers discovered early on that corn steep liquor, a nutrient-rich and inexpensive by-product of cornstarch manufacturing, provided an ideal growing medium for the mold.[9] This advance enabled a thirty-fold increase in penicillin production, making large-scale production possible for the first time.[10]

A second area of improvement involved developing techniques that would allow for submerged fermentation—growing the mold within a solution, rather than just on the surface of the growth medium.[11] Initial US manufacturing operations grew mold cultures in milk bottles, but the process was time-consuming and labor-intensive, requiring thousands of bottles to be emptied and cleaned by hand.[12] And because the mold only grew on the surface, each bottle produced a comparatively small amount of penicillin. It was estimated that using the bottle method to grow enough penicillin for the war effort would require "a row of bottles that stretched from New York to San Francisco."[13] If researchers could figure out how to grow the mold within the solution, this would allow for vastly higher production volumes.[14]

Alexander Fleming's original mold strain and its descendants proved unsuitable for submerged fermentation, so US researchers began to search for other strains that might work. They soon discovered that strains of mold that produced penicillin at all were quite rare. In one researcher's sample of 1,000 strains of penicillium molds, only one other was found to produce penicillin.[15] After looking through tens of thousands of strains, scientists eventually identified the ideal candidate on a moldy cantaloupe. This strain was not only capable of growing while submerged in fermentation tanks that could hold thousands of gallons of solution, but it also produced much more penicillin than earlier strains.[16]

Scientists were able to further improve the cantaloupe strain by cultivating naturally occurring mutants of it and deliberately inducing mutations via X-ray and radiation bombardment. This process ultimately resulted in the Wisconsin Q-176 strain, which produced up to 100 times more penicillin than Fleming's original and became the antecedent to all major industrial strains.[17]

The overall increase in the volume of penicillin produced through these improvements was staggering. According to Emory School of Medicine professor Robert Gaynes, "In 1941, the United States did not have sufficient stock of penicillin to treat a single patient. At the end of 1942, enough penicillin was available to treat fewer than 100 patients.

By September 1943, however, the stock was sufficient to satisfy the demands of the Allied Armed Forces."[18]

Penicillin production reached 80 million units per month in 1943. By early 1944, it had risen by a factor of 200 to over 18 billion units per month. By 1945, manufacturers around the world were producing 5 tons of penicillin annually.[19] And as production volumes rose, the price of penicillin fell. In 1943, a 600-milligram vial of penicillin cost $200; by 1952, the same vial cost just $1.30.[20]

As a result of the widespread availability of penicillin, Allied forces eliminated an estimated 75 to 80 percent of battlefield infection deaths.[21] And the production advancements didn't stop after the war. Between 1950 and 1980, better manufacturing technology boosted productivity by a factor of three, while additional strain enhancement increased productivity by a factor of 16. By 1982, penicillin production had reached 12,000 tons annually—2,400 times the amount produced in 1945. Over that same period, the inflation-adjusted price of penicillin dropped by a factor of 1,000.[22]

These gains in efficiency, along with the subsequent development of other antibiotics, brought about widespread reduction in death and suffering from infection. Prior to the emergence of antibiotics, bacterial illnesses were responsible for approximately 20 percent of all deaths in the US.[23] Between 1936 and 1952, deaths from bacterial illnesses in the US dropped by nearly 70 percent.[24] By some estimates, antibiotics have extended average human lifespan by 23 years.[25]

The initial discovery of penicillin was, of course, necessary to these later achievements. But it was only by making antibiotics cheap and widely available—that is, by producing them efficiently—that this miracle medicine was able to save millions of lives.

### Efficiency and abundance

The story of penicillin is in one sense an outlier: Rarely in human history have such dramatic efficiency gains arrived so quickly. But in another sense, the story of penicillin is simply the story of civilization. For most of human history, producing almost *anything* required an enormous amount of time, effort, and resources. But, as with penicillin, we have over time figured out ways to produce many things much more efficiently, using less labor and fewer materials.

Prior to the past several hundred years, for example, most human effort went into producing enough food to eat. As late as 1500 CE, roughly 75 percent of the populations of England, France, Germany, and Austria worked in agriculture.[26] The continent was perpetually one crop failure away from devastating famine. Europe's Great Famine of 1315 to 1317, brought on by severe summer rains and low temperatures, is estimated to have killed between 10 and 15 percent of the population, depending on the country. And while the Great Famine was extreme, it was by no means unique: Between 1300 and 1400 CE, Italy experienced an estimated 11 years of famine, while Spain experienced 25.[27]

Contrast this meager state with that of the present-day United States. The US began as an agrarian nation, and as late as 1880, 50 percent of the labor force worked in agriculture, a state of affairs not so different from medieval Europe.[28] Today, however, farm employment comprises just 1.2 percent of the workforce, and the US has fewer people working on farms today than it did in 1820.[29] Yet the US produces vastly more food than it did in the 19th century. Between 1866 and 2018, potato production in the US increased by a factor of 6.8, wheat production increased by a factor of 11, and corn production increased by a factor of nearly 20.[30] And modern farming requires not only less labor but also less land: Producing a bushel of wheat requires just 20 percent of the land today that it did in 1866, while a bushel of corn requires just 10 percent of the land. This increase in production has been accompanied by a steep decrease in price. Depending on the crop in question, between 1866 and today, the price has fallen by as much as 50 to 85 percent.[31]

If we look at the rest of the world, we see a similar trend. The massive increase in worldwide agricultural productivity during the 1960s to the 1980s, often called the Green Revolution, is credited with reducing food prices by 35 to 65 percent, increasing food supply by 12 to 13 percent, and increasing caloric intake by 11 to 13 percent in developing countries.[32] If we compare the period after 1990 to the century from 1870 to 1970, worldwide deaths from famine declined by roughly 96 percent.[33]

We can see the same story outside of agriculture. Fabrics and textiles, for example, have been used by humans since before recorded history. Archaeologists have discovered linen fabrics dated as early as 7000 BCE, and farmers began to breed sheep for their wool as early as 3000 BCE.[34] Textiles were an important technology, used for everything from clothing and bedding to containers and sails for ships.

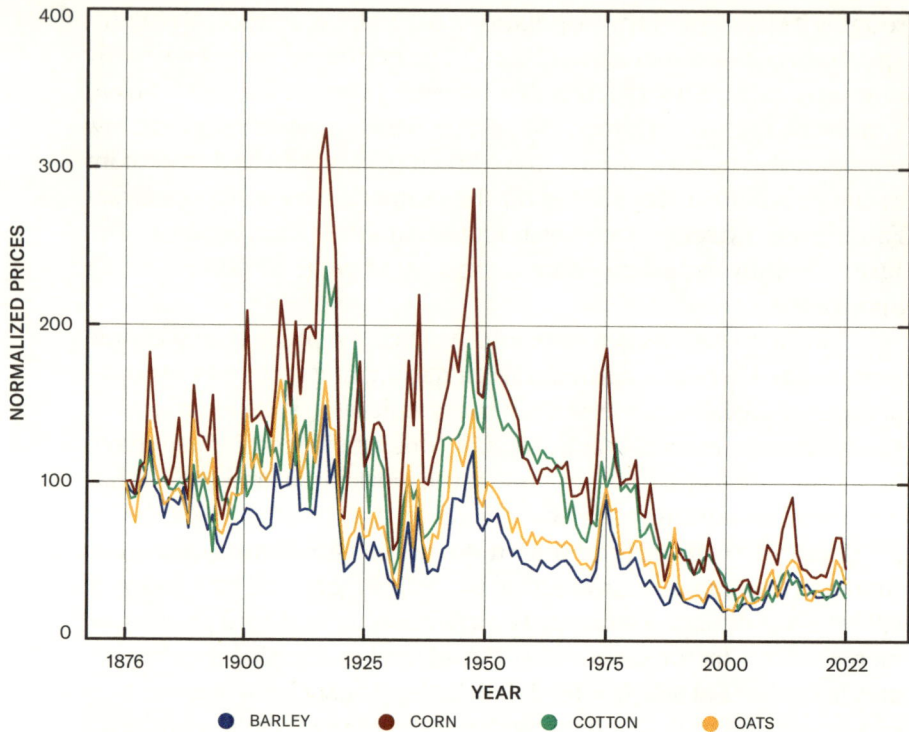

**Figure 1.** Real (inflation-adjusted) price of US crops over time, normalized to 1876 = 100. Data from USDA, *Crop Production Historical Track Records*.

Textiles are made by weaving thread. Thread, in turn, is made by spinning—twisting together numerous short fibers of cotton, wool, flax, or hemp into a single continuous length of fiber. Prior to the Industrial Revolution, textiles were incredibly time-consuming and labor-intensive to produce, largely because they required an enormous amount of thread, which itself was tedious and laborious to make. A 100-square-meter Viking sail, for instance, took about 60 miles of thread, while a Roman toga used about 25 miles.[35] Before the late 1700s, this thread would have been made by hand, using either a drop spindle or, later, a spinning wheel. Even the fastest spinners were only able to spin about 90 to 100 meters of thread an hour on a spinning wheel, while drop spindles produced closer to half that rate. At these production speeds, a Roman toga required roughly 900 hours of labor simply to spin the thread for it and another 200 hours

The Origins of Efficiency

to weave it.[36] To produce the sail for a Viking ship, 385 days of labor were needed to spin the thread, and another 600 days were needed to shear the sheep and prepare the wool.[37]

Preindustrial societies, therefore, devoted an enormous amount of time and effort to spinning thread, with anywhere from eight to 20 spinners supplying thread for a single weaver.[38] As Virginia Postrel notes in her book *The Fabric of Civilization*, "Whether Aztec mothers, orphans in the Florentine Ospedale degli Innocenti, widows in South India, or country wives in Georgian England, women through the centuries spent their lives spinning."[39] In 1770, on the eve of the Industrial Revolution, England had on the order of 1 to 1.5 million spinners, out of a total workforce of around 4 million.[40]

When the Industrial Revolution arrived, textile manufacturing was one of the first industries to be mechanized, resulting in an enormous rise in textile production. Between 1770 and 1841, Britain's output of linen and wool cloth more than doubled, and the output of cotton cloth increased by a factor of over 100.[41] Between 1784 and 1832, the costs of manufacturing cotton thread in Britain dropped by over 90 percent. By 1850, it was cheaper to produce cotton fabric in Britain than almost anywhere else in the world—which enabled the country, which had just 1.8 percent of the global population, to supply half of the world's cotton.[42]

If we fast-forward to the present day, these efficiency gains, and their accompanying abundance, are almost impossible to fathom. As of 2018, the US consumed roughly 35 kilograms of textiles per person each year.[43] Assuming a preindustrial spinner could produce on the order of 100 pounds of fiber a year, supplying enough thread for US consumption using preindustrial methods would require on the order of 230 million people just to spin the thread.[44] Of course, modern thread is not spun by hand but made in highly automated spinning mills staffed by a small number of employees. These mills can manufacture upward of 75,000 pounds of fiber per person per year—750 times what a preindustrial spinner could produce.[45]

Finally, consider a subject dear to both author and reader: the production of books and text. Prior to the mid-1400s, the primary "technology" used to copy a page of text in Europe was manual reproduction. A scribe (often a monk, though later paid copyists became popular) would carefully copy the text from an existing manuscript into a new manuscript by hand. Although there had been some efforts to improve this method, such as the *pecia* system, which broke a manuscript into chunks that many scribes could work on at once, the production and reproduction of text remained time- and labor-intensive.[46] Copying a single page could take a day or more, and increased output could only be achieved by devoting more and more scribes to the task. As a result, books and manuscripts were rare. In Western Europe in the 14th century, manuscripts were produced at an estimated rate of about 20,000 to 30,000 per year, and there were likely fewer than half a million books in all of England—less than a single large bookstore might have today.[47] They were also expensive: For the amount it cost to produce a single volume in 1450, a person could buy two cows, a dozen sheep, or 10 weeks of labor from a farmhand.[48]

As with agriculture and textiles, book production would experience incredible gains in efficiency in the coming centuries. The first dramatic leap occurred sometime around the 1440s, when a goldsmith named Johannes Gutenberg developed a new system for copying text: the printing press. The printer would coat metal blocks shaped like individual letters (cast using a special alloy of lead, tin, and antimony that Gutenberg developed) in an ink made from lampblack and varnish (also developed by Gutenberg), then arrange them on a screw press, a then-common mechanism used for things like producing wine or paper.[49] As the printer turned the screw, the ink-coated letters would press the ink down upon the page.[50]

This system was still slow compared to modern methods: It took Gutenberg nearly five years to make just 180 copies of his first book, the eponymous Gutenberg bible.[51] But even this sluggish rate of production—around 200 to 250 sheets per hour—represented a vast improvement over what had previously been possible.[52] And with the development of powered presses in the early 18th century, that figure climbed to over 1,000 sheets per hour—10,000 times as fast as hand-copying.[53]

The immediate impact of the printing press was a dramatic reduction in the cost of creating a page of text, and a corresponding increase in the number of books produced. By 1500, the rate of book production had increased fiftyfold from the 1300s, to one to two million books per year. By 1600, roughly as many books were printed in Western Europe every five years as had been produced in the entire thousand-year period between 450 and 1450.[54]

The cost of text also dropped precipitously. Though it's hard to get reliable data prior to 1450, it's estimated that by the 1500s the cost of a book was one-eighth of what it had been in the 1300s.[55] By the 1700s, the price of a page of text in England was one-fiftieth of the prevailing price in the Middle Ages.[56] What had previously been a luxury item limited to a small number of universities, monasteries, and the very wealthy became a good a merchant or craftsman might own. And as books and manuscripts became cheaper and more available, literacy rates improved. Across the continent, literacy gradually rose from an average of 12 percent in the late 1400s to 31 percent in the 1700s. In Great Britain, literacy rates increased from 5 percent in the late 1400s to more than 50 percent by the 1600s.[57]

Regardless of where we look, we see the same story—not just in agriculture, textiles, and books but also in steel, watches, shipping, and shoes.[58] For most of history, it took an enormous amount of time, effort, and resources to produce just about everything. In his book on daily life in medieval Europe, Jeffrey Forgeng notes that "it is hard to envision how much effort it takes to produce even the simplest of goods when each piece must be transformed by hand from raw materials to finished product... In all cases the amount of human labor involved was enormous."[59]

Because it took so much time and effort to produce things, most people had very little. In his examination of preindustrial everyday life in Europe, Fernand Braudel observes that prior to the 18th century, the numerous poor "lived in a state of almost complete deprivation." Inventories of French peasants described the sum total of their worldly possessions as "only a few old clothes, a stool, a table, a bench, the planks of a bed, sacks filled with straw."[60]

As the time, effort, and resources required to produce goods and services have fallen, they've become cheaper and more widely available, giving rise to a world of incredible abundance. And it is this abundance that makes modern life possible. Abundant food has improved health, reduced hunger, and nearly abolished famine. Abundant books have enabled mass literacy and education. Abundant steel has made nearly all modern technology possible. More generally, this abundance has freed us from a life of near-constant toil, affording us more time for entertainment, leisure, and creative endeavors. Over the past 150 years, annual hours worked per person has declined by nearly 50 percent in most Western countries.[61]

Without this enormous increase in production efficiency, modern life would be impossible to sustain.

### Where does efficiency come from?

The dramatic rise in production efficiency is partly a story of societies accumulating wealth over time. A business might start out by adopting production processes that are cheap yet labor-intensive, but as it earns profits, it can use those profits to purchase machines and equipment, reducing the amount of labor required. A modern farm in the US, for instance, uses one quarter of the labor of a farm in 1948 but twice as much farming equipment.[62]

But this isn't the entire story. Even if we adjust for changes in inputs—machines, buildings, and so on—there have still been significant reductions in terms of what's required to produce a given amount of output over time. Economists call this measure total factor productivity—the amount of output that can be produced from a given quantity of inputs, where inputs include both labor and capital (such as machines and equipment), as well as materials, energy, and other purchased services. Between 1800 and 2000, US total factor productivity in agriculture is estimated to have increased more than sevenfold, and in manufacturing it has increased more than ninefold.[63] Modern production methods might use more machinery or energy than historical methods, but on balance the input requirements have still fallen massively.

What explains these gains in total factor productivity? Part of the answer involves the development of new and improved technology. Gutenberg's printing press gave us cheap books; the spinning mule gave us cheap thread. More generally, novel inventions make it possibleto produce goods with fewer resources than before. But as we'll see in the chapters ahead, while technology advances are a crucial part of how increased efficiency happens, this explanation is at once too broad and too narrow.

It's too broad because it fails to tell us what specific conditions allow technology to improve efficiency and make things cheaper. In other words, to understand how technology increases efficiency, we need to be able to differentiate between the types of technologies that successfully reduce production costs and those that don't. The world is full of technologies that, despite predictions, failed to make things cheaper. Nuclear power, for instance, was widely expected to greatly lower the costs of generating electricity due to its enormous fuel density.[64] A 1,000-megawatt coal plant burns over 7,000 tons of coal a day; a 1,000-megawatt nuclear plant, on the other hand, uses just 0.07 tons of nuclear fuel (uranium pellets) a day.[65] By essentially eliminating the cost of fuel, nuclear power was expected to produce electricity that was "too cheap to meter."[66] But more than 60 years after the construction of the first civilian nuclear reactor, this hasn't happened. Nuclear power plant construction projects routinely go massively over budget, and since the 1970s, the cost of nuclear power in the United States has gone up nearly tenfold.[67] This is due in part to burdensome regulations that have prevented nuclear power from reaching its potential. But even setting those obstacles aside, nuclear power has trouble producing electricity for less than the coal-fired plants it was predicted to replace.[68]

Another example of a technology that has failed to improve efficiency is factory-built housing. Unlike most industries, construction remains a largely craft-based field, where skilled craftspeople build houses on-site using methods that have changed relatively little in 100 years. For decades, builders have attempted to improve this process by moving construction into the factory, to do for homes what Henry Ford did for cars. But while it's possible to build houses in factories, factory-based construction has consistently failed to yield significant reductions in cost. (We'll look at factory-built housing more closely in Chapter 10.).

Space travel is yet another example. In the 1970s, experts predicted that the development of the space shuttle would reduce the cost of space travel, potentially making space colonization, lunar mining, and space manufacturing common endeavors by the year 2000.[69] But in fact, space shuttle launch costs were more than 10 times as high as those of the Saturn V, the rocket used in the Apollo program. Despite the development of new rockets and better aerospace technology, space launch costs between 1970 and 2010 consistently exceeded those achieved in the 1960s.[70] It wasn't until the rise of SpaceX, which uses a very different launch technology than the space shuttle, that launch costs fell below those of the 1960s.

In other cases, new technology *eventually* makes something cheaper, but only after a long period of time. The first solar photovoltaic panel was built in the late 1800s, but it is only in the past several years, after decades of development, that photovoltaics can produce electricity more cheaply than other sources of power. Similarly, it took nearly 100 years from the invention of the electric arc furnace before it was cost-competitive with existing blast furnace-based methods of steel production.[71]

To state the problem with the new-technology-improves-efficiency explanation more generally, the creation of some new technological capability tells us nothing about whether that capability can be achieved *cheaply*. Saying technology is responsible for production efficiency simply restates the question in a different way rather than answering it. We're still left with the question of what precisely is happening when technology makes something cheaper, and what separates the technologies that make things cheaper from those that don't.

The technological explanation is also too narrow, because significant reductions in production cost often occur in the absence of new technology. In the 1970s, Japanese car manufacturers were more productive than US manufacturers despite often using much older machinery.[72] Similarly, as we'll see in Chapter 3, a wide variety of industrial improvement systems, including design for manufacturing, value engineering, and lean methods, are often able to significantly decrease production costs without technological change.

Arguably, these sorts of changes could be considered technological in a more expansive sense that encompasses organizational structure as a technology. But even so, we're left with the same question: Why is it that some organizational changes reduce production costs, while others don't?

### Explaining efficiency

So, putting aside increasing wealth and advancing technology—which are, at best, partial answers—where do improvements in efficiency come from? What, specifically, is happening on a farm, inside a factory, or within a company when the costs of production fall?

The purpose of this book is to answer that question. As we'll see, there are a limited number of ways to intervene in a production process to make it cheaper. Drawing on historical and contemporary case studies, the mechanics of a variety of industrial improvement systems, and nearly a century's worth of research into production efficiency, this book unpacks each of these interventions and explains when and why they make production processes cheaper.

We begin in the following chapter (Chapter 1) by looking at the basics of production processes, examining what it means to produce a good or service, and identifying the possible points of intervention to reduce the resources a production process requires. The next several chapters each focus on a specific efficiency-improving intervention. We'll start with changing production methods and developing new production technology that requires fewer inputs (Chapter 2). We'll then look at ways of finding cheaper inputs to a process (Chapter 3), including by redesigning the product to be more cost-effective to produce. Next, we'll consider lowering production costs by increasing production volume (Chapter 4), removing unnecessary steps from the production process (Chapter 5), and making a process more reliably do what it has been designed to do (Chapter 6).

After exploring the specific ways producers can decrease the cost of a production process, we'll take a step back and look at how improvements tend to accumulate over time (Chapter 7) and how one improvement often results in additional improvements (Chapter 8). We'll also explore how production processes, despite covering an enormous range of activities, tend to converge on a specific form, known as continuous processes, as they improve along the various axes of efficiency described in previous chapters (Chapter 9). Finally, we'll conclude our investigation into efficiency by reflecting on cases where production costs haven't fallen (Chapter 10) and thinking about how that might change in the future (Conclusion).

As we've seen throughout this chapter, efficiency improvements have transformed daily life, conquered famine, and lifted billions out of poverty. But despite this progress, there's still much work to be done. Many goods and services, including housing, health care, and education, remain expensive, and the impact of many potentially transformative technologies, such as space travel, has been dulled because of their high cost. If we can master the forces of efficiency and find ways to make more and more of our production processes cheaper, we can continue to transform industries and unlock increased abundance. Ultimately, we may be able to create an era of human flourishing as different from today as today is from the preindustrial era.

# 1

# What Is a Production Process?

If we are to understand how a production process becomes more efficient, we first need to understand how a production process works. To do this, let's take a look at a relatively simple process: the manufacture of glass bulbs used for incandescent lights in the late 1800s.

In 1880, Thomas Edison was awarded a patent for his electric incandescent light bulb, marking the beginning of the age of electricity. Although it was the result of thousands of hours of research that took place over decades, the ultimate design of Edison's light bulb was simple, consisting of just a few components: a filament, a thin glass tube in which the filament was mounted, a pair of lead-in wires, a base, and the glass bulb itself.[73]

Until the 20th century, light bulbs were largely manufactured by hand. Workers would run the lead-in wires through the inner glass tube, attach the filament to the lead-in wires, and attach the glass tube to the bulb. A vacuum pump would then suck the air out of the bulb. Initially, this was done by connecting the pump to the top of the bulb, leaving a small tip of glass that had to be cut off. Later, tipless bulbs were developed that had the air removed from the bottom.[74]

Most of this manufacturing process was done in house by Edison's Electric Light Company, but the production of the glass bulb itself, known as a bulb blank, was outsourced. Edison placed his first order for bulb blanks with the Corning Glass Works company in 1880.[75] The process of making the bulb blanks was fairly straightforward: Glassworkers would mix together sand, lead, and potassium carbonate, along with small quantities of niter, arsenic, and manganese oxide, place the mixture in a crucible, and melt it in a furnace into liquid glass. A worker called a gaffer would then gather a blob of glass on the end of a hollow iron tube and place the blob into a mold the shape of a light bulb. While the blob was still attached to the iron tube, the gaffer would blow into it to form the body of the bulb, then open the mold and cut the bulb from the end of the tube.[76]

We can draw this series of steps using a process flow diagram, a visual representation of how a process unfolds. See Figure 2 for an example of what the bulb blank process might look like.

Making bulb blanks is an example of what we'll call a *production process*—a series of steps through which input materials are transformed incrementally into a finished product. Each step in the process induces some change in the input material. The changed material is then passed on to the next step, which makes another change, and so on, until the finished product comes out the other side. In the bulb blank process, sand, lead, and other chemicals are the inputs. These are gradually transformed by heat, chemical reactions, and physical manipulation until a finished bulb blank emerges at the other end.

In turn, this output might be the input to a subsequent process. Bulb blanks, for instance, would then be sent to Edison's factory to be assembled into complete light bulbs. Likewise, the input materials for the manufacture of bulb blanks were themselves the output of some other production process. Potassium carbonate, for example, was mined from potassium ore and then refined using the Leblanc process.[77]

Outside of the small number of things we can obtain directly from nature, all products of civilization are the result of some sort of production process—some series of transformations that take in raw materials, energy, labor, and information and produce goods and services. At first glance, services might seem far removed from the production of physical goods like cars or shoes, but the same basic model applies. A house cleaner, for example, goes through a specific series of steps—cleaning the bedrooms, then the bathrooms, then the kitchen—using various inputs—labor, electricity, cleaning products—to transform an input—a dirty house—into an output—a clean one. These processes might be comparatively simple, such as the production of light bulb blanks, or exceptionally complex, with hundreds or even thousands of steps. One 19th-century watch factory boasted that its watches "required 3,700 distinct operations to produce," while a 1940s Cadillac—a relatively simple automobile by modern standards—required nearly 60,000 separate operations.[78]

Even everyday objects can mask a great deal of production complexity. In his book *The Toaster Project*, Thomas Thwaites disassembles a $7 toaster to find that it contains 404 parts made up of more than a hundred different materials.[79] And if we follow the chain of production further back, to the processes required to make the various input materials (and the processes to make the inputs for those processes, and so on), we find a sprawling mass of complexity for even the simplest products of civilization. In his famous 1958 essay "I, Pencil," Leonard Read notes that a full accounting of the inputs required to make an ordinary pencil—the steel used to make the tools to harvest the cedar, the ships used to transport the graphite from Sri Lanka to

**RAW MATERIAL COMES IN**

MIXTURE

GLASS FURNACE

MOLDS

**FINISHED BULBS GO OUT**

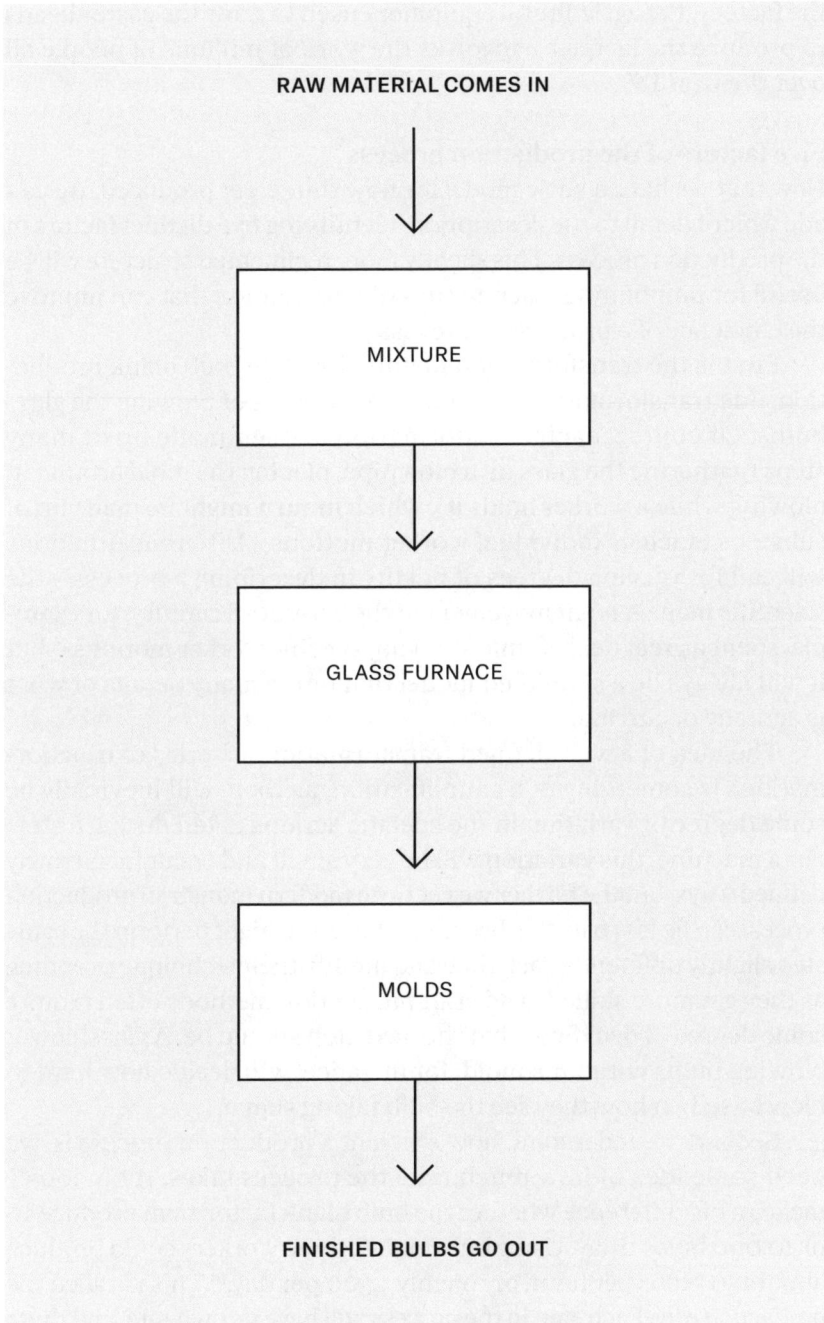

**Figure 2.** Process flow diagram of a bulb blank production process.

the factory, the agricultural equipment used to grow the castor beans to produce the lacquer—involves the work of millions of people all over the world.[80]

### Five factors of the production process

Now that we have a basic model for how things get produced, we can add a bit of detail to the description, identifying five distinct factors of the production process. This slightly more regimented structure will be useful for pinpointing discrete sites of intervention that can improve the efficiency of a production process.

First is the transformation method itself. In bulb blank production, one transformation method is the process of blowing the glass bulbs. Of course, each transformation is itself made up of many steps (gathering the glass on a blowpipe, placing the mold around it, blowing while a worker holds it), which in turn might be made up of substeps (such as individual worker motions). Different situations will call for varying degrees of fidelity in describing a process—the scientific management movement of the early 20th century, for example, spent a great deal of time studying specific worker motions—but it will always be a simplified model that omits many details of what is actually occurring.

The idea of a well-defined transformation or series of transformations is something of a simplification, as there will inevitably be some degree of variation in the specific actions taken during a step. For a machine, this variation will be very small and occur in narrowly defined ways, but the farther we get from modern industrial production processes, the less true this becomes. A person might perform the same step slightly differently each time and modify their technique over time as they get more skilled. And craft production methods often require some degree of deciding what the next step should be. A glassblower blowing bulbs without a mold, for instance, will decide how hard to blow based on how they see the bulb taking shape.

Second, to understand how efficient a production process is, we need some idea of how much time the process takes. It obviously makes a big difference whether the bulb blank factory can produce 10 or 10,000 bulbs a day. Using bulb molds, three workers could produce about 150 bulbs per hour, or roughly 1,500 per day.[81] This is called the *production rate*. Each step in the process will have its own rate, and these rates may differ from those of other steps. For example, filling the glass crucibles might be done just once a week, even though glassworkers were producing bulb blanks daily.[82]

Third, to determine how much a given production process costs, we need to account for all the direct material inputs and outputs to the process. At the furnace step, raw materials go in and molten glass comes out. At the blowing step, molten glass goes in and a bulb comes out. Depending on how detailed we decide to be, we might also include inputs like the coal that fuels the furnace and outputs like the ash and smoke produced by the furnace. There are labor inputs as well. The blowing step, for instance, requires the labor of two or three workers to gather the glass, work the mold, and cut the finished bulb free.[83]

We also need to account for the indirect inputs—things that aren't used directly by the process but are nevertheless necessary. A factory's rent can't be directly attributed to any particular operation within the factory, but the building is still an important input to the process. We can account for this cost by attributing some fraction of it to each step. Similarly, we can assign some fraction of the cost of the equipment, administration, insurance, and any other overhead costs to each step in the process. (The question of how best to assign these indirect costs is an involved area of accounting, but broadly speaking, these costs will be spread over the amount of output we produce.)

Taken together, the costs of the various inputs and outputs allow us to determine the cost to produce a bulb blank. The first blanks from Corning, for example, cost about 3¢ each.[84] For our purposes, the goal of any efficiency improvement is to reduce this cost. Or, to be more precise: An efficiency improvement is anything that is able to lower the cost per unit.

Fourth, to understand whether the process is efficiently arranged, we need to keep track of how much material is in the process at any given time. At any point, some material is actively being worked on and some is waiting to be worked on. In bulb blank production, once the raw materials had been added to the crucible, it might take a while before the glass was gathered by workers and blown into bulbs. If crucibles were filled once a week, there would be about half a week's worth of molten glass waiting to be turned into bulbs at any given time. Any material that isn't currently being worked on is considered to be in a buffer of available material. The total amount of material in the system—that is, the combination of what's in the buffer and what's being worked on—is collectively known as *work in process*.

Fifth and finally, in evaluating a production process we need to make note of how the output of the process varies. While it's tempting to think of a step as producing the exact same output every time, there will inevitably be some variation. At times, the process may simply fail. For example, in some cases, the furnace would produce a batch of glass that was unsuitable for bulbs. In other cases, the crucibles that held the molten glass would crack, spilling the glass before it could be turned into bulbs.

But there will also be more subtle sources of variation. For instance, the composition of the glass and the thickness of the bulbs would differ slightly, perhaps imperceptibly, from bulb to bulb. No two bulbs were exactly alike. This discrepancy can be a natural outcome of the process, the result of a disparity in the inputs, or due to variation in the environment in which the process takes place. The quality of the bulb glass, for example, was greatly dependent on the quality of the chemicals used, how well they were mixed, and the temperature of the furnace.

One simple way of characterizing variation is in terms of yield— the fraction of inputs that are successfully transformed into outputs. A yield of 50 percent would characterize a process that is only successful half the time. An unsuccessful transformation might be a complete failure (a bulb falls on the floor and breaks) or one that is simply outside the range of acceptable tolerance (the glass on the bulb was slightly too thin).

In many cases, however, it will be useful to have a more detailed characterization of the variation in a process. In the production of light bulb filaments, very slight differences in temperature during the carburizing process resulted in the filament producing different amounts of light. Understanding how the resulting filaments varied was, therefore, necessary to determine how many bulbs of a given illumination could be produced. It might turn out that the variation in illumination could be described by a normal distribution with a particular mean and standard deviation, making it possible to track disruptions to the process by looking at whether values fell outside of the expected range. For now, we'll just note that variation is an important factor to consider without worrying about developing a certain measurement for it.

Looking at a single step in the process, we now have five factors that characterize it:

| 1 | The transformation method itself. For example, the act of blowing molten glass into a mold. |
|---|---|
| 2 | The production rate. For example, how many molds the gaffers can fill in an hour. |
| 3 | The inputs and outputs, along with their associated costs. For example, the molten glass, the gaffer's wages, and wear and tear on the molds and blowpipes. |
| 4 | The size of the buffer. For example, how much molten glass is stored in the furnace waiting for the gaffer. |
| 5 | The variability of the output. For example, fluctuations in how fast the gaffer works and the thickness of the bulbs produced, or how often the gaffer drops and breaks a bulb. |

This is, of course, a highly simplified model. For one thing, it omits the complexity of what specifically occurs during each step. For another, it suggests that these factors are steady over time, but in reality they will frequently be in flux. The variation in output may rise when a new worker starts, or at the end of the day when workers are tired, or over a long period of time as workers or managers grow complacent. Alternatively, variation may go down over time as workers gain experience and precision improves.

This model also doesn't include the many possible ways one step may influence another step, beyond how fast the step runs. The temperature of the glass furnace might influence how easy it is to blow the bulb into the mold, for example. Likewise, variation in one process may be a function of variation in some previous process. Bulbs breaking when the mold is removed, for instance, might be a function of inconsistent mixing of the ingredients or uneven temperature of the molten glass.

Finally, this model doesn't include any specifics about what is actually being produced. As we'll see later on, the form of the product and the method of production are intimately connected, and a change in one generally results in a change in the other.

Despite its various simplifications, however, this model gives us a useful way to structure our thinking about production processes and how they can be made more efficient.

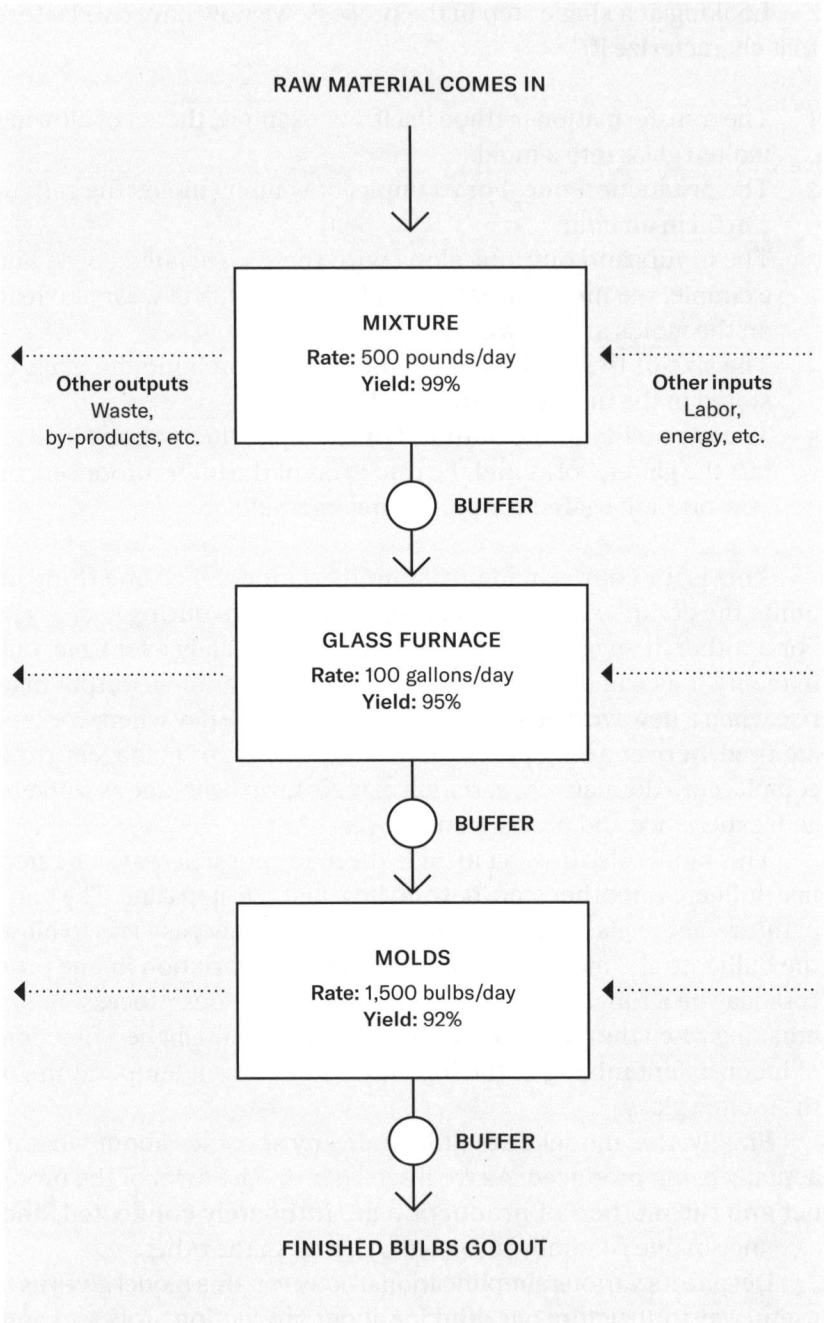

**Figure 3.** Process sketch of a bulb blank process showing inputs, outputs, buffers, production rates, and yields.

**Improvements to the process**

The goal of any efficiency improvement is to minimize the costs of producing something. If we're running a bulb blank factory, we want to figure out how to produce those bulb blanks as cheaply as possible, which means using the fewest, lowest-cost inputs we can. The way to do this is to change one or more of these five factors.

First, we can change the transformation method itself, to one that requires fewer resources. The very first bulb blanks produced by Corning didn't use molds but were produced using a much slower free-hand method, which entailed manually rolling out tubes of glass. Changing the bulb-blowing process to the mold method greatly increased output and decreased the labor required for each bulb, such that workers went from producing 165 bulbs on the first day to 150 an hour.[85]

Second, we can try to improve the rate of production and take advantage of economies of scale—the fact that per-unit costs tend to fall as production volume rises. Glass furnaces in the bulb blank factory ran continuously, because starting a furnace cold took a great deal of time (24 hours or more) and was very likely to damage the crucibles. The furnaces were, therefore, burning coal regardless of whether glass was being blown and bulbs produced.[86] Similarly, the rent needed to be paid whether the factory was producing bulb blanks or not. For these reasons, a factory that manufactured bulbs continuously over 24 hours would have lower unit costs than a factory that only operated for eight hours a day (and, in fact, some glass manufacturers did run continuously for this reason).[87]

Third, we can try to reduce variation in the process. The quality of the glass was dependent on the temperature of the furnace: Variations in temperature would result in glass that would break after a short period of use. Reducing temperature variation would, therefore, result in more glass within acceptable bounds, producing a higher yield.

Fourth, we can try to decrease the costs of our inputs. Replacing the hand-blowing process with the bulb mold process not only reduced the amount of labor required but also enabled the factory to use less expensive labor, since the molding process required less skill.[88]

Fifth, we can try to reduce our work in process by decreasing the size of our buffers. Work in process is material that has been paid for but hasn't yet been sold—it's an investment that has yet to yield a return. If glass furnaces were filled with new glass once a week, on average there would be half a week's worth of glass simply sitting idle in the factory.[89] If the crucibles were instead half the size and filled twice a week, on average there would only be a quarter of a week's supply of glass in the crucibles, reducing work in process by 50 percent.

We also have one more option available to us: We can try to delete an entire step in the process. This will, obviously, remove all of its associated costs. If, for instance, it becomes possible to buy premixed glass powder, we no longer need to perform the mixing step ourselves—our input materials can instead go directly to the glass crucibles.

These are the options available to improve the efficiency of a process. So, what does this suggest about what an extremely efficient process looks like?

- It's a process with no buffers. Material moves smoothly from one step to the next without any waiting or delay, and material tied up in the process is minimized.
- It's a process with no variability. The process works every time and always produces exactly what it's supposed to, at exactly the time when it's needed. More generally, the output of the process is as close to perfectly predictable as possible.
- It's a process with no unnecessary or wasteful steps. Every step is contributing value, and no steps can be eliminated.
- It's a process with inputs that are as cheap as possible and no wasted outputs. Either all inputs are successfully transformed, or the ancillary outputs are repurposed elsewhere.
- It's a process that acts at as large a scale as the technology and market will allow. Fixed costs are spread over as much output as possible, and the process takes maximum advantage of scale effects.
- It's a process that uses transformation methods that require as few inputs as possible, at the limits of what production technology will allow.

This sort of production process is sometimes called a *continuous flow* process—it continuously transforms inputs into outputs without any delays, downtime, waiting, unnecessary steps, or unneeded inputs. A steady stream of inputs goes in, and a steady stream of completed products swiftly and smoothly comes out.

One way of thinking about a continuous flow process is that it's like driving on the highway. In the city, there's the constant stop-and-start of traffic lights and waiting behind other cars. But on the highway, the flow of cars is consistent and uninterrupted, as one car smoothly follows another. In practice, it's often not possible to achieve a true continuous-flow process, just as it's not always possible for traffic to flow perfectly smoothly on the highway. The technology may not allow it, or the size of the market may not justify the cost of the equipment required. There are any number of reasons why continuous flow may not be achievable. But when it *is* possible, it results in the production of enormous volumes of incredibly inexpensive goods.

To see what a continuous flow process looks like in practice, let's look at how the light bulb manufacturing process evolved in the century after Edison.

In 1891, just over a decade after Edison's invention, the US was producing 7.5 million incandescent bulbs per year.[90] By the turn of the 20th century, that figure had climbed to 25 million. But production was still largely manual, and the cost of light bulbs, though falling, was still high. In 1907, a 60-watt light bulb cost $1.75, or about $54 in 2022 dollars.[91]

In 1912, Corning introduced the first semiautomatic machine for blowing light bulbs, called the Empire E machine. Though it still required workers to manually gather the molten glass, the machine could produce bulbs at a rate of 400 per hour, over twice as fast as the manual mold method.[92] This was followed by General Electric's fully automatic Westlake machine, as well as Corning's Empire F. In 1921, a Westlake machine could manufacture over 1,000 bulbs an hour. By the 1930s, improved Westlake machines could produce 5,000 bulb blanks an hour.[93]

The Westlake machine, though speedy, was largely a faster, mechanized version of the existing method for hand-blowing bulbs. It consisted of a large rotating drum with a series of iron tube arms mounted to it; as the machine rotated, the arms would lower into a glass furnace, gather a glob of molten glass, and swing it into a mold, after which air would be blown into it to form the bulb. Then, in 1926, a new type of machine for manufacturing bulb blanks was introduced: the Corning ribbon machine. Unlike previous machines, which largely duplicated the manual bulb-blowing process, the ribbon machine used a different mechanism for forming the bulbs. Instead of gathering a blob of glass on an iron rod, molten glass was poured onto a conveyor belt, which produced a continuous ribbon of molten glass (giving the machine its name). The glass would sag through holes in the belt, forming a bowl shape. As the conveyor moved, a mold attached to a second conveyor below would snap shut around the bowl-shaped glass and air would be blown in from above, forming the shape of the bulb. The formed bulbs were then released and carried away by conveyor belt.[94]

What had previously been a process with many small stops and starts became an uninterrupted, continuous flow. Glass poured onto the conveyor, sagged through the holes, and was repeatedly transformed into finished bulbs, one after the other, without any delays or waiting. Every step was perfectly synchronized.

The ribbon machine was extraordinarily complex and required constant intervention to keep operational.[95] But it could produce bulb blanks in truly staggering volumes. The first ribbon machine produced 16,000 bulbs an hour—over three times faster than the Westlake machines. By 1930, an improved ribbon machine could produce 40,000 bulbs an hour.[96]

The ribbon machine represented the final evolution of incandescent bulb blank production. It produced bulbs in such enormous quantities that by the early 1980s, fewer than 15 ribbon machines were needed for the entire world's supply of light bulbs. By then, machine improvements had increased the production volume to nearly 120,000 bulbs an hour, or 33 bulbs every second.[97]

Similar improvements took place in the rest of the light bulb manufacturing process, though none were quite so dramatic as the ribbon machine. In the late 19th and early 20th centuries, machines were developed to attach the inner tube to the outside bulb, mount the filament to the tube, make and then insert the lead-in wires, and seal the bulb.[98] Enhanced vacuum pumps were developed to evacuate bulbs much more quickly—Edison's original pumps took five hours to produce a vacuum in a bulb—and they did so automatically.[99]

By the 1920s, most steps in the bulb manufacturing process had been automated, but they were largely performed by separate machines. Large volumes of in-process bulbs would accumulate between workstations, creating severe storage problems.[100] Starting in 1921, these steps were rearranged into groups, or cells, so one machine would smoothly feed another at synchronized rates.[101] Work in process was greatly reduced, storage requirements fell, and output per worker nearly doubled.[102] By 1930, the major manufacturing innovations were complete, and by 1942, finished bulbs could be produced by a work cell at a rate of 1,000 per hour.[103]

As a result of these improvements, the cost of a light bulb plummeted. By 1942, the cost of a 60-watt bulb had fallen to 10¢. Over this same period, bulb efficiency, the amount of light emitted per watt, also improved, nearly doubling from 1907 to 1942. Combined with cheaper electricity, the cost per lumen dropped 98.5 percent between 1882 and 1942.[104]

Other parts of the light bulb-making process benefited from the same types of improvements: new production technology that required fewer inputs, increased economies of scale, reduced variability, minimization of buffers, and the elimination of unnecessary steps. As with bulb blanks, these processes gradually evolved toward a continuous, uninterrupted transformation of material.

Of course, such gains are not restricted to light bulbs. Any production process that can be described as a series of sequential steps can be made more efficient in the exact same ways. As we'll see throughout this book, these types of improvements have resulted in increased efficiency in everything from steelmaking to cargo shipping. Over the next several chapters, we'll take a closer look at each of the five factors of a production process and how they can contribute to increased production efficiency.

# 2

# New Processes

In the previous chapter, we looked at the five factors that can be used to characterize a production process: the transformation method, the production rate, the cost of inputs, the degree and type of variation, and the buffer size. Of these, the transformation method itself—the physical steps a process employs to produce something—is perhaps the most important, since it's a major determinant of most of the other factors. When penicillin manufacturers switched from bottle plants to submerged fermentation, that changed the production rate (the new, larger tanks produced much more penicillin), the inputs required (the new process involved less labor and eliminated the need for milk bottles), what other steps were necessary, and everything else about the process. So, understanding how a process gets more efficient first requires understanding how the production methods used in that process change over time.

### Production method change and nails

To better understand how production method changes drive efficiency improvements, let's look at the evolution of the process for making another relatively simple manufactured good: nails.

Until the late 1700s, nails were forged by hand. A smith would start with a long rod of wrought iron, known as nail rod, heat it until it was red-hot, then hammer the end into a tapered point.[105] They would then cut the tapered end from the rest of the rod and place it in a heading device, an iron mold with a nail-shaped hole that left the top of the nail sticking out. Then the smith would hammer the top of the nail to form the head. Different types of nails would have different types of points and heads.[106]

This method of making nails drove every other factor of production. For instance, it created a certain amount of variability in the output: Because each nail was made by hand, each would be slightly different. And because it required a smith to hammer the nail into shape, the process could only produce nails so quickly. Even the best smith could make only a few thousand nails per day; making more nails meant hiring more smiths.[107] Likewise, the inputs the process required—the nail rod, the smith's tools, fuel to heat the forge, the amount of labor—were determined by the nature of the process. A nail made in a different way, for example by pouring molten iron into nail-shaped molds (a less common process that was sometimes used), would require different inputs: molten iron instead of nail rod, molds instead of heading devices, and so on.

Each step in this process was slow and laborious, and other than the development of the slitting mill for making nail rod (first invented in England in the 16th century), the process had changed little since ancient times. Indeed, similar hand-forged nails were produced by the Romans.[108] Because of the time and effort required to make them, nails were expensive. In fact, they were so valuable that colonial buildings were occasionally burned down to recover the nails. As late as 1810, nails made up an estimated 0.4 percent of the United States' GDP— roughly as large a fraction as computers do today.[109]

But starting in the late 1700s, a new type of nail began to appear: the cut nail. The cut nail was made using a different process than a hand-forged nail. While hand-forged nails were produced by a smith hammering a rod of iron, cut nails were produced by a machine that used a reciprocating blade to cut a nail-shaped piece of iron from an iron sheet, and a press to squeeze the top of the nail to form the head. These machines were initially manually operated, but later they would be driven by waterwheels or steam engines.[110]

The cut-nail process was much faster than the hand-forging process. Though both methods required manual labor, the cut-nail process replaced a long sequence of steps (repeatedly striking the nail body with a hammer) with a single step (a single motion of a blade). Early manual cut-nail machines could produce 10,000 nails a day—faster than any smith—and by the 1870s, steam-powered cut-nail machines were producing 30,000 nails per day, with a single worker overseeing multiple machines.[111] Because cut nails were made by a repetitive mechanical motion, each nail was nearly identical, resulting in much less variation in size and shape than hand-forged nails. And the repetitive mechanical action meant the rate of production was much more uniform than a smith working by hand.

The cut-nail process also greatly reduced input requirements. The machines eliminated the need to slit iron sheets into nail rods and, eventually, the need to heat the iron before working it, which eliminated all of the inputs those steps required: fuel for heating, tools to do the work, and the like.[112] The process also entailed less labor per nail, both in terms of time (since the worker time per nail was greatly reduced) and exertion.

These reductions were, of course, partially offset by the cost of building or acquiring the machines, as well as the cost of powering them. But on balance, the input costs were much lower. As a result, the price of nails fell by more than 50 percent in real terms during the 19th century, about half of which can be attributed to the more efficient process. (The remainder was largely the result of the falling cost of iron.)[113]

The cut-nail process, in turn, was eventually superseded by another process: the wire-nail process. Here, a machine would take steel wire, cut a nail-sized length of it, straighten it, and point the end. And, as with the cut-nail process, the change affected every other factor of production. It required different inputs—steel wire instead of iron sheets—and could run at a much faster rate. By the end of the 19th century, wire-nail machines could produce hundreds of thousands of nails a day, more than a threefold increase in the production rate compared to cut nails.[114] Because it made nails so quickly, and because a wire nail required roughly half the metal of a cut nail, the wire-nail process resulted in a further reduction in the cost of nails. Between the late 19th and early 20th century, the price of nails once again fell by approximately 50 percent, half of which was due to improved process efficiency.[115]

As the manufacture of nails illustrates, efficiency improvements and changing production methods go hand in hand. Nails got cheaper because new production methods required fewer inputs, ran more reliably, and could operate at a higher rate. Conversely, an improvement to any of our axes of efficiency—production rate, input costs, buffer requirements, variability—will often warrant a change to the production method itself.

### Production methods, technological change, and S curves

When we talk about production methods, we're really talking about *technologies*. Cut-nail machines, penicillin fermentation tanks, and the Corning ribbon machine are all production technologies, and a production process can be thought of as a large collection of different technologies strung together to accomplish a particular goal. Production method change, then, is a specific example of the more general process of technological change.

For individual technologies, progress tends to follow an S-shaped curve, with time or effort on the horizontal axis and technological performance on the vertical axis.[116] Early on, the technology performs extremely poorly, if it works at all. The phenomenon at play in the technology may have only recently been discovered and as a result is poorly understood. It might not yet be clear how it behaves under different conditions, or what arrangement of components can best take advantage of it, or even for what purposes it might be harnessed.[117] Fixing one problem with a nascent technology tends to simply reveal more problems, so significant time and effort might be invested without any noticeable increase in performance.

But over time, as scientists, engineers, and tinkerers explore different ways of implementing it, the technology's characteristics become better understood. As people begin to figure out what works and what doesn't, the search space of the technology is narrowed, and it attracts more talent and funding. As attention converges on the most promising avenues for advancement, performance improves more quickly. This often occurs as a series of small enhancements or tweaks to the technology that individually may not appear noteworthy but which collectively result in a significant performance boost. This gradual refinement that leaves the basic nature of the technology unchanged is often called incremental or evolutionary improvement. During this period, the technology might converge on a dominant design: a specific way of implementing the technology that can be easily adapted to serve the needs of many potential users.

Eventually, a technology's performance approaches some natural limit: the maximum level of performance that the given effect or principle at work or the structure of the dominant design can achieve. As the technology approaches this limit, gains in performance are harder and harder to achieve, and the rate of improvement slows.[118] Further performance improvements often come from the development of a new technology that has similar functionality but relies on a different mechanism or design with a higher performance limit. This is sometimes called revolutionary, breakthrough, or disruptive improvement. The new technology will similarly advance in an S-shaped fashion until it gets replaced by yet another technology with an even higher performance limit. The result is that technological progress often looks like a series of overlapping S curves as a technology is invented, improved, and ultimately replaced by some successor technology.

### S curves and electric light
As an example of technological S curves, let's return to the electric light bulb. The first electric light bulbs relied on the mechanisms of

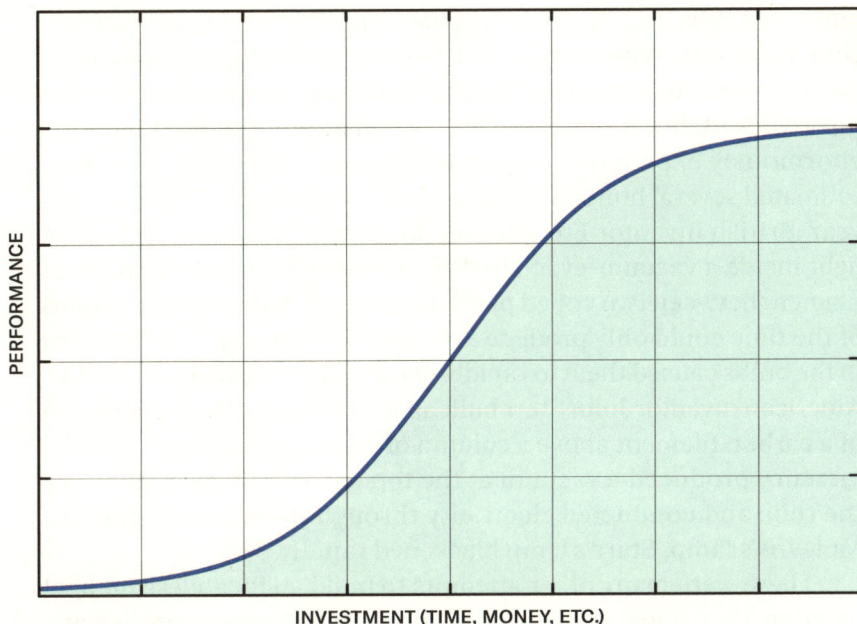

**Figure 4.** Schematic illustration of a technological S curve.

incandescence and electrical resistance—the observations that a material heated to a high temperature will emit light, and that this heat can be generated by running electricity through a resistive material. This method of producing incandescence was first demonstrated in 1761, when English scientist Ebenezer Kinnersley showed that running electricity through a wire could heat it until it was red-hot. As early as 1802, British chemist Humphry Davy had created an incandescent light by connecting a thin strip of platinum, chosen for its high melting point, to a battery.[119]

Davy demonstrated that an incandescent light was possible, but his rendition didn't shine brightly enough or last long enough for it to be practical. Over the next several decades, various inventors worked on turning incandescence into a practical source of lighting. Though they all used the same basic principle—electricity running through a resistive filament—they implemented it in a variety of ways, each searching for an arrangement of components that would work most effectively. In 1840, William Grove, a Welsh judge and scientist, built an incandescent light that consisted of a coiled platinum wire surrounded by glass, placed upside down in a bowl of water to minimize the platinum's exposure to air. Grove's light worked, but it only gave off a small amount of light and was enormously expensive to operate (the cost of electricity was an estimated several hundred dollars per kilowatt-hour). The following year, British inventor Frederick de Moleyns built an incandescent light inside a vacuum-evacuated glass bulb, which used a charcoal filament between two coiled platinum wires. But the vacuum pumps of the time could only produce a partial vacuum, and air remaining in the bulbs caused them to rapidly blacken. Four years later, in 1845, American inventor John Starr built an incandescent lamp consisting of a carbon filament above a column of mercury in a long tube. The mercury produced a vacuum at the top of the bulb as it fell down the tube and conducted electricity through the filament. But like Moleyns's lamp, Starr's lamp blackened rapidly.[120]

There were many other attempts to build an incandescent light using an assortment of filament materials and shapes (platinum and iridium wire, thin platinum sheets, graphite rods, thin strips of gas carbon) and an array of bulb atmospheres (vacuum, air, nitrogen, hydrocarbon) arranged in a variety of ways.[121] The goal was clear: to produce an incandescent light that lasted a sufficiently long time, emitted a reasonable amount of light, and that didn't require too many expensive components or was otherwise costly to operate. But the way to achieve it was far from obvious. Early attempts used expensive components like platinum, burned out too quickly, or resulted in bulb blackening caused by the hot filament evaporating.[122]

Edison himself tried many different lamp designs, including early iterations that used platinum filaments and thermostats to automatically break the circuit if the platinum got too hot. Over the course of multiple trials, he eventually arrived at a particular arrangement of elements that worked effectively. Edison's lamp used a thin, long filament of carburized cotton thread, which had a high resistance and a high melting point. The filament was placed in a high-vacuum atmosphere, made possible by newly available Sprengel mercury pumps, which used falling drops of mercury in a thin tube to pump out air. Edison's lamp was designed in concert with a centralized system of electricity distribution: Lamps would be powered by remote power stations rather than on-site generators. Because the distribution system used low current to minimize transmission losses, Edison's bulb used a high-resistance filament.[123]

Beyond this basic arrangement of elements, there were many other issues to work out, and solving one problem tended to reveal another right behind it. For instance, producing a high enough vacuum required Edison's team to build custom vacuum pumps that could produce a higher vacuum than any other pumps in the world.[124] Once a high enough vacuum could be produced, they discovered that gases within the filament would escape when heated, spoiling the vacuum. They solved this by continuing to run the vacuum pumps while heating the filament and only then sealing the bulb.[125] And once high-vacuum bulbs could be produced, they found that the bulbs tended to leak around their seals, destroying the vacuum. Solving this problem required inventing a new type of seal: a double seal separated by an air bubble.[126]

Once Edison and his team finally arrived at a working arrangement of elements, light bulb performance improved rapidly as the various components were refined and enhanced. Edison continued to search for better filaments, eventually landing on carburized bamboo in 1880. A series of improved filaments—asphalt-coated bamboo, flashed carbon, metallic carbon, osmium, tantalum—followed, each producing more light from a given amount of electricity (a measure known as luminous efficiency) than its predecessor. This rapid series of advancements culminated in the ductile tungsten filament in the early 1900s.[127]

Metallic filaments were more efficient than carbon, but they were more susceptible to evaporation, depositing a residue on the interior of the bulb. This new problem was eventually solved by switching from a vacuum atmosphere in the bulb to an inert gas—initially nitrogen and then argon.[128] Inert gas, however, caused a cooling of the filament, requiring further design alterations in the form of a double-coiled filament, which reduced problems of heat loss.[129]

By the mid-1930s, manufacturers had converged on a dominant design for the incandescent bulb: a double-coiled ductile tungsten filament in an inert gas atmosphere.[130] Bulbs had reached a luminous efficiency of about 16 lumens per watt, which was difficult to improve on; even a typical modern incandescent lamp is around 17 lumens per watt.[131]

Further increases in luminous efficiency resulted from new types of lamps that used different mechanisms to generate light. Fluorescent lamps, for example, began to appear in the late 1930s; instead of incandescence, fluorescents used electricity to excite a mercury vapor, which emitted ultraviolet light, causing a phosphor powder to glow. This method of generating light had a much higher luminous efficiency than incandescent lamps. Early fluorescents had a luminous efficiency of 50 lumens per watt—three times that of modern incandescents—and by the 1990s had reached nearly 100 lumens per watt.[132] Fluorescents were then followed by light-emitting diodes, or LEDs. These used yet another mechanism for producing light, called electroluminescence, in which electrons flowing through a semiconductor recombine with electron holes. LEDs can achieve even higher luminous efficiency than fluorescents—theoretically up to 300 lumens per watt.[133]

Here we see the basic overlapping S-curve pattern. Edison's incandescent lamp was preceded by many decades of unsuccessful attempts to create an incandescent light, and Edison himself had to overcome many problems before the technology worked well enough to commercialize (the initial, flat part of the S curve). But once it did, it improved rapidly, increasing luminous efficiency by a factor of eight over the next 50 years and displacing other lighting technology such as gas (the slope of the curve). But as improvements became harder and harder to find, performance flattened (the top part of the curve), and the incandescent lamp began to be replaced by new lighting technology such as fluorescents, which performed better. By 2012, 70 percent of light in living spaces in the US was generated by fluorescents.[134] These followed their own S curve, steadily improving over the second half of the 20th century before starting to be supplanted by LEDs. Between 2010 and 2021, LEDs went from representing 1 percent of global lighting sales to an astounding 48 percent.[135]

## Other technological S curves

We find this same pattern when we look at other technologies. Inventors attempted to produce powered aircraft for decades prior to the Wright brothers, without much success. British inventor John Stringfellow flew an unpiloted steam-powered monoplane as early as 1848, and the second half of the 19th century saw dozens of attempts to build a heavier-than-air flying machine.[136] But it was only after the Wright brothers' success that aircraft performance began to improve rapidly. In just 36 years, maximum aircraft speed increased from

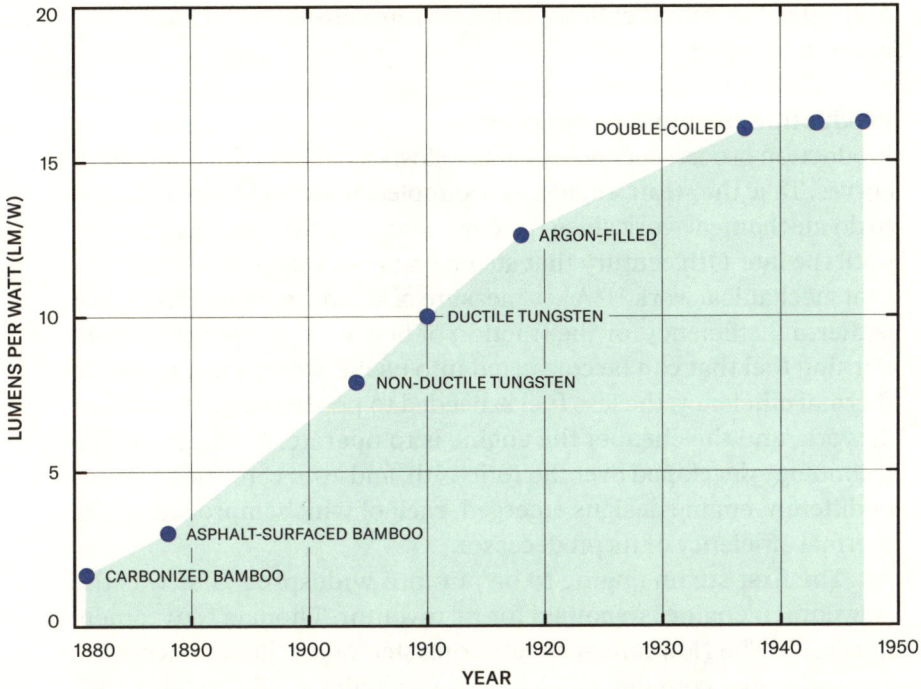

**Figure 5.** Luminous efficiency of incandescent bulbs over time. Data from Bright, *Electric-Lamp Industry.*

6.8 mph (achieved by the Wright Flyer in 1903)[137] to 466 mph (achieved by the Heinkel He 100 in 1939).[138] But these aircraft all used the same propulsion technology—a propeller driven by an internal-combustion piston engine—which is limited in how quickly it can move an aircraft, so further increases in airspeed became difficult to achieve. In 1969, 30 years after the He 100, piston-driven aircraft had reached a speed of 483 mph, less than a 4 percent improvement from 1939.[139] Today, the speed record for a piston-driven aircraft stands at 554 mph, achieved in a modified World War II-era P-51 Mustang.[140] Additional spikes in aircraft speed came from a shift to a new technology, the jet engine, which has a higher performance ceiling. Jet-powered aircraft would eventually exceed speeds of 2,000 mph.[141]

Examples of technological S curves abound. Similar overlapping curves have been observed for a variety of other technologies, including permanent magnets, vacuum tubes, hard drives, semiconductor lithography, ship propulsion, particle accelerators, turbine blades, deep space communications, dynamic random-access memory, and vacuum pumps.[142]

## Production technology S curves

Production process technology also follows a pattern of overlapping S curves. Take the steam engine, for example. Steam had been exploited to do mechanical work since the days of ancient Greece, but it wasn't until the late 17th century that steam engines could perform significant mechanical work.[143] A key measure of steam engine performance is thermal efficiency, or the fraction of heat energy generated from burning fuel that can be converted into useful work. The higher the thermal efficiency, the less fuel is needed to perform a given amount of work, and the cheaper the engine is to operate. As steam engine technology developed over the 18th, 19th, and 20th centuries, a series of different engine designs emerged, each of which improved on the thermal efficiency of its predecessor.

The first steam engine to be put into widespread use was the Newcomen engine, so named for its inventor, Thomas Newcomen, in 1712.[144] The Newcomen engine consisted of a boiler connected to a cylinder that contained a piston. The cylinder would be filled with steam and then cooled with water, a process that condensed the steam and created a partial vacuum in the cylinder. The surrounding atmospheric pressure would then drive the piston down, and the action of the piston would raise a beam connected to it by a chain. Steam would be let back into the cylinder, breaking the vacuum, and the weight of the beam and the equipment attached to it would raise the piston back up, starting the process over again. The raising and lowering of the beam could then be used to do mechanical work, such as pumping water.

Early Newcomen engines were incredibly inefficient, operating at less than 1 percent thermal efficiency.[145] But by the second half of the 18th century, efficiency had doubled thanks to a variety of improvements (largely implemented by civil engineer John Smeaton), including more accurately boring the cylinders to prevent the loss of steam, insulating the cylinder to prevent heat loss, and changing the length of the piston stroke.[146] However, even with these and other improvements, a Newcomen engine could achieve a thermal efficiency of less than 5 percent at best.[147] For that reason, the engine was mainly used to pump water out of coal mines, where fuel was cheaper, since coal from the mines could be used without having to transport it and otherwise useless slack coal (pieces of coal too small to be sold commercially) was readily available. Of the 100 Newcomen engines built between 1712 and 1733, 89 of them were used at coal mines.[148]

The Newcomen engine's inefficiency was largely a function of its design. The cylinder would cool down when water was sprayed into it to condense the steam, so a significant amount of energy was wasted on continuously reheating the cylinder. In 1776, the Watt steam engine, which used a separate condenser for cooling the steam and thus avoided the problems of cylinder reheating, addressed this issue. As a result, the Watt engine tripled the thermal efficiency of Newcomen engines.[149]

Like the Newcomen engine, the performance of the Watt engine improved over time. For example, efficiency was doubled by closing the valve letting steam into the cylinder before it had fully raised and allowing the steam to expand in the cylinder.[150] But as with the Newcomen engine, performance eventually plateaued, and the Watt engine was superseded by another type of steam engine: the compound engine, which used higher-pressure steam and multiple expansion chambers to achieve even higher thermal efficiencies. Compound engines were eventually surpassed by steam turbines, which used yet another mechanism—steam passing over airfoils attached to a rotating shaft—to further boost thermal efficiency.[151]

We see a similar pattern in other production process technologies. Cut-nail machines also underwent a series of small improvements in the decades following their introduction. The first cut-nail machines could only cut the body of the nail; the head still had to be made manually by putting the nail in a heading mold and hammering it. Soon, machines were developed to create the heads mechanically, but these were initially separate machines requiring the smith to manually move nails between machines. By the early 1800s, one-step machines that could both cut the nail and form the head began to appear. Similarly, the nail plate initially had to be manually flipped over after each cut, but eventually machines could do this automatically. The earliest machines also lacked the power to cut nails without first heating the nail plate, which both added a step to the process and reduced the blade life, but later machines were powerful enough to cut the nail plate without heating it first. Each improvement reduced the time and effort required to produce nails, and as the cost of cut nails fell, they rapidly displaced hand-forged nails. The cut-nail machines were then superseded by wire-nail machines, which improved along their own S curve.

We can find many other examples of the same pattern. We've already seen how manual bulb blank-making was superseded by automatic bulb-making machines, which were in turn replaced by the ribbon machine, which used a completely different mechanism of operation. Similar production process improvements occurred in electrical generators, gasoline production, steel manufacturing, glass manufacturing, alkali production, cement production, and aluminum production.[152] In every case, the pattern remains largely the same: A new production method is developed, improves until it approaches some limit, and eventually gets replaced by another production method with a higher limit.

## Complications to the S curve pattern

Overlapping S curves give us a basic model for understanding technological progress and how production processes improve over time. But there are a number of important complications to this pattern. First, technologies often have many relevant axes of performance that can't be captured by a single S curve. Second, the process method will be linked to some degree to the design and functionality of the product, which can make it difficult to substitute new production technology. Third, transporting a production technology (from one facility to another, or from a pilot plant to a full-scale plant) requires adapting it to a new context, which can be expensive and time-consuming. Finally, a successor technology does not always appear, and it's not always easy to determine the natural performance limit of a given production method. We'll discuss these complications in more detail in the remainder of this chapter.

### Multiple axes of performance

An S curve tracks technological performance along a single metric, such as luminous efficiency.[153] But adoption of a technology is frequently a function of many different measures of performance that can't easily be compared along a single axis. When it comes to lighting, for example, beyond luminous efficiency we also care about the color and quality of the light, the up-front cost of the system, the size of the bulb, its lifespan, whether it can be dimmed, and more. Improved luminous efficiency resulted in fluorescent lights displacing incandescent lights in commercial settings, where operating cost was a driving concern, but incandescents remained in use in residential settings, where other performance metrics, such as color, mattered more. Similarly, low-pressure sodium lamps were even more efficient than fluorescents, but the yellow light they emitted made them unsuitable for commercial buildings, so their use was largely limited to outdoor and industrial settings.[154]

Adoption of production technology is also often determined by multiple relevant performance considerations. For instance, the early cut-nail process produced a nail in which the grain of the metal ran perpendicular, rather than parallel, to its body. This meant that cut nails, unlike forged nails, couldn't be clinched—that is, when the nail was hammered through a piece of wood, it couldn't be bent back on the other side to act like a staple. So hand-forged nails continued to be used in cases that required clinching, such as producing door battens.[155]

Similarly, adoption of the steam engine involved multiple performance considerations. Despite being more efficient and having lower fuel costs than the Newcomen engine, the Watt engine had a higher up-front cost, likely due to the additional components and fabrication precision it required. Newcomen engines were also thought to be easier to maintain than Watt engines, since the precision components of the latter required more frequent intervention.[156] These considerations—along with the hefty license fee charged by Watt—meant that Newcomen engines remained popular well after the Watt engine was introduced, despite their low thermal efficiency and high fuel consumption. As of 1800, Newcomen engines outnumbered Watt engines three to one in Britain.[157]

At any given time, for any given production process or transformation, there might be a palette of available technologies that can be used, and the best option may be a function of many different considerations: what material is available, what expertise the technology requires, the quality requirements of the product, the expected production volume, and more. A 1980 taxonomy of manufacturing processes listed more than 100 available processes for shaping or joining metal, including 10 kinds of casting, seven kinds of forging, six kinds of brazing, and eight kinds of soldering.[158]

Advances in production technology often take the form of additions to this palette rather than substitutions. Most of the metal-shaping options described in 1980 are still in use more than 40 years later, but today they've been joined by a variety of additive manufacturing and

3D printing methods, such as selective laser sintering and wire arc fabrication.[159] Complicating this picture further is the fact that these technologies are all improving at different rates, and the best method for satisfying a particular production problem will change over time. Wire nails made of iron had been manufactured since the early 1800s, but they didn't begin to replace cut nails until they could be crafted from cheap steel, which was made possible by the invention of the Bessemer steelmaking process in 1856.[160]

In his book *The Innovator's Dilemma*, Clayton Christensen notes that multiple axes of performance often play an important role in the development and spread of new technology. An S-curve pattern means that early on, a new technology often performs significantly worse than an established technology along the most important measures of performance, even if its theoretical performance ceiling is much higher. But the new technology is often superior along some other axis of performance that, while not important to mainstream users of the technology, is important to some particular set of customers, giving the technology a foothold in the market. As the technology improves, it eventually surpasses the existing technology on other performance metrics as well.

Take mechanical excavation technology. In the mid-20th century, excavators largely used a system of steel cables to manipulate the digging arm and bucket. The primary market for these excavators were sewer line and water pipe diggers, open-pit miners, and builders digging basements, who all tended to measure performance by the reach of the digging arm and the amount of earth that could be lifted in a single scoop. In the 1940s, excavators that used hydraulics rather than cables to manipulate the digging arm began to appear. These new excavators initially had meagre scoop capacities one-twentieth the size of cable-articulated excavators and, therefore, weren't appealing to most existing users.[161] However, their slight size, speed, and maneuverability made them attractive to residential contractors for digging small trenches and excavating foundations—projects that had previously been too small to justify a large excavator and had instead been dug by hand. Over time, various improvements increased the capacity of hydraulic excavators, to the point where they had as much scoop capacity as the cable-articulated excavators. And because hydraulic excavators were more reliable than cable-articulated excavators, they eventually displaced them.[162]

In a similar vein, technological development has often been funded by the military, which has historically often been willing to tolerate much higher costs than could be justified commercially in pursuit of greater performance. Semiconductors, jet aircraft, nuclear energy, interchangeable parts, and numerically controlled machine tools all relied on defense funding for their early stages of development.[163] In each case, the technology was initially expensive and was, therefore, of limited commercial use, but it had capabilities that were worth the cost to the military. Interchangeable parts allowed firearms to be repaired more quickly in the field, while numerically controlled machine tools enabled the production of airfoils for high-performance aircraft and helicopters. Only later did these production methods find commercial applications.

*Coupling between production method, product design, and functionality*

Another complication of technologies often having multiple axes of performance is that a new production method may not be an exact substitute for an old method. That is, it may not perform the exact same transformation. Rather, it may perform a similar transformation that nevertheless differs in meaningful ways.

To return to our nail-making example, the cut-nail process wasn't simply an improved hand-forging process that performed the exact same steps more cheaply. It used a completely different series of operations, and nails produced by the cut-nail process were not identical to handwrought nails. They were similar—both were a tapered length of iron—and had comparable performance in many ways, but there were key differences. Cut nails couldn't be clinched, for example, while handwrought nails could. Similarly, wire-nail machines used a different series of steps than cut-nail machines, which meant there were important differences between cut nails and wire nails. Most notably, a wire nail wasn't as strong as a cut nail.

Whether these differences matter depends on the design of the product and the function it's intended to fulfill. A given production process reflects a particular product design: The material a product is made from, its size and shape, the arrangement of its parts, and how those parts attach will all determine the necessary production methods. In turn, the design of the product will reflect a certain set of functional requirements: an array of things the product should do or properties it should have. Because functional requirements determine design and design determines production methods, switching to a new production method might require changing the design of the product, and perhaps even its functionality.[164]

Automotive paint provides an example. The color of a car is perhaps not its most important characteristic, but it's something people care about. All else being equal, car buyers prefer to be able to choose from a variety of color options. When Ford first began to produce the Model T in 1908, it offered the car in a range of different colors depending on the body style. But at the time, colored automotive paint took weeks to apply and required up to 14 coats, with sanding and drying in between each coat. Black paint, however, could be applied much more quickly. As Ford increased production of the Model T and storing huge numbers of cars for painting became an issue, the company decided to paint every Model T black, believing that the low price enabled by the more efficient production process was worth the trade-off.[165] (And it was, for a while.) Offering cars in a spectrum of colors, as General Motors did, meant sacrificing production efficiency.

In the 1920s, DuPont developed a quick-drying pyroxylin-based paint called Duco that made it possible to paint a car with colored paint in a single day. General Motors started to use Duco-colored paints in 1923, and Ford adopted pyroxylin paint for the Model T in 1926.[166] But while Duco made it possible to manufacture cars in different colors efficiently, it came with its own functional trade-offs. Duco paints were much less lustrous than previous colored paints, and car buyers would have to accept much more muted colors than had originally been available.[167] As automotive paint technology changed, functionality—in this case, the color and lustrousness of the car—was forced to change with it.

Similarly, the introduction of mass production technology was predicated on consumers being willing to accept a homogeneous product that could be easily produced by specialized machines. American consumers proved much more inclined to accept uniform goods than British consumers, who demanded products customized to their individual needs. This constraint stifled the use of mechanized manufacturing in Britain. As economic historian Nathan Rosenberg writes, "English observers often noted with some astonishment that American products were designed to accommodate not the consumer but the machine."[168]

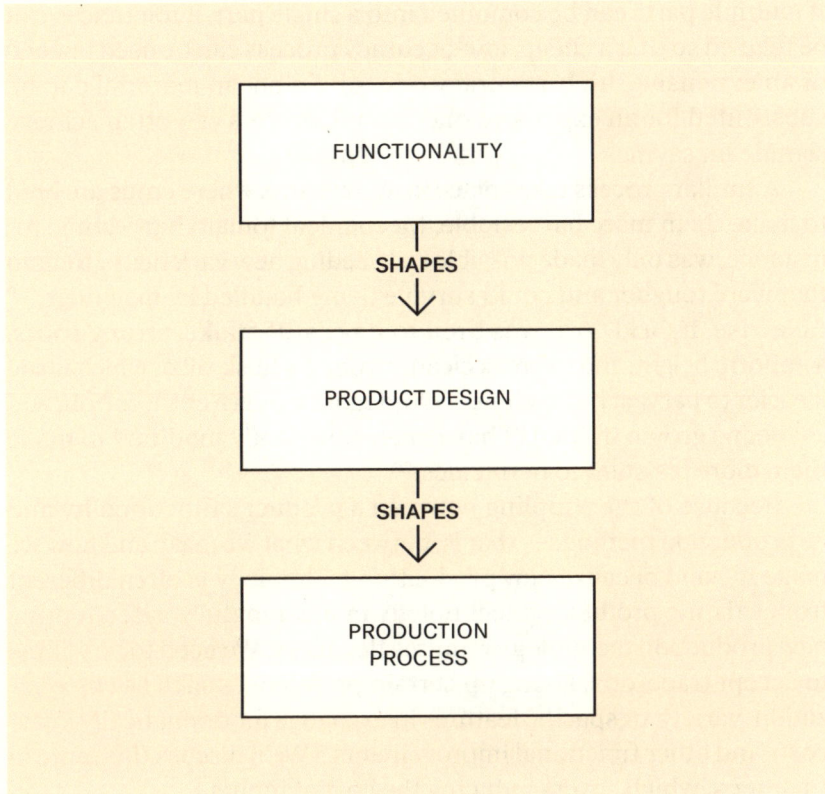

```
┌─────────────────────────┐
│                         │
│      FUNCTIONALITY      │
│                         │
└─────────────────────────┘
         SHAPES
            ↓
┌─────────────────────────┐
│                         │
│     PRODUCT DESIGN      │
│                         │
└─────────────────────────┘
         SHAPES
            ↓
┌─────────────────────────┐
│                         │
│       PRODUCTION        │
│        PROCESS          │
│                         │
└─────────────────────────┘
```

**Figure 6.** Linkage between a product's functionality, its design, and the process used to manufacture it.

As these examples demonstrate, adopting a new production method often requires some flexibility in both the design of the product and the functional requirements the product is built to meet. The less flexibility in the design of the product or the more tightly coupled functionality is to the production method, the more difficult it is to change how it's made. If part of a product's appeal is that it's handmade, for instance, that precludes the use of any sort of automatic machinery to manufacture it.

The flip side is that it's often possible to greatly reduce an item's production cost if you can adapt its design to the requirements of a particular production method. In manufacturing, this is known as design for manufacturing, or DFM, and typically involves redesigning a product to require fewer, less expensive operations and materials.

If multiple parts can be combined into a single part, if tolerances can be relaxed so that a cheap, low-accuracy process can be used instead of an expensive, high-accuracy one, or if a cheap material can be substituted for an expensive one, manufacturers can often achieve significant savings.

A similar process takes place in agriculture, where crops are bred to make them more harvestable. Mechanical tomato harvesting, for instance, was only made possible by breeding new varieties of tomato that were tougher and could survive being handled by machines.[169] Likewise, hybrid corn was bred to have stiff stalks, strong roots, a uniform height, and to break cleanly from the husk, all of which made it easier to harvest by machine.[170] And most modern corn, cotton, and soybeans grown in the US have been genetically modified to make them more resistant to herbicides.[171]

Because of the coupling between a product's functionality and its production methods— that is, between what we make and how we make it—and because new production technology is often different from existing production technology in meaningful ways, adopting new production technology requires flexibility. We need to be willing to accept trade-offs, giving up certain preferences such as customization, variety, or specific features in exchange for dramatically lower costs and other functional improvements. (We'll discuss this more in Chapter 3, which covers reducing the cost of inputs.)

*Lack of portability*
Another complication to the basic S-curve model is that even when new production technology is straightforwardly superior, adopting it is far from straightforward. Production methods are often surprisingly unportable: There tend to be significant barriers to transferring a production process and using it effectively in a new location.

Moving a production process doesn't only require relocating machines and equipment. It also means transferring the skills, expertise, and patterns of behavior necessary to operate the new process to a new group of workers and a new organization. Historically, this has proved difficult. In the 19th century, English gun manufacturers couldn't successfully adopt American gun-making machinery unless they also employed Americans to operate it. Even Samuel Colt, the most successful private arms manufacturer in the US at the time, failed with his firearms factory in London, partly because the British workers were simply unable to operate the American machines.[172] Similarly, in the late 1700s, Americans were aware of British textile technologies such as spinning jennies and water frames, and knew roughly how such technology worked, but their own attempts to build British-style textile mills failed. They would only succeed in building a working mill when they obtained the assistance of Samuel Slater, who had worked for more than 10 years in British textile mills.[173] Likewise, when DuPont tried to get into the dye manufacturing business in the early 1900s, it suffered heavy financial losses for several years, despite building an exact replica of an existing dye factory, due to its lack of know-how of dye manufacturing. (Its first attempt at manufacturing indigo resulted in a green, rather than blue, dye.) It wasn't until DuPont poached several German chemists with the necessary expertise that its dye manufacturing operations became profitable.[174]

Similar difficulties continue to occur today. In a 1990 study of 32 firms transferring manufacturing technology to a new location, six failed to achieve profitability at the new location and were eventually shut down, and another four experienced severe productivity problems.[175] In a 1992 study of seven manufacturers that launched continuous improvement efforts (modeled after lean production methods), only three had successfully implemented the new methods after two years.[176]

These sorts of difficulties can also be observed across a single industry, as there is often a surprising amount of variation in plant productivity, and the best firms in a particular industry will produce far more efficiently than the worst. A 1967 study of 102 manufacturing industries concluded that "in no industry was productivity (value added per man-hour) in the top quartile of plants less than 25 percent greater than the industry-wide average. In the typical industry it was 65 percent greater, and in 16 industries it was more than double the average."[177]

These challenges can even exist within a single company. Famously, General Motors was unable to replicate the success of its New United Motor Manufacturing, Inc. (NUMMI) joint venture with Toyota at its other plants, nor was it able to replicate the success of Saturn practices in its Oldsmobile division.[178] Similarly, TSMC's semiconductor fabrication facility in Oregon was not, after 25 years, able to match the productivity of its facilities in Taiwan.[179]

Even when new production methods are successfully adopted,

the process often takes many years. The transfer of manufacturing capability of the jet engine for the F-104 fighter from the US to Japan, while successful, took five years to complete, and the transfer of shipbuilding capabilities from the UK and Japan to South Korea took place over more than 15 years.[180] In *What Machines Can't Do*, industry scholar Robert Thomas provides several case studies of firms adopting new manufacturing technology, from flexible machining cells to continuous aluminum casting to robotic assembly, where the process also took many years.[181] Similarly, practitioners of lean production suggest that fully adopting lean production methods, which are almost entirely organizational and require little to no new equipment, often takes five to 10 years.[182]

Several factors are at work to make moving production technology from one place to another so difficult. For one thing, there will often be substantial differences between production facilities, even if they're in the same industry and producing the same goods. Different facilities may have different production volumes, target different markets, or have different product designs. (Aston Martin and Toyota both make cars, but their markets, production volumes, and designs vary dramatically.) They may have chosen other options from the palette of available production technologies, or they may utilize separate standards, such as imperial versus metric units or 120-volt versus 220-volt electricity. They may be operated by organizations with different strategic goals, organizational structures, or work cultures. They may be subject to individual regulatory requirements or have different resources available to them, such as what skills can be found in the local labor pool.

In short, a production process, like any technology, doesn't stand on its own but is always embedded in a particular context—a surrounding environment of resources, knowledge, assumptions, and relationships. Moving a production method or system from one location to another requires adapting it to a new context, which takes time and effort.

One example of this sort of adaptation is the use of engineering drawings, which convey production information—what operations to perform, what parts and materials are needed in what quantities, and so on—to the factory floor. Different countries and industries have developed distinct engineering drawing styles, and the transfer of a product or process to a new location with another drawing style might require recreating all the relevant drawings, often at substantial expense. During WWII, for example, some Ford plants were repurposed to manufacture the B-24 bomber. But when Ford received the drawings from the original designer, Consolidated Aircraft Corporation, the two companies found that "they spoke different languages."[183] The Consolidated drawings used different symbols and dimension styles and provided different levels of explanation; Consolidated drafters assumed the drawings would be interpreted by an experienced foreman as opposed to factory workers, and the drawings were much sparser than what Ford was used to. Ultimately, Ford engineers were forced to redraw the 30,000 drawings from Consolidated, turning them into 60,000 drawings in a format their autoworkers could understand.[184]

Similar adjustments, both technical and cultural, were needed when transferring jet aircraft manufacturing technology from the US to Japan in the 1950s. Not only did additional drawing information need to be provided, which took "a considerable fraction of the US technical assistance efforts," but the Japanese quality control inspectors also needed to be given promotions so they would have sufficient standing to "criticize" other workers.[185]

Another common context change that can create difficulties is variations in scale. For example, transferring a chemical manufacturing process from a lab to a commercial facility is complicated by the differences between lab-scale and commercial-scale operations. Economist Gary Pisano observes:

> The chemical process may behave and perform differently in a small-scale versus a large-scale vessel or in a glass versus a stainless steel vessel. The chemists may stir the process very differently from the way it would be stirred by large, computer-controlled rotators. For these reasons, the researcher can expect that [commercial performance] will not be equal to [lab performance]. Yields may be lower; costs may be higher; different impurities might be present. Even tests in a pilot plant can be quite different than what might be expected in a full-scale plant.[186]

The supporting infrastructure a production technology relies on may also vary in different locations. When steel buildings began to be constructed in India in the 1960s, builders had to account for the fact that the steel would be transported to the jobsite in pushcarts rather than the trucks typical in the US. So they adapted their building designs to use smaller, more easily movable pieces of steel.[187] Similarly, bringing the first Newcomen engine to the US required a large supply of spare parts, as well as British technicians to operate it, because the necessary skills to build and repair the engine didn't exist in America.[188]

A related difficulty is that a production facility may be oriented around a previous production technology and can't easily adapt to an updated one, or can't take maximum advantage of it. For example, factories designed to use steam power tended to be tall and narrow, with a single large steam engine delivering power to individual machines via axles and belts. The introduction of electric power made it so that each machine could have its own motor sized specifically for that machine and allowed the machines to be spread out over a single floor, which made moving material significantly easier. So factories built around steam power couldn't take maximum advantage of electric power because of constraints on floor space and factory arrangement.[189] Similarly, older UK shipyards had difficulty adopting more modern methods of ship construction, where ships would be assembled block by block inside a single large building.[190]

Another major barrier is the transfer of knowledge required to operate the process. While much of the needed knowledge may be in the form of drawings, specifications, documentation, or physical objects that are (comparatively) easy to transfer, much of it might also exist in the form of tacit knowledge that is hard to articulate or express. It's hard to explain how to ride a bicycle; the skill can only really be acquired through practice. Successfully operating a production process likewise often requires significant tacit knowledge. As an example, nuclear weapons manufacturing in the 1980s relied on skilled machinists who could, among other things, observe when a layer of oxide had been completely removed from a piece of metal they were working on. Attempts to use instruments for the same task proved inferior to a skilled set of eyes for "detecting the change to a shiny, then slightly hazy, appearance that indicates a clear surface."[191] During the Industrial Revolution, some considered it best practice to start children working in textile mills as early as age 10 to allow them to develop the instinctual knowledge of fiber behavior that was necessary to successfully operate spinning mules as an adult.[192]

In a study examining problems that cropped up after new machines were installed in a factory, only two of 27 problems could be resolved without engineers physically inspecting the machines, because the problems resisted being communicated in other ways. As one user explained, "We tried the telephone, the tube [electronic mail]... It was endless. You just had to be there and see it." The engineer concurred: "I just don't know what is going on until *I see* the problem. To look and know what I see is exactly what I get paid for!"[193] Following WWII, the British encountered similar difficulties when they tried to transfer German manufacturing technology by way of written reports, which were quickly found to be insufficient. As the Association of British Chemical Manufacturers observed in one heated letter, "We have not spent all the time and trouble in organizing investigating teams merely to produce a row of reports on the shelf... Firsthand investigation would eliminate a great deal of the usual trial and error in setting up a plant here... Much of the 'know-how' is impossible to put into words."[194]

Beyond encompassing skill-based know-how, tacit knowledge might also be embodied in the working relationships among the staff of a plant or firm. In other cases, tacit knowledge may be explicit but invisible to the organization at large—for instance, kept in the "little black books" of tricks foremen and operators devise to keep operations running smoothly but conceal from higher-level supervisors.[195]

Transferring tacit knowledge often requires the equivalent of an apprenticeship, where workers from a new facility spend an extended period of time at an existing facility (or vice versa) to absorb the relevant production knowledge. Just such a method was used to transfer shipbuilding knowledge from British and Japanese firms to South Korea when the nation developed its shipbuilding capabilities in the 1970s and 1980s.[196] Likewise, fast-food franchises, when opening a new location, will often have new employees work side-by-side with existing employees from other locations for several weeks.[197] This makes transferring the knowledge a production process requires expensive: In a study of 26 projects where production processes were installed at new locations, the cost of transferring knowledge comprised on average 19 percent of the entire project cost, and in some cases it was as high as 59 percent.[198]

Another barrier when transferring production technology is that it may be imperfectly understood. There may be some degree of causal ambiguity in what makes it function, or there may be unknown variables that affect the process but aren't considered until they manifest in a new location. For example, the Bessemer steelmaking process initially only produced quality steel if the input iron was low in phosphorus. Henry Bessemer was unaware of this; he'd only succeeded because he had, by chance, used some of the only ore in Britain that was low in phosphorus. When the more common higher-phosphorus ores were used, the resulting steel was of poor quality, and Bessemer was forced to refund most of the initial licensees of his process.[199] Similarly, the float process for producing plate glass—which involves floating molten glass on a bath of molten tin to produce a smooth surface—is dependent, among other things, on nearly imperceptible material properties of the tank lining. When the British company Pilkington first developed the process in the 1950s, they had inadvertently used a tank lining with the exact composition needed to produce good results. Pilkington only became aware of the importance of the tank lining when licensees of the process had problems using superficially similar but microscopically different tank linings.[200]

Yet another difficulty is that as production tolerances become tighter, slight variations caused by different environmental conditions have greater effects, making transferring production technology more fraught. Semiconductor manufacturing, for instance, relies on extremely pure silicon doped with just the right amount of chemical impurities. The slightest environmental change can upset the balance. Early semiconductor fabrication was found to be influenced by such factors as the phase of the moon (higher tides were thought to raise the groundwater and increase the humidity in the plant), female workers' menstrual cycles (which slightly changed the amount of oil on their hands), and whether male workers had recently visited the bathroom (which deposited microscopic droplets of urine on their hands).[201] Because there were so many factors to take into account, many of which were not always obvious, Intel developed its Copy Exactly technological transfer methodology, which required every new semiconductor fabrication facility to copy an existing fabrication process down to the smallest detail, including the color and supplier of the paint used on the building.[202]

Finally, there can also be social or political barriers to transferring production technology. A change in production method often means new roles and patterns of behavior for the people who operate, manage, or are otherwise involved in it. To the extent that this creates "winners" and "losers," there may be constituencies who will work against adopting a new process. Automation and other labor-saving technologies have often been resisted by labor unions, trade guilds, and other worker constituencies worried about job loss. During the Industrial Revolution, weavers smashed power looms and other textile-making technologies.[203] Longshoremen and other maritime unions resisted the introduction of the shipping container in the 1950s and 1960s and, more recently, the use of autonomous ships and automated container handling equipment.[204] This sort of resistance has been with us for as long as people have been building machines. As economic historian Joel Mokyr notes:

> As early as 1397, tailors in Cologne were forbidden to use machines that pressed pinheads. In 1561, the city council of Nuremberg, undoubtedly influenced by the guild of red-metal turners, launched an attack on a local coppersmith by the name of Hans Spaichl who had invented an improved slide rest lathe. The council first rewarded Spaichl for his invention, then began to harass him and made him promise not to sell his lathe outside his own craft, then offered to buy it from him if he suppressed it, and finally threatened to imprison anyone who sold the lathe. The ribbon loom was invented in Danzig in 1579, but its inventor was reportedly secretly drowned by orders of the city council... In 1299, an edict was issued in Florence forbidding bankers to use Arabic numerals. In the fifteenth century, the scribes guild of Paris succeeded in delaying the introduction of printing into Paris by 20 years. In the sixteenth century, the great printers' revolt in France was triggered by labor-saving innovations in the presses.[205]

New technology or processes can change the balance of power, control, or influence within an organization, so employees who stand to be disadvantaged by the change may resist. In one case, a donor facility, frustrated that it was being forced to share its newly developed technology with another facility, sabotaged the transfer to the recipient.[206] In other cases, facilities have opposed the introduction of methods that were "not invented here" or that conflicted with short-term goals. A need to meet production quotas, for example, might inhibit the adoption of methods that are likely to temporarily reduce output.[207] Summarizing the difficulties one company experienced with several plants that shared a process technology, Kim

Clark and Robert Hayes note, "The transfer of relevant information appeared to be limited by the organizational difficulties associated with coordinating and communicating engineering knowledge, by the desire of each plant to protect its proprietary knowledge, and by each plant's reluctance to assimilate superior techniques developed at other plants." In some cases, they wrote, a plant "balks at the thought that others might be able to do it better."[208]

Adopting new production technology isn't like buying a new car, where you simply replace the old one and get a better ride without any muss or fuss. It's more like trading in your car for an airplane: You have to learn how the new technology works and develop the skills required to operate it, many of which take experience to learn. It means adapting your habits to the new technology's strengths and weaknesses. And it means successfully navigating an entirely new context that surrounds its operation.

*Performance ceilings and successor technologies*
The S-curve model makes it seem as if it's obvious when a technology has begun to approach its performance ceiling. And for many technologies, particularly those based on simple physical phenomena or chemical reactions, it can be. The theoretical maximum amount of light that can be produced from a given amount of energy via incandescence, or the minimum amount of energy it takes to fix nitrogen for synthetic fertilizer production can be calculated based on the laws of physics and knowledge of chemistry.

But more often, determining a technology's performance ceiling is difficult, if not impossible. Indeed, Clayton Christensen has argued that most technology is simply "too complex" to accurately predict its development.[209] A technology thought to be at its limit will often continue to improve, in many cases due to pressure from a new, supposedly superior technology, while a technology that's promising on paper may fail to pan out in practice. In response to competition from Edison's incandescent light bulb, gas lighting companies developed the gas mantle, which greatly increased the efficiency of gas lighting, making it so competitive with early incandescents that Edison nearly went bankrupt.[210] Similarly, wind energy experts in the 1990s sometimes argued that wind turbines had reached their maximum efficient size, but continued advancements in blade manufacturing, among other things, allowed turbines to get larger and more powerful, which has been a major driver of increased turbine efficiency.[211] And in semiconductor manufacturing, optical lithography techniques continued to improve in performance despite predictions that the technology's natural limit had been reached.[212]

This lack of predictability makes new technological development fraught. It's hard to predict what the ultimate performance of a new technology will be or even if it will outperform existing technology, and the transition from one technology to a successor can be chaotic. In some cases, betting on the wrong production technology has cost companies dearly. Starting in the 1960s, Alcoa spent 20 years and millions of dollars in an unsuccessful attempt to develop an alternative to the Hall–Héroult process for smelting aluminum, which was derailed in part by continued improvements to the Hall–Héroult process.[213] Similarly, in the 2010s, Intel opted to further develop immersion lithography technology rather than try to adopt extreme ultraviolet (EUV) lithography for semiconductor fabrication, which later caused strategic challenges when EUV proved much more successful.[214] Famously, the Manhattan Project attempted to guard against the unpredictable nature of technological development by simultaneously developing several different methods of enriching uranium—gaseous diffusion, electromagnetic separation, and, later, thermal diffusion—in hopes that at least one of them could be made to work.[215] A similar tactic was taken during the Covid-19 pandemic, when the US government simultaneously funded the development of several promising vaccine variants.[216]

One final complication to the basic S-curve model is that there's no guarantee that a successor technology will appear at all. The wire-nail process remains the standard way of making nails more than 100 years after it was invented. The machines for crafting them continue to undergo marginal improvements, but the fundamental mechanism remains the same, and nails stopped getting cheaper early in the 20th century.[217] Similarly, no technology for light bulb blank production has succeeded the ribbon machine, which remains in use nearly a century later.

If a technology has a natural performance ceiling, its performance will simply continue to approach it. As with nails and bulb blanks, the Hall–Héroult process remains the standard method for smelting aluminum today more than 100 years after it was invented. Its performance, as measured by energy required per pound of aluminum produced, has gradually approached the theoretical minimum the chemical reaction requires. Improvements beyond this will presumably require a new process that uses a different chemical reaction.[218] Likewise, the Haber–Bosch process used to produce ammonia for synthetic fertilizer remains the primary method for making ammonia more than a century after its invention, and continues to approach the theoretical minimum energy consumption required for nitrogen fixation.[219]

When discussing successor technologies, it's worth briefly noting that determining what counts as a new technology worthy of a separate S curve is, to some degree, a question of definition. Does the tungsten light bulb filament constitute a fundamentally new technology, or is it a continuation of incandescent technology? Was Watt's steam engine a fundamentally new development or a refinement of the Newcomen engine? We can try to come up with a classification criteria or rubric, but it is perhaps more useful to simply remember that the S curve is merely a tool for simplifying and organizing thought and condensing information—a map that shouldn't be confused with the territory of the real world. The S-curve model gives us a basic framework for thinking about technological progress and understanding how production processes evolve over time. By understanding the limits of this model and the ways it smooths over and simplifies the roiling complexity of the real world, we can better understand the evolution of production technologies without being led astray.

## Mechanization

Before wrapping up this chapter, it's worth devoting some time to one specific type of technological progress that has had an outsize effect on making production processes more efficient: mechanization, or the replacement of manual labor with machines.

Economists sometimes talk about general-purpose technologies, or technologies that can be broadly applied to solve problems or accomplish tasks in a wide variety of industries or situations. The steam engine, the semiconductor, and the laser are all examples of general-purpose technologies. One way of thinking about manual labor is that it's the original general-purpose technology: a set of capabilities for sensing the environment, processing information, and moving physical objects around that can be applied to solve an extensive range of problems. Human senses like sight, hearing, and touch provide a rich picture of the surrounding environment and how it's changing; a human brain can take in this data and, given enough time, accomplish any type of information-processing task.[220] The number of physical actions a human can perform is extremely broad: A human arm, which is able to move at the shoulder, elbow, and wrist, has seven degrees of freedom, while a hand has over 30 different ways of grasping objects depending on their size, shape, and weight.[221] And a person can vary the amount of force they apply depending on the situation, from carefully manipulating an egg without cracking it to lifting and tossing a bowling ball. This flexibility has made human labor a crucial component of nearly every production process in history. But human capabilities also have limits. A person can only move so fast, lift so much weight, or control their motions so precisely. By replacing the human with a machine, we can achieve much higher levels of performance and much lower costs.

We've already seen examples of mechanization in the production of nails, bulb blanks, text, and thread. In each case, manual labor was replaced by machines that could move faster and more precisely. A team of glass bulb blowers could produce, at most, 150 bulbs an hour, but ribbon machines could produce that many bulbs every five seconds—roughly 700 times as fast. Wire-nail machines could produce as many nails in 10 minutes as a smith could in an entire day. And the introduction of the printing press increased the production of books from roughly 20,000 to 30,000 per year to 1 to 2 million per year, while a single modern mechanized air-jet spindle can produce roughly 20 times as much cotton thread as the fastest preindustrial spinners.

When a manual process is limited by how fast a person can work, output is limited by the number of workers available. Producing twice as many handwritten manuscripts, for example, requires hiring twice as many scribes. And because a person needs to be paid some minimum amount and the costs of that labor will be spread over a relatively small volume of output, the cost of a manual process will necessarily be high. Mechanized processes that reduce the amount of human labor can, therefore, produce things much more cheaply. Recall that the mechanization of spinning caused the cost of cotton thread to fall by 90 percent between 1784 and 1832, while books in the 1500s cost an eighth of what they did in the 1300s thanks to the printing press. Because most production processes have historically relied heavily on manual labor, and because labor is expensive, mechanization has been one of the main avenues for efficiency improvement in the modern era.

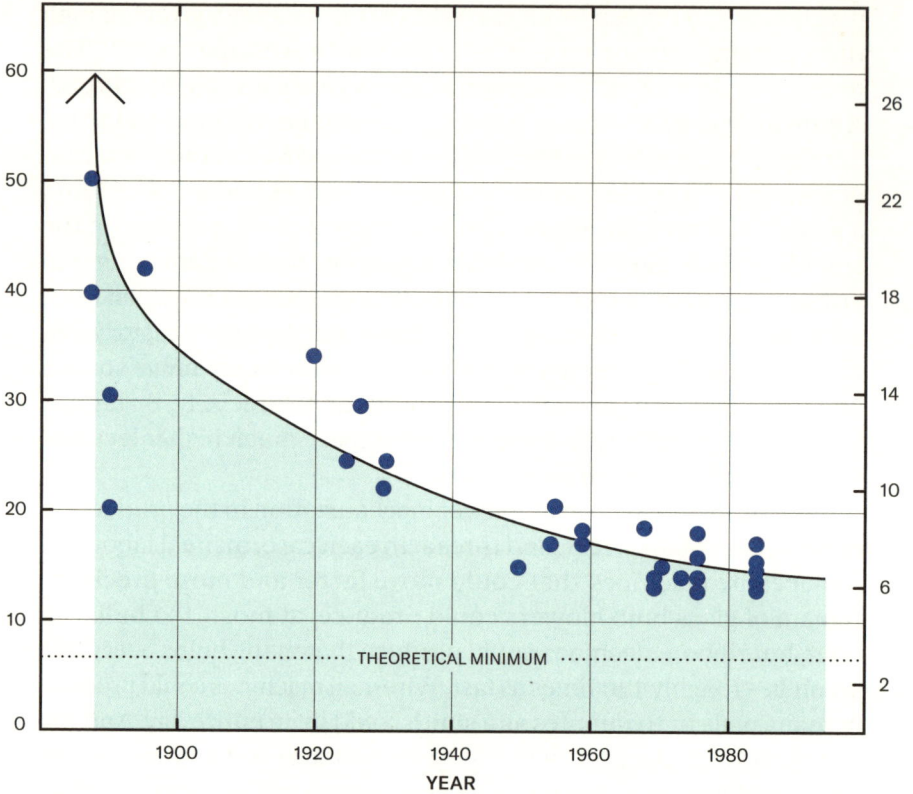

**Figure 7.** Energy consumption in kilowatt-hours per kilogram (kWh/kg) and kilowatt-hours per pound (kWh/lb) of the Hall–Héroult process to produce aluminum from 1890 to 1990. Reproduced by permission from McGeer, "Hall–Héroult," fig. 1.

Earlier, we discussed how production technology is seldom an exact substitute for previous technology: The new technology often does something fundamentally different, and the production process must adapt accordingly. This is also the case with mechanization. Machines built to perform a certain task generally don't perform the exact same actions a person would but instead use different actions to achieve the same (or similar) results. This is partly because it's often difficult to build a machine to mimic human motions precisely, and partly because, as with other cases of technological progress, higher performance is often achieved by finding a different, superior mechanism. To move items horizontally, we don't build machines with mechanical legs; we build them with wheels. A wheel is not only much simpler to build than a mechanical leg, but it can also move much faster. Likewise, mechanizing nail production didn't involve creating a mechanical hammering device that duplicated the movements of a smith, but instead involved creating an entirely different process—a reciprocating blade—that produced similar results.[222] And mechanizing text production didn't require the creation of a mechanical hand that could write script but rather used an entirely different method of putting text on a page.

As with other advances in production technology, then, mechanizing a production process often requires a change in the design of the product. Mechanically manufactured nails were different from hand-forged ones, and text from a printing press was different from text that had been copied by hand. Similarly, the mechanization of light bulb assembly required a significant change to the design of the bulb itself. Instead of a long length of wire draped over hooks, a difficult operation to mechanize due to the flexibility and delicacy of the wire, the filament was redesigned as a shorter, rigid coiled filament that was easier for a machine to manipulate.

More specifically, since machines usually can't respond flexibly to their environment the way a person can, mechanization requires products to be made by a series of specific, repetitive motions planned out in advance. While there wasn't a specified number or type of hammer strikes needed to produce a hand-forged nail (because the smith could simply keep hammering until the nail was done), cut-nail production required planning out the exact motions that would produce the desired nail, then building a machine to perform them.

Often, reorganizing a production process to use a series of well-defined, repetitive motions precedes actual mechanization, since it enables a manual division of labor. By fully specifying each step in the process, it becomes possible to divide the work among multiple workers and arrange the steps to avoid any wasted effort or time. With each step in the process planned out in advance, it then becomes possible to improve individual steps. Thus, many processes were "industrialized" well before a machine was built to perform it. Both Ford's assembly line and Adam Smith's pin factory are examples of industrialized processes that were still performed manually even as they dramatically cut labor costs.

When mechanization does occur, it often proceeds in stages, starting with the simplest or most labor-intensive tasks and gradually spreading to the rest of a process. In cut-nail manufacturing, the initial machines only cut the body of the nail. The rest of the process—heating the nail plate, inserting the nail plate at the correct angle, moving the body of the nail from the cutting station to the heading station, adding the head—was still done manually. But over time, more and more steps were mechanized. First, separate heading machines were developed, followed by machines that could cut the nail and add the head in a single machine, followed by machines that could also insert the nail plate at the correct angle, and so on, until by the 1840s nearly every step was performed mechanically.

We see a similar progression with light bulb assembly. Around 1920, assembly was almost entirely manual. Workers would start by making the stems, the small glass tubes that connected the wires to the filament. They would then insert the wires into the stems and attach the filament. Then they would affix the exhaust tube to the top of the bulb blank (to remove the air), then fasten the bulb blank to the stem assembly. Next, they would pump air out of the bulb and test-light the lamp. Following a successful lighting, the base was added, then the bulbs were labeled, tested, and packed for shipping. There was little mechanization involved. Most tools were hand-operated, except for the exhausting machines and test-lighting machines, which ran automatically but required manual loading and unloading. With this process, 10 workers could assemble about 200 bulbs per hour.[223]

By 1925, much of this work had been mechanized. Manually controlled machines were replaced with semiautomatic machines for stem production, wire insertion, sealing, and adding the base. These machines still required labor to operate, but they had predetermined settings and were much simpler to use. Indexing machines, which moved partially completed bulbs to the next workstation at a preset production rate, were added to the process to smooth out the production rate and eliminate work in process. The steps of sealing the exhaust tube, mounting the stem, and labeling the bulbs were still performed manually, as were the loading and unloading of bulbs from the machines. Using this production process, five workers could assemble 300 bulbs per hour.

By 1936, even more of the process had been mechanized. There were now automatic machines to make the stems, insert the wires, exhaust the bulbs, test-light them, and add the labels. Mechanical arms would transfer bulbs from machine to machine. A single machine would seal and exhaust the bulbs, and another would install the base, light the bulbs, and label them. Mounting the stems was still done manually. Now, seven workers could assemble 700 bulbs an hour. And by 1955, both stem-making and mounting had been fully automated. The only manual operations that remained involved loading the sealing, basing, and testing machines, as well as packing the bulbs. Four operators could assemble 1,350 bulbs per hour.

Today, the few remaining incandescent light factories are almost completely mechanized. Every step in the process is performed by a specially designed machine, and the few workers who remain mostly load and unload the machines and perform quality control. The machinery can produce nearly 2,000 bulbs an hour.[224]

Most processes won't be automated to the level of nail and light bulb production, however. The highest levels of automation have traditionally been limited to high-volume production of highly uniform products—things like nails and light bulbs, as well as bulk materials like cement, chemicals, and petroleum. Lower-volume production of higher-variability products tends to have lower levels of automation, with a significant amount of the work done by humans who can respond flexibly to changing conditions and make adjustments on the fly. And, of course, there are things that humans remain better at than machines and places that mechanization hasn't yet touched (at least, as of this writing). Humans couple a broad range of movement with senses and information-processing capabilities that afford them the flexibility to chain together sequences of actions. A smith can adjust the force and location of their next hammer strike based on the results of the previous one and can select a new tool or take some other action (such as reheating the iron) if needed. A glassblower can adjust the amount of air they blow or how they spin the molten glass depending on how the glass responds.

A manual process is often more resilient to environmental variation and doesn't need to be specified below a certain level of fidelity. The person performing the work can, to some extent, fill in the blanks, and the more expertise and skill they have, the more blanks they can fill in. A machine, on the other hand, lacks many of these capabilities. While duplicating one specific physical movement is usually possible, it's much more difficult to build a single machine that can recreate the broad range of a human's physical capabilities. And machines have lagged well behind their human counterparts when it comes to integrating these movements with sensing and information-processing capabilities. We have yet to create a robotic hand with the tactile sensor capability that a human hand has. While machines are getting better and better at reacting to their environment—a modern industrial control system can now monitor and make adjustments based on thousands of different variables—they still have much narrower capabilities than humans. And many sensory-dependent tasks that are comparatively easy for a human—driving a car, folding a towel, picking a strawberry—remain difficult to fully mechanize.[225]

One aspect of this information-processing problem is that objects that are difficult to predict the behavior of, such as fabrics and other deformable materials, are often challenging to manipulate mechanically. As Swiss physicist Aude Billard notes, while "robots are adept at handling rigid objects, they still struggle with flexible materials—such as fruits and vegetables or clothing items—that differ in size, weight, and surface properties. Manipulations that produce a deformation (e.g., inserting, cutting, or bending) are particularly difficult, as accurate models of the deformations are needed."[226] Despite investing millions of dollars in mechanization, the textile industry still relies on human labor to piece together fabric and feed it into a sewing machine.[227]

Successful mechanization has thus historically required reducing or otherwise limiting the amount of information processing that must be performed and the environmental variation that must be considered. Mechanical harvesting machines, which have been steadily replacing farm labor since the 18th century, often require no information processing at all. Instead, these machines operate via simple filter mechanisms. An ear of corn is wider than the stalk it attaches to, so corn harvesters work by forcing the stalk to pass through an opening that's too narrow for the ear. As the combine passes over the cornstalk, the ear of corn pops off and makes its way through the rest of the combine. Other harvesting equipment functions in a similar way. Potato harvesters work by scooping up the soil the potatoes are planted in and running it over a grate that separates the dirt and debris from the potatoes. The cotton gin operates by pulling cotton fibers through a fine mesh that cotton seeds can't pass through, eliminating the laborious work of removing them by hand.[228] In this way, a crop harvesting machine can ignore almost all the information a human would use to harvest a plant. It doesn't need to know exactly where the ear of corn is, or how big it is, or how firmly it's attached to the stalk.

The machine just needs to have a few pieces of information about the physical properties of the crop it's harvesting built into the mechanism so that, with some sufficiently high degree of probability, it only lets the right part of the plant pass through.

As we've seen with the evolution of light bulb and cut-nail production, mechanizing a process rarely eliminates labor completely. Instead of performing the task itself, humans oversee or operate the machines executing it. Even for a task done almost entirely by a machine, worker expertise remains valuable in keeping the machine operational and responding to variation the machine can't handle. Mechanizing cotton spinning, for example, didn't eliminate the need for worker skill and experience, as Earle Buckingham noted in his 1942 book on manufacturing:

> The story goes that one operator had very little trouble with the breaking of thread on his spinning frames while all the other operators were continually in difficulties. His actions were carefully watched in the attempt to discover how he was able to prevent this breakage of threads, but without success. As far as could be seen, his actions and methods were the same as those of the other operators. Finally he was asked how much payment he wanted to disclose his secret. At first he denied that he had any. After much persuasion, he named his price, the guarantee that he should have a pint of beer every day as long as he lived. The bargain was struck, then he said: "Chalk your bobbins."
>
> This operator carried a piece of chalk in his pocket. Whenever he had to change a bobbin, which was made of wood, he rubbed his hand on the chalk in his pocket and then rubbed the bobbin. The chalk dust was sufficient to prevent the fibers of the thread from catching on the fine splinters on the surface of the wooden bobbins.[229]

The same phenomenon can be observed across a number of mechanized processes. The operation of 18th-century spinning mules required a keen sense of how fibers would behave; the invention of machine tools required operators to develop a new set of skills to properly work them, even as they eliminated the labor required by previous manual methods; the operation of automated bulb blank machines required constant intervention by skilled workers. As a former ribbon machine operator put it, "Operating the machine was at times an art form... These machines required constant maintenance, love, and care to keep production continuous for days at a time."[230]

Ultimately, mechanization proceeds like any other type of technological progression: Machines start out working poorly, but with time and effort they can become capable enough to replace the existing "technology" of manual labor. This progression is subject to all of the difficulties of any new technological development: predicting the performance of the mechanized process, transferring the technology, failing to displace the previous manual methods in certain cases (though it's perhaps rarer for manual labor technology to improve in response). Progress then continues apace, with new and better technology eventually superseding the new machines.

## Summary

The S-curve model, along with its various refinements and complications, paints a picture of how production technology changes and how production methods evolve over time to become more efficient. A production process is created and gradually improves along various axes of efficiency: over time, its output increases; it requires fewer, cheaper inputs; it cuts out extraneous process steps; and so on.[231] As the low-hanging fruit is picked clean and it becomes harder to wring out additional performance gains, the improvement slows down or stops. Eventually, a new technology comes along that uses a different design or functional mechanism, unlocking higher performance. As the new technology improves, there is a chaotic and unpredictable period of transition. The new technology may not achieve the expected performance level; the old technology may continue to squeeze out surprising performance gains; or the new technology may require changes to the product or process that aren't easily accommodated. But in many cases, the new technology supplants the old technology, though the process might take years or decades (and even then, the old technology will often still be used in certain niches where it maintains an advantage over the new one). The whole process then repeats in a messy, lurching progression.

The processes we've looked at in this chapter skew toward the simple. Many real-world production processes are far more complex, involving dozens or hundreds of varied technologies. Likewise, an individual technology is often made up of many subtechnologies, which are in turn made up of sub-subtechnologies, and so on, until you reach the level of simple raw materials. This process of technological progression might be taking place at any level within the hierarchy. But regardless of which level of organization we look at, we see the same basic process at work.

Of course, new technology alone does not improve efficiency. Advancements along the various axes of efficiency—higher rates of production, greater reliability, fewer and cheaper inputs—are what make efficiency improvements possible. We'll start to look at these in the next chapter.

# 3

# Reducing
# Input Costs

One way to think about a production process is that it starts with a particular bundle of inputs—raw materials, purchased components, energy, labor, equipment, facilities—and concludes with a particular good or service. Making handwrought nails, for example, required blacksmith tools (tongs, hammers, anvils), nail rod, a forge and fuel, a heading mold, and a certain amount of relatively unskilled labor. The lower the cost of this bundle of inputs, the lower the cost of the finished nail.

So how does one go about finding a cheaper bundle of inputs? In some cases, it may be possible to simply swap out one input for a different, lower-cost one without changing anything else. If a smith can find a cheaper source of nail rod, that can reduce the costs of production while keeping the overall process the same. But more commonly, finding cheaper inputs involves some sort of rearrangement or redesign of the production process. It's rare that any combination of processes, inputs, and organizational structure is unique in its ability to generate a given product. Even in the absence of new technology, many possible arrangements can produce the same product or a functionally similar one, many of which will be cheaper than others. By exploring the landscape of possible ways to produce a good or service, we can often find approaches that require fewer, cheaper inputs—different recipes that make the same thing with cheaper ingredients.

In general, reducing input costs will involve redesigning the product or production process, changing the organizational structure or the location in which the product is made, or increasing the value of the outputs. In this chapter, we'll look at each of these methods in turn, as well as their trade-offs and implications.

### Redesigning the product

One important way to reduce the cost of inputs is to change the design of what is being made. Redesigning a product to use different materials or parts can often significantly lower the cost of the product.

This won't always be possible, of course. Some products have a very close coupling between their functionality (what a product does) and their design (how a product is built), making it difficult or impossible to make even slight changes. A pharmaceutical manufacturer can't simply modify the chemical formula for the drug it's making, because even a single atom's difference can alter a drug's behavior. Similarly, a steel manufacturer will have limited (but not zero) ability to change the makeup of the steel it produces, since the chemical composition of a particular grade of steel is governed by certain technical standards.

But other products have many more degrees of freedom and much less coupling between functionality and design. When it comes to manufactured goods, there are often many possible designs and arrangements of components that will accomplish a particular goal. Even something as simple as a screw can be produced in almost limitless varieties: McMaster-Carr, a supplier of hardware and industrial materials, lists over 50,000 types of screws on its website.[232] For something as complex as a car, which has thousands of parts, the number of possible varieties vastly exceeds the number of atoms in the universe.[233]

Manufacturers have long pursued this strategy of cost reduction. One of the early improvements to the Newcomen engine in the mid-18th century involved replacing the expensive brass cylinder with a cheaper cast-iron cylinder.[234] Similarly, in the 15th and 16th centuries, gunsmiths began to replace expensive bronze with cheaper cast iron in the production of cannons. And in the mid-19th century, machine toolmaker Joseph Whitworth redesigned his machines to require less time and labor to manufacture by designing their parts to be easily produced by machine tools.[235]

These types of improvements became even more important toward the end of the 19th century, when small-volume and craft production gradually gave way to industrialized mass production. With large production volumes, even small reductions in per-unit manufacturing costs would result in large savings, so manufacturers increasingly looked for ways to make their products cheaper to produce. Much of the early work on the incandescent light bulb focused on finding ways to reduce the amount of expensive platinum it required, which, as of 1890, made up one-third of the cost of a light bulb. The platinum clamps that held the filament were eliminated in favor of attaching the filament directly to the lead-in wires, and the platinum lead-in wires were replaced with wires made of copper or other materials, leaving just a small amount of platinum at the point where the wire entered the bulb. Because platinum has the same coefficient of thermal expansion as glass, a rare property among materials, designers had trouble finding a suitable substitute for this last bit of platinum. But it, too, was eventually removed and replaced with a cheaper composite of iron, copper, and brass first developed in 1912.[236]

Similarly, when designing the Model T, Henry Ford strove to use components that were as simple and easy to manufacture as possible. For the Model A in 1927, he allegedly went so far as to specify how to design the shipping boxes for the batteries so they could be reused within the car itself. Ford specified that the box

> must be made of a certain type of wood, be reinforced in places, and contain some rather peculiar "vent" holes in most of the pieces... As the first Model A of the new series was coming down the line, he called all his department heads... to the factory floor. The battery box was knocked down to get at the battery, and the battery was installed under the floor at the driver's feet. Henry then picked up the pieces of the box and fitted them above the battery, exactly forming the floorboards on the driver's side of the car. The holes in the boards were for the brake, starter switch, etc. The screws that held the box together were then used to bolt the floorboards down.[237]

Though most manufacturers didn't go so far as to optimize the designs of their boxes, by the early 20th century the importance of refining the design of a product to minimize manufacturing costs was well understood. Books on designing for mass production or interchangeable manufacture began to appear that included recommendations for minimizing manufacturing costs.[238] Large-volume manufacturers, such as National Cash Register, often included a step in their product development process for making initial prototypes easier to manufacture.[239] Similar suggestions were made by the draftsmen of early 20th-century British car manufacturers.[240]

During WWII, the urgency of war production resulted in numerous examples of products designed to be made as quickly and cheaply as possible. The Liberty cargo ship, built in large numbers during the war, was originally based on a British design that assumed the work would be done by skilled shipbuilders.[241] During the war, it was redesigned so it could be built quickly by unskilled labor, replacing riveted construction with cheaper, faster welded construction.[242] Similarly, firearms used at the start of the war were succeeded by lower-cost, easier-to-produce models that eliminated complex mechanisms, used cheaper stampings in place of expensive forgings, and had parts that required a minimal amount of machining. The American Thompson submachine gun was eventually supplanted by the cheaper M3 submachine gun, and the German Luger P08 pistol and MG-34 machine gun were replaced by the cheaper Walther P38 and MG-42, respectively.[243]

Following WWII, these strategies for minimizing manufacturing costs began to coalesce into more formal methodologies. One of the earliest of these was value analysis, today also known as value engineering. A vice president at General Electric, Harry Ehrlicher, noticed during the war that many of the product changes made in the name of material economy resulted in products that performed better than the original.[244] In 1947, he assigned a subordinate, Lawrence Miles, the task of finding better value for materials. Miles's work grew into what would become known as value analysis. Within 12 years, GE employed 120 value analysts, whose work collectively saved the company millions of dollars. Word spread, and the methodology began to be discussed in trade publications. Companies like RCA, Westinghouse, and IBM created their own value analysis departments. By the late 1950s, the Society of American Value Engineers was formed (it continues to promote value engineering today as SAVE International), and government agencies began to add contractual clauses requiring value engineering on their projects.[245]

Value analysis or engineering works by systematically investigating the design of a product. For each component, the value analyst asks what it does, how much it costs, and whether something else might do the same job for cheaper. In *Techniques of Value Analysis and Engineering*, Miles describes the methodology applied to the design of an electrical switch:

> One of the first parts to come up for review was a wire clip, which held the cover on. It was made of phosphor bronze and cost $7,000 a year. The value specialist asked, "What is the function? How does phosphor bronze contribute to that function? It is a material that is commonly used when a part has to flex millions of times. But this clip is flexed only when the cover is removed for servicing— an average of six times in the life of the product." By changing to spring brass, the same job could be done for $3,000 a year, that is, for less than half.
>
> The cover itself cost 4 cents, an expenditure of $40,000 a year. Its function was to keep extraneous material out of the small mechanism. The entire control was mounted inside another

enclosure. Using a plain piece of laminated plastic reduced the cost from 4 cents to 1.5 cents each, and out went $25,000. Many similar findings were made as the analysis of the complete device proceeded.[246]

Value engineering remains popular today and is widely applied in manufacturing, construction, and government projects. Departments of transportation, for instance, often have value engineering guidelines that projects must follow to improve quality and reduce costs. A study of Georgia Department of Transportation projects found that $1 spent on value engineering saved anywhere from $23 to $300 in construction costs.[247]

A similar development was the creation of guidelines for producibility, later known as manufacturability: designing products and parts to make them easy to manufacture. Producibility guidelines, such as Roger Bolz's *Production Processes: The Producibility Handbook*, first began to appear in the 1940s and became increasingly popular over the following decade, with organizations like GE and the US Army developing their own.[248] Producibility guidelines aimed at providing specific, process-by-process recommendations for designers, and took into account how easy or difficult it was to produce a particular element of the design. "As part designs are created, developed, and detailed," Bolz writes, "the designer is actually selecting the process or processes by which the parts will be manufactured. Conversely, too, he may be limiting the usable methods to those least desirable."[249]

Manufacturers could achieve significant savings by tailoring the design of a product to the constraints of a specific process. In one version of his manual, Bolz notes that it is difficult to cut a recess that has a radius tangent to the surface (that is, a perfect semicircle) into a piece of metal without accidentally nicking the sides, which would then require expensive reworking. Changing the design of the recess to make the angle shallower and eliminate the tangent, he writes, could reduce the costs of this operation by 66 percent or more.[250]

Producibility and manufacturability recommendations often focused on designing parts that were easy to manufacture, but this sometimes came at the expense of making the product more difficult and complex to assemble. Making parts simple to manufacture often results in a product with a greater number of parts, which then takes significant time and effort to put together. Conversely, combining several different parts into one adds complexity to fabrication but makes assembly much easier and faster, because with fewer parts there are fewer assembly operations. Designing a product to be easy and cheap to manufacture, therefore, requires attending to both the ease of fabrication and assembly.

In the 1960s, researchers at Amherst College began to study how products could be designed so their parts could be handled mechanically. They found that certain part shapes and features were much easier than others for machines to manipulate. Symmetrical parts without small features, for example, were much easier for automatic machinery to handle than asymmetrical parts with small features.[251] The researchers behind this early work, notably Geoffrey Boothroyd and Peter Dewhurst, then began to develop more general assembly recommendations for both humans and machines, and integrated these recommendations with considerations of producibility and manufacturability. The combined methodology became known as Design for Manufacturing and Assembly, or DFMA, and was widely adopted by companies in the 1970s and '80s, including Ford, IBM, Xerox, GE, Westinghouse, and DEC.[252] Additionally, some organizations created their own systems for analyzing ease of assembly, such as Hitachi's Assembly Evaluation Method (AEM), developed in the late 1960s and early '70s, and the Lucas method, developed by Lucas Engineering in the '80s.[253] Today, DFMA and its descendants remain standard tools in manufacturing design.

Like producibility guidelines, DFMA guidelines are based on adapting a product to what is easiest for various production processes. But DFMA also includes guidelines for minimizing a product's assembly cost. Since each separate part requires an assembly operation, the main mechanism for this cost reduction is to minimize the number of parts by combining as many separate parts as possible. DFMA also includes recommendations for making assembly operations easier and less likely to be done incorrectly. Making a part symmetric, for instance, means the assembler will spend less time orienting the part while also often preventing the part from being inserted the wrong way.

Similar strategies for designing a product around the production method have been adopted beyond the manufacturing context. In agriculture, crops are often specifically bred to make them easier to harvest by machine, thereby reducing production costs. And in restaurants, a chef might choose menu options based on how easily and quickly they can be prepared in the kitchen. In *Kitchen Confidential*, Anthony Bourdain described his process:

> I'm fine-tuning the specials in my head: Grill station will be too busy for any elaborate presentations or a special with too many pans involved, so I need something quick, simple, and easily plated... I'm thinking grilled tuna livornaise with roasted potatoes and grilled asparagus for fish special. My overworked grill man can heat the already cooked-off spuds and the preblanched asparagus on a sizzle-platter during service; the tuna will get a quick walk across the grill, so all he has to do is heat the sauce to order... Appetizer special will be cockles steamed with chorizo, leek, tomato and white wine—a one-pan wonder; my garde-manger man can plate salads, rillettes, ravioli, confits de canard while the cockle special steams happily away on a back burner... Both leg of venison and some whole pheasant are coming in, so I opt for the pheasant. It's a roasted dish, meaning I can par-roast it ahead of time, requiring my sous-chef simply to take it off the bone and sling it into the oven to finish, then heat the garnishes and sauce before serving, easy special. A layup.[254]

Whether in a restaurant kitchen or on the assembly line, designing a product so it's as easy as possible to produce reduces the cost of making it.

**Redesigning the production process**
In addition to redesigning a product to be cheaper to manufacture, it's often possible to redesign or rearrange elements in the production process itself to reduce costs, even if no changes are made to what is being produced. Ford's assembly line, for instance, improved efficiency by rearranging workers and how they did their jobs. Many of these strategies fall under the heading of operations research, or OR, a field of study that uses scientific and analytical methods to find ways to optimally allocate resources. OR makes it possible to navigate a vast space of potential production arrangements algorithmically to quickly find the most low-cost options.

An early commercial use of OR occurred in the paper industry. Paper was produced in wide rolls, which were then cut into narrower widths for resale. The producer could save money if it was able to minimize trim loss, or leftover paper too narrow to be used, when cutting the rolls into narrower widths. By formulating the problem in terms of a linear mathematical model, paper producers were able to use an algorithm (such as the simplex method) to find an arrangement of roll cuts that would minimize trim loss.[255]

Another example comes from rail travel. On a single-line railway in India with trains running in both directions, whenever two trains running in opposite directions would meet, one would have to stop to let the other pass. This reduced the distance a train could travel in a single day. A mathematical analysis revealed that the number of crossings rose faster than the square of the number of trains, so a small number of additional trains meant many more stops and much longer delays. Combining multiple trains together, thereby reducing the number of individual trains, could, therefore, increase daily travel distances. In the seven months after the scheme was implemented, daily travel distances on the railway increased by over 30 percent, lowering transportation costs and allowing people and cargo to get where they were going faster.[256]

OR methods can also be used for tasks like determining the optimal amount of inventory to hold to ensure service, minimizing transportation costs by finding the shortest possible route between two points, decreasing the time spent by products or people waiting in line, and defining the maximum amount of goods that can move over a transportation network connecting two points.

Outside of OR, productivity researchers have developed other methodologies to redesign production processes to make them less expensive, often focused on specific industries or types of operations. For example, the single-minute exchange of die (SMED) methodology was developed by Japanese engineer Shigeo Shingo in the 1950s and '60s as a method for reducing the time spent on setups, or operations that need to be done in preparation for the actual transformation a process performs. The name comes from the practice of changing out a die, a specially shaped insert placed in a press to cut or stamp metal into a particular shape. Historically, die changes took several hours, but Shingo found that the time could often be reduced to less than 10 minutes, and that the same methodology could be applied to other types of machine setups. Reducing setup time increases the amount of time the machine can perform useful work, since time spent on setups is time when the equipment isn't producing. It also increases production flexibility, since it's easier to use the same machine to produce many different parts if every changeover doesn't require a time-consuming setup.

A setup can be thought of as a smaller production process within a larger one: taking some inputs (labor, parts, equipment in a particular orientation) and transforming them into outputs (the equipment in the proper configuration) via a series of operations. Much like DFMA, SMED is about finding the fewest and least expensive operations that will produce a given output: a properly inserted die, for example.[257] SMED guidelines include things like replacing bolts, which are time-consuming to tighten, with clamps, which can be attached and removed much faster; installing jigs, guides, and fixtures so parts can be placed quickly; and modifying machines so no adjustment is needed for commonly used settings. Here is a description of the SMED process applied to a machining operation:

> In one plant, a limit switch was used to set the end point in machining shafts. Since there were five shaft lengths, the switch had to be moved to five different locations. It could not be positioned correctly without as many as four trial adjustments every time the setup changed. By installing limit switches at the five sites, each equipped with an electrical switch that is supplied with current independently of the other switches, this problem was eliminated. Now, setup is performed by flipping a switch.[258]

SMED also focuses on when the setup takes place. It's often possible to perform much of the setup *before* the actual changeover, while the machine is still running. For example, Shingo notes that although a new die can be attached to a press only when the press is stopped, "the bolts to attach the die can be assembled and sorted while the press is operating."[259] This is known as converting an *internal* setup to an *external* setup. While this doesn't reduce the time and effort required to perform the setup, it does reduce machine downtime.

Pharmaceutical manufacturing provides another example of how a process can be redesigned to be more efficient. Pharmaceuticals are produced through a sequence of chemical reactions called a route, but the route developed during the initial R&D process is often not appropriate for large-scale commercial production. It may require expensive reagents, have a large number of steps, result in low yields, or require the expertise of PhD chemists to perform the operations.[260] For these reasons, scaling up a process for production typically involves what is known as route design or route selection, in which chemists explore possible synthesis routes, searching for a sequence of reactions that will, among other criteria, produce the desired product with a small number of steps and few expensive reagents.[261] An early route for one antipsychotic drug required raw materials that cost over $2,300 per kilogram, mostly because it required a large amount of an expensive aldehyde and involved a reaction step that had only a 65 percent yield. Eventually, chemists identified an alternative route that required significantly less aldehyde, reducing raw material costs by nearly 50 percent.[262] Identifying the most effective route can reduce the overall manufacturing costs of a drug by orders of magnitude, making it possible to produce cheap, widely available drugs for a variety of conditions.[263]

This sort of process exploration is also a key part of work simplification, a descendant of the theory of scientific management that became popular in the mid-20th century. Like scientific management, work simplification sought to find the best, simplest way to perform a job, but it expanded beyond time and motion studies to analyze entire processes, breaking them down into individual components and studying each part. For each step, and for the process as a whole, the analyst must ask *what*, *why*, *how*, *who*, *where*, and *when*. Here's how one engineer describes the method, as applied to rewinding electric motors:

> What motors should be rewound? Is it economically feasible and sound to rewire all of them?... Why are so many motors being rewound? Is there a recurring failure?... How should the motors be rewound?... Are the rewound motors as good or better than the original manufacturers' motors?... Who should rewind the motors?... Can part of the work be assigned to helpers or other lower paid men?... Where are the motors being rewound? Is the location now being used the best one available consistent with economic standards?... When is the job being done? Is overtime being used to rewind motors that are not actually needed at the time they will be completed?[264]

Once the process had been analyzed, improvements could be made. In the case of rewinding motors, for example, it was discovered that one operation (cutting off the existing coils) could be reduced from 30 minutes to 30 to 60 seconds using a specialized piece of equipment.[265]

**Changing the organizational structure**

Another way of rearranging a production process to minimize the costs of inputs is to reconsider the organizational boundaries around the process. In some cases, the cheapest input may be one that is produced by another company and purchased on the open market. In other cases, it may be cheaper to produce it internally. This is known as the make-versus-buy decision.[266]

On one end of the spectrum, every step could take place within a single firm. For nail making, this would involve a firm mining the iron ore and the coal, smelting the iron in a blast furnace to produce pig iron, turning that pig iron into wrought iron, then cutting the wrought iron into nails. In fact, in the late 19th century, nail production did take place in highly integrated companies that controlled every step of the process. Cut nails were made in large ironworks with dozens of nail-making machines (as well as equipment for making other steel products), which were fed with nail plate made in on-site blast furnaces and rolling mills. Other steps in the production process were similarly integrated: Companies that made nails also often owned coal mines, keg factories, riverboats, and even railroads.[267]

On the other end of the spectrum, it's possible to have a separate firm handle each step of the production process and allow a market mechanism to manage the coordination between them. Building construction, for instance, is typically done by a large team of separate contractors. The developer who funds the project will hire a general contractor, who will then hire subcontractors to perform specific tasks: pouring the foundation, building the structural framing, installing the electrical systems, and so on. These subcontractors will obtain their materials from building supply stores, which will, in turn, be supplied by many different material manufacturers making narrow families of products.

In addition to making or buying, there are also more complex arrangements. Firms can have a variety of relationships beyond arm's-length market transactions, such as strategic partnerships, part ownership, or supplier of choice.[268] Similarly, an integrated firm's operations might be marketlike, such that internal divisions might need to bid against outside suppliers for any given piece of work. Firms might also obtain some of a given input internally but secure the rest from outside (known as tapered integration), use outside suppliers only during periods of especially high demand, or structure their operations in any other number of ways to minimize costs.[269]

Where and how it makes the most sense to draw the organizational boundaries around the process depends on many factors. An outside producer may have lower costs because they operate at a larger scale or have cheaper labor costs. On the other hand, as economist Ronald Coase points out in "The Nature of the Firm," there are transaction costs associated with interacting with the market. It takes time and effort to do things like write and negotiate contracts and vet suppliers. Outside suppliers, whose incentives are less aligned, may decide to prioritize other more important customers. And suppliers may have high profit margins that will add to the costs of purchasing an input.[270] Where transaction costs are low—if what's being purchased is an undifferentiated commodity with many suppliers in a competitive market—it's more likely to be cheapest to buy. But if transaction costs are high—if the input is specialized and requires supplier-specific equipment or knowledge—it might be cheapest to make.

Similarly, there may be various risk factors associated with obtaining a particular input. If a plant requires the input to operate, then it might be worth making it in house simply to ensure a steady supply, even if cheaper inputs are currently available on the open market. Doing so also protects the company from being at the mercy of its suppliers: If a company has adapted its process to a specific input and can't easily drop it, the supplier may leverage that dependence to charge higher prices. Making some input internally and maintaining the capabilities to do so might also be important for strategic reasons, such as being able to develop new products in the future or to maintain competitiveness.[271] In other cases, relying on the open market may be a way to minimize risk. If an industry is in flux such that the basic form of a product or the equipment used to produce it has an uncertain future, obtaining inputs from the open market may be less risky than buying production equipment that could shortly become outdated. Similarly, if production capabilities are difficult or expensive to develop, it might be less risky to buy them.

Projected production volume, available time, physical location, firm assets and capabilities, and even accounting systems can also influence whether it's cheaper to make or buy an input, both in the short term and in the long, risk-adjusted term.[272] As a result, different firms in the same industry may decide to draw their organizational boundaries in different places and outsource individual aspects of the production process depending on the specifics of their business and what they perceive the risks to be. As an industry evolves, the location of these boundaries often shifts.[273]

As an illustration of the complex dynamics of the make-or-buy decision and how it influences costs, consider the US automobile industry. In the early 1900s, car manufacturers were largely assemblers who purchased parts—axles, wheels, engines, and the like—that may have originally been developed for other purposes.[274] Indeed, as economist Susan Helper notes, "In 1903, Ford's entire contribution to the manufacture of the cars that bore the company's name was to install bodies, wheels, and tires on completed chassis made by the Dodge brothers, owners of a Detroit machine shop."[275] Buying inputs made sense in the early days of the automobile because the market was small and the form of the car was still taking shape, with steam, electric, and gasoline models vying for dominance.[276] Moreover, the concentration of the industry in the Midwest made it easy to source and purchase components.[277] But as the automobile industry grew dramatically (from selling 4,192 cars in 1900 to 181,000 in 1910 to 1.9 million in 1920), firms began to take advantage of the possibilities of vertical integration.[278] Once car manufacturers reached a certain production volume, it became cheaper to produce parts in house than to buy them from suppliers. While early cars were made mostly from parts built by outside suppliers, by 1922 outside suppliers were responsible for only 55 percent of the wholesale value of American cars. By 1926, that figure was 26 percent.[279]

Ford's business model, which involved producing just a single model of car, the Model T, in enormous volumes and selling it at a low price, inspired Ford to go further than any other manufacturer with respect to vertical integration. In its effort to continually reduce the cost of production, by the 1920s Ford had extended its operations backward into acquiring and refining raw materials. Ford owned forests, sawmills, iron mines, and coal mines, and even explored drilling its own oil, mining its own lead and magnesium, and operating its own rubber plantations.[280] Ford continued to use outside suppliers for raw materials (the iron mines and timber supplies could provide only a portion of what the company used), but Ford-owned resources acted as insurance against supply interruptions or suppliers raising prices.[281] When a railroad couldn't ensure timely transportation to Ford's enormous Rouge facility in Dearborn, Michigan, Ford purchased it and completely reworked its operations, speeding up distribution and reducing the volume of parts in transit.[282] These savings were large enough to completely offset the purchase price of the railroad.

Ford's Rouge manufacturing facility, which employed 75,000 people in 1928, was a paragon of vertical integration. Limestone, coal, iron ore, and timber were brought to the Rouge on Ford's railroad and on Ford-owned lake freighters. The facility included coke ovens and blast furnaces for smelting iron, open-hearth and electric arc furnaces for turning that iron into steel, and foundries for turning the iron and steel into parts.[283] Excess slag from the blast furnace was turned into cement, which was used in the construction of additional buildings, with the balance sold on the open market.[284] Ford-owned forests supplied the factory with lumber, with sawmill waste turned into charcoal (later sold under the brand name Kingsford).[285] The Rouge also included power generation facilities, a glass plant, a tire plant, and a paper mill that turned scrap wood into cardboard boxes.

This level of vertical integration enabled unparalleled reductions in cost. In 1911, a Model T Runabout cost $680. By 1924, the price had dropped to just $260—a 75 percent reduction when adjusted for inflation—which no competitor could match.[286]

But Ford's enormous production machine was built specifically for the Model T and couldn't easily adapt to the evolving car market, where models changed yearly. When Ford decided to switch to a new model, the Model A, in 1927, the Rouge had to be shut down for six months to retool at a cost of $100 million (the equivalent of $1.7 billion in 2022), not including sales lost during the shutdown, estimated at another $250 million.[287] Following the introduction of the new model, Ford began to purchase more from outside suppliers, including wheels, bodies, shock absorbers, storage batteries, piston rings, and much else. Today, more than 1,400 firms provide the company with parts.[288]

Other car manufacturers were less vertically integrated than Ford, so they had an easier time adapting to a changing market. General Motors, for example, had a policy of having full control over only 33 percent of the parts and accessories for the cars they produced. Chrysler produced even fewer of its own components, which proved to be a benefit during the Depression, when fixed costs were lower and strained suppliers were willing to sell parts extremely cheaply.[289] As a result, Chrysler's share of the auto market increased from 6 percent in 1927 to 25 percent in 1933.[290] Chrysler didn't increase its vertical integration until after WWII, when high demand forced it to acquire a parts manufacturer to ensure reliable supply.[291] Over the course of the 20th century, the Big Three of Ford, GM, and Chrysler would continue to pursue different internal production strategies. By 1965, GM had emerged as the most integrated manufacturer, with 48 percent of the wholesale value of GM cars purchased from outside suppliers, compared to 60 percent at Ford and 73 percent at Chrysler.[292]

We can observe similar ebbs and flows of integration in other industries. In the late 19th century, steel production became highly vertically integrated. The two largest steel companies, Carnegie Steel and Illinois Steel, built high-volume steel facilities, then purchased sources of raw materials, such as iron and coal mines, to ensure a steady flow of input material. When makers of steel products such as wire, hoops, and bridge girders began to integrate backward and produce their own steel, Carnegie responded by integrating further downstream, building a large plant that could produce wire, nails, and other steel products more cheaply than anyone else. This culminated in the formation of US Steel, a merger of Carnegie and several other steel and steel product companies into one enormous, integrated producer. Its immense size gave US Steel very low production costs, although it failed to pass these savings on to the consumer.[293] Then, in the second half of the 20th century, large, integrated steel mills gave way to much smaller mini mills, which produced small batches of steel using a different technology: steel scrap melted in an electric arc furnace. Mini mills, in turn, often integrated downstream and began to produce other products like steel decking and metal buildings.[294]

Similar variations in integration occurred in other industries in this period. In general, as production and transportation technologies evolved over the 19th and early 20th centuries, there was a shift from small firms and independent contractors that coordinated via market mechanisms to larger, highly integrated firms that coordinated material flows internally.[295] Toward the end of the 20th century, with the rise of competition from low-priced foreign inputs enabled by low-cost labor, the opposite occurred, and US firms largely began to outsource aspects of production that weren't strategic or a core competency.[296]

As a result, between 1970 and 1997, the average size of a US firm shrank by 20 percent.[297]

## Changing the location

Another way to rearrange a production process to reduce the price of inputs is to change the location in which production takes place. The cost and availability of inputs are often a function of geography: Many things are more expensive or harder to obtain in some places than others because moving them from place to place is either impossible or incurs large transportation costs.

Labor is the classic example of this phenomenon. The cost of labor differs from place to place, and labor-intensive production operations will often be located where labor is cheaper.[298] Garment manufacturing, for example, has proven resistant to automation and remains a labor-intensive industry. As a result, garment manufacturing operations tend to continuously relocate to sources of low-cost labor (which are often, not unrelatedly, places with poor working conditions). By the late 19th century, New York City had become one of the largest garment manufacturing centers in the world, such that in 1890, New York made 44 percent of the ready-to-wear clothes produced in the US.[299] But by the 1920s, the industry had begun to move to where labor was cheaper still—first from Manhattan to Brooklyn and New Jersey, then to New England, then, as the interstate highway system developed, to the South, and, finally, overseas.[300] Similarly, Nike began importing footwear from Japan in the 1960s, then began manufacturing its own shoes in Japan in the 1970s. As labor costs in Japan rose, Nike shifted production operations to Taiwan and Korea, then China, then Vietnam and Indonesia.[301] There have been similar shifts in other labor-intensive industries, such as shipbuilding (which moved from the UK to Japan, then to South Korea, then to China), as well as toy manufacturing (which moved from the US to Japan, then to Hong Kong, then to China).[302]

The phenomenon of relocating operations to reduce input costs can also be observed on a smaller scale by looking at land costs in urban areas. In the late 19th and early 20th centuries, US cities became much more densely populated, and land within these cities became much more expensive. As transportation technology improved and electrification incentivized the use of wide, spread-out factories, manufacturing operations moved away from central urban areas to places where land costs were cheaper.[303]

Likewise, energy-intensive industries are continually moving to wherever electricity is cheapest. Iceland and Norway have disproportionately large aluminum smelting industries because aluminum smelting is electricity-intensive and electricity costs in Iceland and Norway are low. In Iceland, electricity costs roughly 30 percent less than in the US. As a result, Iceland smelts nearly as much aluminum as the US does, despite having roughly one nine-hundredth the population.[304]

Historically, the relative cost of inputs in different places has had a major influence on the type of production system that develops in a region. In the 19th-century United States, land was abundant and cheap while labor and capital were scarce. So the US developed agricultural practices that made best use of this relative abundance and scarcity, with early US farmers often eschewing yield-improving but labor-intensive practices like crop rotation and fertilizing and instead increasing output by cultivating more land.[305] Similarly, early US iron and steel production methods were tailored to local material availability. While steel manufacturers in England switched from charcoal (made from timber) to coke (made from coal) as a blast furnace fuel in the late 18th century, in part due to the increasing difficulty of obtaining timber supplies, America's vast forests enabled it to continue to use charcoal as a furnace fuel for much longer.[306] And US iron ores tended to be low in phosphorus, which made them more suitable to be refined into steel by the open-hearth process rather than the Bessemer process. As a result, the open-hearth process became the most popular steelmaking method in the US in the early 20th century.[307]

More broadly, manufacturers often select production locations to minimize transportation costs. Prior to the development of the railroad and the automobile, moving goods over land was incredibly expensive compared to moving them over water, so 19th-century factories were often located in port cities where materials were cheaper. Chauncey Jerome, a New England clockmaker who located his business in the port of New Haven in the 1840s, speculated that the availability of cheap fuel and other inputs allowed clocks to be made in New Haven for "less than one-half of what they could be made for in any part of [Europe]."[308]

We see similar forces at work today. In aluminum production, the first step is to refine mined bauxite into alumina (aluminum oxide), which then gets smelted into metallic aluminum. To minimize bauxite transportation costs, alumina refining is often located near bauxite mines. The bauxite is transported from the mine to the smelter on conveyor belts that can be up to 50 kilometers long.[309] Likewise, to minimize the cost of transporting mobile homes over long distances, mobile home manufacturers have many small production facilities spread throughout the US, each one serving customers within roughly a day's drive.[310]

### Increasing the value of outputs

When considering how to reduce input costs, we also have to pay attention to the other side of the ledger: the value of the outputs. Many production methods will generate by-products. Machining operations that cut metal into a particular shape also produce large volumes of metal shavings; iron and steel refining operations, such as blast furnaces and Bessemer converters, emit exhaust gases and leave behind waste products such as slag; crop production yields edible fruits and vegetables but also inedible plant parts such as stalks, roots, and leaves. These extra outputs need to be disposed of, and disposal contributes to the cost of the production process. If we can find a way to reduce or reuse these extra outputs, we can decrease or offset production costs.

Consider the manufacture of iron, for example. Iron is made by putting coke (coal reduced to a mass of nearly pure carbon), limestone, and iron ore in a large blast furnace. In the heat of the furnace, the iron ore (largely iron oxide) reacts with the carbon and other impurities to form a liquid, high-carbon iron known as pig iron. The impurities float on top as slag that is skimmed off. Over time, producers have reduced the cost of making iron by making better use of the by-products of the process. In the 1850s, British mechanical engineer Edward Alfred Cowper developed the Cowper stove, which used waste heat from the furnace exhaust gases to preheat the incoming air, decreasing the furnace's fuel consumption. Similar systems were developed in the late 1800s to reuse the exhaust from coke ovens, lowering the energy requirements for coke production.[311] Likewise, the slag produced by the blast furnace was originally considered a "nuisance of little intrinsic value."[312] In the US, railroads often hauled away the furnace slag for no charge, but in the early 20th century they began to charge for this service, inspiring steelmakers to find uses for it. They soon discovered that slag could be used as a cement alternative in concrete and asphalt, and by the 1940s, more than 60 percent of blast furnace slag was sold commercially.[313] As of 2019, the market for slag produced in US blast furnaces was valued at $470 million per year.[314] Similarly, paraffin, a by-product of coal, was originally viewed as worthless. But it was eventually found to be an excellent material for making candles, and low-priced paraffin candles went on to dominate the industry.[315]

Improved technology or process knowledge often makes it possible to increase the use of process by-products. For example, advances in chemistry enabled the creation of synthetic dyes from coal tar, a waste product of coal gas production.[316] And in the 19th-century cotton printing industry, a simple chemical treatment was devised to allow companies to recover additional dye from what had previously been harmful residual matter.[317]

In some cases, the waste output even becomes more valuable than the primary output. The Leblanc process used to produce soda ash in the 19th century emitted hydrogen chloride gas as a by-product, which was vented directly into the atmosphere or dissolved in water and dumped into rivers. But in the 1860s, chemists developed a process to convert hydrogen chloride into valuable bleaching powder. As the process for producing soda ash became more efficient, this bleaching powder became the primary product to come out of the Leblanc process.[318] Gasoline, too, was considered a waste product in the late 19th century and was thrown away during the production of kerosene. But following the development of the automobile, gasoline became much more valuable than kerosene.[319]

When calculating the cost and value of waste products, the other considerations we've discussed in this chapter play a key role. For example, since by-products are often bulky and impractical to transport very far, their value may be a function of how vertically integrated the company is, since vertically integrated operations may be better able to make immediate use of them.

### Trade-offs and coupling

Even though there are well-understood ways to reduce the cost of a bundle of inputs, choosing the cheapest one can be a thorny problem because numerous trade-offs exist between different production processes. As we saw in Chapter 2, a palette of different production technologies is typically available for any given operation, and each will have its own advantages and drawbacks.

Consider the Bessemer converter versus the open-hearth furnace, the two primary methods for making steel at the end of the 19th century. The Bessemer process was faster and required no fuel as an input, but it was limited in the kinds of ores it could use and the amount of scrap steel that could be added to the mix. In contrast, the open-hearth process was much more flexible. The Bessemer process also exposed the molten steel to much more air than the open-hearth furnace, resulting in more nitrogen impurities and lower-quality steel. Which of these processes was considered superior would depend, among other things, on the availability of steel scrap, the relative prices of various ores, and what the steel would ultimately be used for. Bessemer steel initially dominated for steel rail production, but open-hearth steel began to displace it for other uses such as axles and bridge components, which demanded a quality of steel that Bessemer converters often struggled to produce.[320]

We can observe similar trade-offs in other processes. For example, low-volume cutting operations are often done via water-jet cutting or laser cutting. Water-jet cutting uses a stream of high-pressure water, often mixed with an abrasive, whereas laser cutting uses a beam of light that melts or vaporizes the material. Water-jet cutting is less accurate and requires more cleanup, but it can cut thicker materials and doesn't heat up the part being cut, which is important in applications like aerospace manufacturing. Laser cutting is more accurate and doesn't put any force on the parts being cut, making it easier to use on smaller parts, but it's limited in the thickness of material it can cut.[321]

Often, these trade-offs appear as differences in up-front costs and marginal, or per-unit, costs. Many production processes have high up-front costs but low per-unit costs, and thus become less expensive than other methods with high enough production volumes. Glass bottles, for instance, are sometimes still manufactured using methods similar to Edison's early bulb blanks: A glass blower will manually blow glass into the proper shape, sometimes with the aid of a mold. But very high-volume glass bottle production uses industrial machinery that automatically pours molten glass into a mold and then blows it into shape. This machinery has very low unit costs, but the up-front costs involved in designing and fabricating the molds and other equipment are enormous. Hand blowing, on the other hand, has high unit costs, since each bottle takes a comparatively large amount of manual labor, but very low up-front costs, since it doesn't require any complex or expensive equipment. So hand blowing continues to be used for small-volume goods such as art pieces or custom tableware, while machine blowing is used for large-volume goods like wine bottles.[322]

Different methods also often trade flexibility for speed or cost. A dedicated, special-purpose machine will often be faster and have lower unit costs than a general-purpose machine, but a general-purpose machine can be used for a wider variety of operations. In the early 20th century, Ford achieved high throughput in Model T production by using a large number of special-purpose machine tools, often designed by Ford engineers. These kept production costs low and could be operated by relatively unskilled workers, but they also limited the plants' flexibility; recall that adapting the Rouge for Model A production required shutting the facility down for six months for retooling. Another, more modern example is the 3D printer, which has higher unit costs than injection molding but can be adapted to produce different parts at very little cost, because a new part design can be created using software and fed directly to the machine. Injection molding has lower unit costs, but each new part design requires the creation of an expensive new mold. For this reason, 3D printing is much more useful in prototyping work, where the design of the part or product is likely to change frequently.

Similar trade-offs exist when considering the manufacturability of a product. As we discussed earlier, designing a product often involves a trade-off between ease of manufacture and ease of assembly. A design with a larger number of simple parts might be easier and cheaper to manufacture than a design with fewer, more complex parts, but it will likely also have much higher assembly costs. Early efforts at creating manufacturability guidelines sometimes failed to consider assembly costs, so the recommendations would increase rather than decrease production costs by recommending simple-to-make but difficult-to-assemble part designs.[323] Likewise, ease of production at one step may require complicated, expensive operations at a different step. For example, early auto factory engineers sometimes minimized their machining costs by specifying very tight tolerances for raw material, but the added expense of obtaining the correct raw material wound up more than offsetting the savings from the reduced machining.[324]

Because of these trade-offs, cost minimization can't be done by considering each item in isolation. Achieving the lowest overall cost for the process often requires that certain portions be made more expensive.[325] For instance, using a higher quality but more expensive material may decrease overall costs if the increased quality reduces downtime in the factory caused by delays from having to rework the product. One automotive factory opted to produce flywheels from expensive steel rather than cheaper cast iron because the cast-iron flywheels often caused delays due to material irregularities.[326] Another factory opted to add a seemingly unnecessary machining operation to a steel pin to make it symmetrical. The additional machining expense was more than offset by the reduced labor cost, because the symmetrical pin could be inserted without needing to be examined first.[327] And, of course, many cost-saving efforts, such as automating labor or installing equipment to recover waste material, require up-front capital expenditures that only pay for themselves over time and are, therefore, only cost-saving in the context of a particular production volume.

More broadly, there will typically be various degrees of coupling between different steps in the production process, as well as in the design of the item being produced. Decisions made at one step of the process determine the available options and performance of other steps in the process. In the case of cast-iron manufacturing, the choice of fuel used in the blast furnace affected the silicon content of the iron,[328] which affected the material properties of the iron, which affected the type of mold that could be used for casting. In the early 1700s, British ironmaster Abraham Darby was able to produce cast-iron pots much more cheaply by changing his blast furnaces to burn coke instead of charcoal, which yielded a higher-silicon iron that could safely be cooled much more quickly, allowing for the use of cheaper unheated molds in lieu of more expensive heated ones.[329] And in cut-nail manufacturing, whether or not the nail plate needed to be heated before being cut depended on the power applied to the machine, which depended on the power source used; steam could provide more power than waterwheels or manual labor. This coupling means that the space being explored to find the least expensive recipe isn't fixed but changes based on other decisions made about the process.

In some cases, this coupling is somewhat loose. For a product made of modular components that only connect to each other through well-defined and well-understood interfaces, the coupling is limited to the interface and the overall physical geometry of the component. Changes that don't affect those aspects can be made somewhat freely.

For example, modern light bulbs with different light-generating mechanisms (say, incandescents versus LEDs) can be used in the same lamps, so long as they have the same shape and use the same screw base. In other cases, the coupling is extremely tight, and even a slight process change can have a substantial impact on the rest of the process. In pharmaceutical manufacturing, even a single atom's difference in output will change the drug being produced. Similarly, in semiconductor manufacturing, the slightest modification to a process can introduce impurities or cause process failures, irreparably damaging the chips being produced.[330]

## Technological development and explore-exploit

In this chapter, we've discussed the idea of exploring the landscape of possible options to find the cheapest bundle of inputs. But, as we discussed in the previous chapter, new technology—that is, new production methods and processes—can often perform an operation, or a functionally similar one, using fewer resources. This can be thought of as expanding the available landscape of possible options.

These two ideas represent an explore-exploit dynamic: Do we explore new territory by trying to develop new technology, or exploit existing territory by trying to make the best use of the technology, materials, and systems we already have?

Explore-exploit is generally thought of as a binary choice representing two divergent paths. But the distinction between the two is somewhat fuzzy. Instead, we can think of these different approaches as existing on a single spectrum of how well we understand the territory we're operating in.[331]

On one end, we have a well-understood and well-characterized landscape. At this end of the spectrum, for a given set of materials and technology, practitioners have a strong understanding of performance, the domain they operate in, best practices for use, and interactions with other systems, so their behavior is highly predictable. For example, to minimize the cost of constructing the Empire State Building, engineers developed a somewhat novel system of steel framing.[332] An extra steel beam was attached to the edge of the building, outside the floor beams, and used to directly support the limestone cladding. This was a departure from the more common method of cladding support, which used steel angles attached to the floor beams. The extra beam allowed construction to proceed much faster because the steel could be designed without having to account for the specific design of the cladding, and the cladding panels could be set without having to install and adjust steel angles, a time-consuming process. Well-understood materials (steel beams) were being used in novel ways (a unique arrangement to support the cladding) but the result was largely predictable, and the engineers didn't need to learn anything new about the behavior of steel beams or limestone panels to make this system possible. Making innovative use of standardized, well-understood materials enabled the Empire State Building to be built at unprecedented speed, taking less than a year to construct what was, at the time, the tallest building in the world.

On the other end of the spectrum, there's the unknown landscape of new technological development, where the behavior and performance of systems or materials isn't known and interactions between different elements are difficult to predict, often requiring a painstaking process of trial and error. Consider Edison's light bulb. Its development required exploring many different materials and how they would respond to electrical current flowing through them. Edison's lab ran thousands of experiments on hundreds of potential filament materials. Over the course of the development process, Edison and his colleagues discovered how the elements would behave: The filament in a vacuum would emit gas; inert gas in the bulb would cause the filament to cool; a heated filament would emit electrons. All these phenomena needed to be understood before a successful bulb could be created.

Between these two extremes are shades of variation. We may understand certain aspects of an element's performance but not others. Or we may be able to partially predict an element's behavior but still encounter a great deal of unexplained variation. We may be exploiting existing technologies that haven't been used together before, which might have unexpected interactions, or we may be taking an existing technology and pushing it beyond its current performance limit.

Consider, for example, the development of flush riveting in aircraft manufacturing in the 1930s. Riveting, which consists of driving a metal pin through two sheets of metal, was used to fasten together metal panels to create the body of an aircraft. Prior to the 1930s, the head of the rivet was left protruding above the surface of the metal, which caused extra drag and a loss of fuel efficiency. In the 1930s, aircraft manufacturers began to develop flush riveting, in which the rivet head was flush with the surrounding surface, to reduce drag. Flush riveting required placing a small divot, or dimple, in the metal sheets, into which the rivet head would sit.[333] But even though the ideas behind flush riveting were either well known or conceptually obvious and the process used existing materials and technology, getting it to work was complicated. As one aircraft engineer put it, "At first glance, nothing seems simpler than putting a dimple into a sheet of metal, yet it has proved to be no simple problem for the best engineering talent in the industry."[334] Successfully and reliably flush riveting two sheets of metal required understanding how the metal would be affected by the size of the hole, the thickness and angle of the head, the shape of the dimpling die, how hard the die was pressed, the composition of the metal used, and the orientation of the joint. This understanding had to be worked out through trial and error, and different manufacturers ultimately developed different methods of flush riveting.

A collection of well-understood mechanical movements can be combined in an infinite number of ways to make an infinite number of machines, much like a fixed collection of letters can be combined to form an infinite number of texts. Also like text, the possibility space of potential machines is so vast that finding a valuable place within it—a machine that does something useful—is difficult.[335] The individual elements may be known, but understanding each one individually doesn't mean we understand how a large collection of them working together will behave. For that, some exploration is required. Engineers, designers, and production workers gradually explore this vast landscape of partially understood possibilities as they make small tweaks and push technologies and processes to work in slightly new ways.

Technological development often involves exploring the adjacent possible: the set of possibilities outside but near current possibilities. Over time, the adjacent possible grows ever larger as people create new elements and combine them in new ways. There are hundreds of mechanical movements, thousands of materials, and millions of products that can be combined in as-yet unexplored ways.

### A dynamic landscape
The ever-changing technological landscape means that the lowest-cost production method today might not be the lowest-cost method tomorrow. New processes are constantly being developed, and older ones are getting cheaper. In the mid-20th century, the ability to manufacture liquid oxygen on an industrial scale opened the door for a

modification of the Bessemer process for steelmaking: injecting pure oxygen rather than air into the steel. This basic oxygen process went on to replace the previously dominant open-hearth process. In 1960, the open-hearth process produced nearly 90 percent of steel in the US, but by 1970, that share had fallen to 36 percent, with 48 percent made by the basic oxygen process.[336] The basic oxygen process, in turn, was challenged by electric arc steelmaking, which had been steadily improving since its invention in the late 1800s. In 1970, electric arc furnaces produced around 15 percent of the steel in the US. Today, they produce more than 70 percent.[337]

Changing technology will sometimes make an input that was once expensive cheap and vice versa. Wire-nail machines had been used in the US since the early 1800s, but wire nails didn't displace cut nails until the Bessemer process made steel wire cheap. The automobile made gasoline, once a worthless waste product, one of the most valuable petroleum products, and oil companies would spend many years and millions of dollars developing cracking technology to produce more of it from crude oil. The landscape of production methods is constantly being reshaped by Schumpeter's gale of creative destruction, the "process of industrial mutation that continuously revolutionizes the economic structure from within, incessantly destroying the old one, incessantly creating a new one."[338] Even raw materials taken directly from the earth aren't immune from this shift. In the early 2000s, the structural capacities of southern pine lumber had to be downgraded because modern tree farms grow trees faster, making them less strong at harvest than trees had been historically.[339]

Forces beyond technology are reshaping this landscape, too. Changing consumer tastes, for example, might require new ways of organizing production. Ford's vertical integration worked in the auto market of the 1910s but was less well suited to the market of the 1920s. In other cases, shifting political winds might impact costs or upheave the competitive landscape. In the 1920s and '30s, many rural US farms used small wind turbines to generate electricity, a market that was completely eliminated when the Rural Electrification Administration built transmission lines to rural areas that delivered low-cost electricity.[340] Fifty years later, thousands of new wind turbines were built following a spate of government subsidies that made constructing wind farms an attractive prospect for investors, resulting in an enormous drop in the cost of wind-generated electricity as manufacturers fell down the learning curve.[341] (We'll talk about learning curves in Chapter 7.) On the other hand, pollution controls on coal plants resulting from environmental legislation enacted in the 1970s greatly increased the cost of coal-generated electricity.[342] No part of the production landscape stays still for long.

The challenge, then, is that finding the lowest-cost bundle of inputs requires some degree of predicting the future. Every cost is really an expected cost predicated on certain assumptions about the future—that certain materials will continue to be available, that a certain level of demand will continue to exist, that competitor behavior will continue along a certain trajectory. The low-cost option might lock us into a particular course of action, whereas a higher-cost option might provide more flexibility. Spending money on a high-volume, low-unit-cost production operation like a new injection molding might be the cheapest fabrication method, but only if our projections for product demand are correct. Minimizing inventory levels might reduce costs, but only if we can ensure reliable supply. After the Fukushima nuclear disaster in Japan, Toyota, famously known for minimizing inventory using just-in-time production, opted to stock an inventory of certain critical components, including semiconductors. As a result, the company was able to sustain production much more successfully than other automakers in 2020, when supply chains frayed during Covid-19.[343] When we acknowledge that the future is uncertain, minimizing expected costs in the long run might look different than simply choosing what's cheapest today.

To better understand how strategies like DFMA can reduce production costs and how they intersect with the shifting landscape of production technology, let's look at a specific example: Tesla's move from steel stampings to large aluminum castings.

*Case study: Tesla and large castings*
Historically, car bodies have been made from thin sheets of stamped steel, bent into shape by machines and welded together to form what's known as the body in white—the structural frame of the car that the rest of the components (engine, subassemblies, trim) attach to. This method creates a strong, light structure, but it also requires hundreds of individual parts, each with its own separate production process, and thousands of joining operations to attach them together.

When electric car manufacturer Tesla began to produce its Model 3 in 2017, it followed existing manufacturing practice and built the body in white from many small stamped-steel parts (along with other materials such as aluminum). But since then it has gradually developed an alternative method of car construction, using large aluminum castings to replace dozens or hundreds of stamped-steel parts. The front and rear underbody of the Model Y, introduced after the Model 3, are made up of two large castings.[344] Later versions of the Model 3 were also redesigned to use large aluminum castings instead of steel stampings.[345]

By redesigning car bodies to use large pieces of cast aluminum, Tesla eliminated hundreds of parts, along with the machines to produce them and the robots and labor to attach them.[346] According to one estimate, replacing stamped-steel parts with large castings could reduce labor costs by 65 percent and factory floor space by nearly 50 percent. Large castings also generate significantly less waste metal than steel stampings, because excess aluminum can simply be placed back into the furnace, melted, and reused.[347] Altogether, the redesign is estimated to have reduced the cost of producing a car body by 20 to 40 percent. Many other automakers, including Volvo, Mercedes, and Volkswagen, have since begun investigating using large aluminum castings in their car bodies.[348]

The new cast-aluminum elements were different from the stamped-steel parts they replaced. A stamped-steel body is made from many thin pieces of steel attached together, but aluminum casting (more specifically, high-pressure die casting, or HPDC) struggles to produce very thin parts.[349] Whereas a stamped-steel part can be as thin as 0.7 millimeters, cast-aluminum parts can't be much thinner than 2 millimeters; any thinner, and the molten aluminum will have trouble flowing through the casting mold before it solidifies.[350] More generally, a cast-aluminum part must be designed so that it can be produced by rapidly pouring molten aluminum into a metal die and cooling it quickly. Smooth corners are preferable to sharp angles; large areas of solid metal should be eliminated because they might cause voids in the center as the metal cools; blind walls where the flow of metal needs to reverse course are best avoided.[351] Tesla's cast parts were specifically shaped so the molten aluminum flowed smoothly, "like a river."[352] Cast-aluminum parts must also be designed to be easily ejected from the steel mold, and design elements like undercuts, where the part is wider at the top of the mold than at the bottom, must be avoided.[353]

Tesla's large aluminum castings were made possible thanks to several technical developments that occurred in parallel. The most notable was the creation of Giga Presses, large HPDC machines that had nearly twice the capacity of previous casting machines.[354] Prior to the development of the Giga Press in 2018, aluminum castings were limited to around 100 pounds of metal. In contrast, Giga Presses could handle nearly 200 pounds.[355] Doubling the capacity of existing machines required innovations to the HPDC process, such as a new metal injection system capable of moving the large amount of molten metal quickly into the die before it cooled.[356]

Tesla wasn't the first car manufacturer to use cast-aluminum parts, but historically such parts had been much smaller in size. Other car manufacturers were often reluctant to increase the size of their castings because of the time and effort required to develop the dies. A die might require millions of dollars to produce and might need to go through many iterations as the company refined the part design. Tesla, however, was able to rapidly and cheaply iterate on its casting designs by making test molds with binder jetting, a type of 3D printing that builds up the molds using thin layers of sand. This allowed new molds to be made in a matter of hours and reduced the time it took to validate a die, or to ensure the part being made was correct and met requirements, by up to 75 percent.[357] To further reduce the costs of using cast aluminum, Tesla developed a novel aluminum alloy that was extremely strong, had good castability, and didn't require heat-treating or additional coatings, cutting out more production steps and potential sources of error.[358]

Tesla's use of aluminum castings illustrates the basic benefits of DFMA and similar cost reduction strategies: By redesigning a product to use a different production technology (in this case, large aluminum castings instead of steel stampings), a manufacturer can dramatically lower its production costs. Fewer parts means fewer assembly operations and eliminates the labor, factory space, and other inputs those operations require. And while some of Tesla's input costs increased (because aluminum is more expensive than steel), costs still fell on balance. Making this change required modifying the design of the car body; the new cast components weren't identical to the stamped-steel parts they eliminated, but they were functionally similar. Demonstrating similarity, however, was far from straightforward. Crash tests were required to ensure the castings would perform properly in a vehicle collision, for example.[359] And the car bodies weren't the only thing that needed to be redesigned—using Giga Presses also required changing the production facilities, adding heavy foundations, and ensuring facilities had high enough roofs to fit the machines.[360] The challenges of adapting existing facilities to use Giga Presses is one possible reason why Tesla has been faster to adopt them than other manufacturers.[361]

This cost-reducing redesign was made possible by the constantly evolving landscape of production technology. While HPDC had been in use for decades, it was only with the development of large die casting presses, 3D printing for rapid mold iteration, and even things like faster computers and better simulation technology that Tesla was able to push aluminum casting into previously unexplored areas of size and function. Some of these explorations, such as the development of new aluminum alloys and testing to ensure crash safety, were instigated by Tesla itself, while others were driven by broader technological developments. Often, production processes move forward not from a single technological advancement but from a combination of many separate improvements.

## Summary

Society often celebrates inventors and inventions—the breakthroughs that unlock new possibilities and change the way we live. Many more people know about Alexander Fleming's discovery of penicillin than the subsequent painstaking efforts to scale up its manufacture, and the telephone is more closely associated with Alexander Graham Bell's 1876 invention than with the enormous capital investments by AT&T and other companies that made using it practical.[362] Inventors are like explorers, finding previously unknown locations on the map of technological possibility, and they're celebrated as such.

As we saw in the last chapter, inventions and new technologies are crucial mechanisms for improving efficiency. But what comes after the invention is often as important as the invention itself, if not more so. Finding an undiscovered location on the map of technological possibility is important, but it's only useful if we can find an efficient way of getting there—if we can develop a sequence of steps that can realize the new machine, chemical, or drug at an affordable cost. And just because we've found one successful route doesn't mean there aren't better, faster, and cheaper routes out there.

As this chapter outlines, complex manufactured goods exist in enormous, multidimensional landscapes with an effectively infinite number of possible ways of producing them. What's more, this landscape is constantly changing. As technology creates new possibilities, market demand shifts and the cost of inputs rise and fall. Because of this, there's immense opportunity in finding new, better routes through this landscape—cheaper ways of making the same product—and locating them is a crucial tool for efficiency improvement.

# 4

# Production Rate and Economies of Scale

One common phenomenon in production processes is that as the production rate rises, production costs fall. The more you make of something, the cheaper making it becomes.

Take iron production. Iron is made by heating a mixture of iron ore and carbon (such as charcoal derived from wood or coke derived from coal) in a blast furnace, producing a liquid iron known as pig iron.[363] Though the blast furnace has been in use in Europe for more than 800 years, modern blast furnaces are far bigger and produce far more iron than historic furnaces. A blast furnace in 16th-century England was a truncated stone pyramid perhaps 20 feet high, capable of making 250 to 300 tons of pig iron a year.[364] By contrast, a modern blast furnace is a steel shaft more than 100 feet high capable of making more than 3 million tons of pig iron a year—10,000 times as much as a medieval furnace.[365] This surge in size and output, most of which occurred during the late 19th and early 20th centuries, resulted in a dramatic decrease in the cost of producing iron. Between 1870 and 1910, the price of pig iron in the US declined by roughly 40 percent in real terms, in part due to the rising scale of blast furnace operation.[366] In the second half of the 20th century, increasing blast furnace size was the single biggest factor in the improving economics of iron making.[367]

Relationships between decreasing production costs and increasing production volume are known as economies of scale, and they have historically been one of the most important mechanisms behind falling production costs. As business historian Alfred Chandler Jr. notes in *Scale and Scope,* in the late 19th century producers of everything from smelted metals and light machinery to refined chemicals and packaged food and drink all greatly reduced their production costs by building large, high-volume facilities that had sizable economies of scale.[368] More recent studies have confirmed the ubiquity and importance of economies of scale. A 1967 study of 221 industrial plants across 14 different industries found economies of scale in more than 80 percent of them, while a 1992 study of 147 Canadian manufacturing industries found economies of scale in roughly 60 to 70 percent of them. Another study of seven US manufacturing industries found that within each industry labor costs at the biggest plants were generally 60 to 90 percent lower than at the smallest plants, and that overall costs in the smallest plants were two to seven times as high as in the largest plants.[369]

Economies of scale are often discussed in terms of a scaling exponent, where production costs are modeled by the equation $C \sim x^{\wedge}b$. According to this equation, unit costs ($C$) are proportional to total production volume ($x$) raised to some exponent ($b$).[370] If $b$ is less than 1, production has increasing returns to scale, and unit costs fall as production rises. If $b$ is 1, there are no economies of scale, and the 10th unit costs just as much as the 10,000th to produce. If $b$ is greater than 1, production has decreasing returns to scale, or *dis*economies of scale, and unit costs rise as production rises.

What factors and contexts determine this relationship? In this chapter, we'll dig into the mechanisms that produce economies of scale. We will also discuss why, despite their importance, it's sometimes best to try to eliminate scale effects.

## Mechanisms behind economies of scale

There are, broadly speaking, five potential mechanisms that can cause production costs to fall as production volume increases: fixed cost spreading, geometric scaling, statistical scaling, influence scaling, and learning curve effects. There is also one closely related mechanism—network effects—that, while not itself an economy of scale, facilitates such economies. Let's take a look at each in turn.

### *Fixed cost spreading*

Fixed cost spreading occurs when a production process has some costs that are variable (that is, they rise or fall with production output) and some costs that are fixed (that is, they're largely independent of production volume). The greater the production volume, the more thinly the fixed costs are spread across each good, causing per-unit costs to fall.

Consider iron production in the blast furnace. Inputs like iron ore, coke, and limestone for flux are all variable costs; doubling the amount of pig iron produced requires twice as much of each. However, other costs of the operation are fixed. The land the furnace sits on doesn't cost any more if a furnace produces more iron. Similarly, the costs of much of the furnace structure itself—the foundations, the steel shell—are independent of production volume. (The brick furnace lining, on the other hand, has a lifespan proportional to the amount of iron the furnace produces and is more properly considered a variable cost.)[371] And in a modern, highly mechanized blast furnace, much of the labor to operate it is also close to a fixed cost, because furnace output can often be increased significantly without requiring any additional labor. For example, when a 1923 mechanized blast furnace was rebuilt to double its production volume, the per-unit labor costs fell by 50 percent.[372]

We can also observe fixed cost spreading in the transportation of goods via container shipping. Fuel for the ship is a variable cost that increases with the size of the ship and the distance it travels. The cost of loading and unloading the ship is similarly variable; the more containers a ship is carrying, the more it will cost to unload. The cost of the ship itself, however, is a fixed cost for the ship owner. Paying for the ship is independent of how much cargo it carries or how many trips it takes, just like a car loan payment is the same regardless of how much a person drives the car (although, of course, different-sized ships will cost different amounts to build). The more cargo the ship transports, the more the fixed costs of the ship's construction are spread out.[373]

What are sometimes referred to as economies of scope are also often a type of fixed cost scaling. Economies of scope occur when manufacturers reduce unit costs by increasing the variety of goods a process produces. In the late 19th century, German dye makers decreased their unit costs by using large plants to produce hundreds or thousands of different dyes from the same raw materials and intermediate chemical compounds.[374] By producing more dyes in a single plant, the fixed costs of production could be spread over a larger volume of output (although this strategy likely also exploited other scaling mechanisms, such as geometric scaling, which we'll discuss shortly). Likewise, economies of speed—increasing the production rate of a piece of equipment or facility—are also a form of fixed cost scaling. The more a given piece of equipment or facility can produce in a given amount of time, the more goods its fixed costs can be spread over, lowering unit costs.

Fixed cost effects can be found at many different levels of a production process, from the actions of an individual worker to the operation of an entire factory. In a blast furnace, for example, fixed cost spreading occurs at the plant level: Increasing furnace output will spread the fixed costs of the entire operation over more iron produced, lowering unit costs. But it can also be found at smaller scales. Before furnace operations were mechanized, getting pig iron out of the furnace (a process called tapping) was a labor-intensive procedure that required eight to 10 men to drill a hole in the side of the furnace.[375] Because the time and effort required to drill the hole was independent of how much iron was drained, drilling the tap hole was effectively a fixed cost. The more iron that could be drained on each tap, the more thinly that fixed cost could be spread, and the lower the per-unit cost of iron making would be.

The same phenomenon is at work in the transportation of goods in large container ships. Fixed cost spreading occurs at the level of the entire ship, where it's advantageous to spread the fixed costs of the ship itself over as much cargo and as many voyages as possible. But we also see fixed cost spreading at the level of an individual container. It takes roughly the same amount of time and effort to handle a container regardless of how full it is, and shipping rates for containerized cargo are generally per container.[376] The greater the number of goods over which you can spread the fixed costs of handling the container itself—that is, the more cargo you can fit into it—the lower your per-unit transportation costs. Shippers, therefore, have an incentive to stuff their containers as full as possible.

We can get even more granular. Consider a factory worker who uses a variety of tools to do their job—wrenches, hammers, screwdrivers, and so on. The time it takes the worker to grab a tool and position themselves to use it is a fixed cost—that is, it's independent of how long they use the tool for. The more the fixed costs of grabbing the tool can be spread out—the longer the worker uses the tool without having to switch to another—the lower the labor costs of a particular task will be. This sort of cost spreading is part of the reason why increasing the division of labor often improves production efficiency. Dividing an operation into many separate tasks, each performed by an individual worker, means that each worker spends less time switching between tasks, so the fixed cost of any given task (for example, the time it takes to get the tools and get in position) is more thinly spread. More generally, fixed costs often take the form of setups, or operations that must be done at the beginning of a process before it can start producing. The greater the number of goods over which the fixed costs can be spread—that is, the longer a process can run uninterrupted—the lower the unit costs will be. (This sort of cost spreading by avoiding setups can also be thought of in terms of eliminating a step in the process, which we'll discuss in Chapter 5.)

Different production technologies will vary in their relative proportions of fixed and variable costs, even for processes that produce similar outputs. For example, as discussed in Chapter 3, injection molding has high fixed costs (due to the time and effort required to make the mold) and low variable costs. By contrast, stereolithography, a type of 3D printing, has much lower fixed costs: Instead of a physical mold, all that's needed is a digital file of the part. Its variable costs, however, are much higher than injection molding. For that reason, injection molding is most suitable for very large production volumes where the fixed costs of mold making can be widely spread, while stereolithography is much more suitable for low production volumes, such as prototyping.[377]

We can see similar variation in fixed and variable costs in other types of production, such as electricity generation. Nuclear power plants have high fixed costs due to the time and effort required to construct the plant and its various safety features. But their variable costs, such as fuel and maintenance, are comparatively low. Natural-gas turbine plants, on the other hand, are much cheaper to build than nuclear plants, giving them lower fixed costs. But their variable costs, primarily the cost of the gas itself, are higher.[378] These different cost structures influence how different power plants get used. Because electricity demand fluctuates over time, plants are dispatched, or turned on and off, based on their variable costs of production. Low-variable-cost plants, such as nuclear plants, run constantly to supply baseload power, while high-variable-cost plants, such as natural gas turbines, are turned off and on frequently as demand rises and falls.[379]

Before moving on, it's important to note that fixed costs are often fixed only over a certain range. A 40-foot shipping container, for example, has a capacity of just under 2,400 cubic feet. Up to that volume, the costs of shipping via container are relatively fixed; shipping 400 cubic feet of goods will cost basically the same as shipping 800 cubic feet because everything will still fit in one container. But going beyond 2,400 cubic feet requires a second container, which will cause shipping costs to jump significantly. The limited range of fixed costs often creates a series of cost cliffs, where costs suddenly jump when the capacity of some process, such as a machine's output or a container's volume, is exceeded and more equipment needs to be purchased.

*Geometric scaling*
The second mechanism behind economies of scale is geometric scaling. Geometric scaling occurs when making production equipment physically larger causes costs to fall. In container shipping, for

example, a larger ship is cheaper to build on a per-container basis than a smaller one. One 2016 study estimated that the per-container cost for a ship with a capacity of 12,000 containers was roughly half that of a ship with a capacity of 1,500 containers.[380] Similarly, a 1967 study of the iron and steel industry found that a 50,000-ton blast furnace was just 53 percent more expensive to build than a smaller 25,000-ton furnace, reducing capital costs per ton of iron by 23 percent.[381]

Geometric scaling is typically the result of area-volume relationships. The volume of a sphere rises in proportion to its radius cubed, while its surface area rises in proportion to its radius squared. Doubling the radius of a sphere will, therefore, increase the volume much more than it increases the surface area. (This is known as the square cube law.) A similar relationship holds for any hollow shape: cylinders, cubes, boxes.

Industrial equipment often consists of tanks, pipes, and containers that can be thought of as a collection of hollow shapes. The capacity of such equipment often rises with its volume (how much material it can hold), while its cost rises with its surface area (how much material is required to enclose that volume). As a result, higher-capacity equipment requires less material to build per unit of capacity, and thus has proportionally less cost than lower-capacity equipment. Of course, real-world material relationships are more complicated than this simplification—doubling the size of a container, for example, might require thicker walls, more supports, or other additional material beyond just the increase in surface area—but it offers a reasonable model of how costs fall as equipment gets larger.

Similar geometric relationships exist in container shipping. A ship's cargo capacity is a function of its volume, while the cost to build the ship is closer to a function of its surface area.[382] Larger container ships also require proportionally less fuel than smaller ships.[383]

This kind of geometric scaling occurs across a wide variety of production equipment, including trucks and aircraft, steam turbines and wind turbines for generating electrical power, engines and motors, and chemical and petroleum processing plants.[384] Cost estimates for processing plants and industrial equipment often follow a six-tenths rule, where the scaling exponent is assumed to be 0.6: Each doubling of size increases total costs by $2^{0.6}$, which works out to unit costs falling by around 25 percent for every doubling of capacity.[385] In his textbook *Chemical Engineering Economics*, Donald Garrett lists

| CONTAINER SIZE (TEUs) | COST/TEU |
|---|---|
| 0–999 | $ 23,065.11 |
| 1,000–1,499 | $ 20,606.62 |
| 1,500–1,999 | $ 19,215.59 |
| 2,000–2,999 | $ 16,436.43 |
| 3,000–3,999 | $ 16,255.45 |
| 4,000–5,099 | $ 14,672.54 |
| 5,100–7,499 | $ 13,912.16 |
| 7,500–9,999 | $ 11,491.36 |
| 10,000–13,299 | $ 11,234.63 |
| 13,300+ | $ 9,298.82 |

**Table 1.** Cost per TEU (twenty-foot equivalent unit, the standard unit of measurement for cargo containers) of different-sized container ships. Reproduced from Murray, "Container Ship Costs."

scaling exponents for 50 types of chemical equipment and nearly 250 types of process plants, which have an average scaling exponent of 0.65 for the former and 0.67 for the latter.[386] These scaling exponent values are, unsurprisingly, almost identical to the values you get if you assume costs rise in proportion to surface area and capacity rises in proportion to volume.[387]

The existence of geometric effects often incentivizes producers to make production equipment as large as possible to minimize unit costs. However, this typically isn't as simple as making every part of the equipment twice as big. As a piece of equipment increases in size, the way it operates or performs tends to change: Fluid or material might flow through it differently, or reaction conditions might differ. As we saw in Chapter 3, the small, bench-scale equipment used for producing early samples of pharmaceuticals often behaves significantly differently than the larger, production-scale equipment used to produce drugs for the mass market. Similarly, in the 1960s, large coal power stations were found to behave very differently from their smaller counterparts. One utility executive complained, "We hoped the new machines would run just like the old ones we're familiar with. However, they sure as hell don't."[388]

This means that scaling up production equipment often requires redesigning it.[389] Nuclear power plants, for example, required significantly different safety systems as they got bigger. One potential failure mode of a nuclear plant is a core meltdown, a loss of coolant that causes the nuclear fuel to become hotter and hotter until it melts through the reactor itself. Early nuclear plants were small enough that even if a core meltdown occurred, the molten nuclear fuel wouldn't escape the containment building, a large concrete structure surrounding the reactor designed to prevent the escape of radioactive material. But as nuclear plants got bigger, they reached the point where molten fuel was hot enough that it could melt through the containment structure itself. This required a radical redesign of emergency cooling systems to ensure they would be operational even in the worst-case scenarios.[390]

Because the behavior of production equipment tends to change as it's scaled up, the best approach is often to make slow increases in size, solving new problems as they emerge. Nineteenth-century blast furnaces, for example, were enlarged gradually to ensure they wouldn't behave much differently from existing furnaces.[391] And when wind power began to be heavily developed in the 1970s and '80s, efforts to immediately build large turbines failed, as the large turbines proved to be unreliable. Instead, larger turbine sizes had to be achieved gradually, by slowly increasing turbine size over time.[392]

*Statistical scaling*
Statistical scaling occurs when a larger production volume is better able to damp the negative effects of random variation in output.

One example of this sort of scaling can be seen in electric power generation. Because electricity can't be easily stored and usage varies over time, a power plant must be able to accommodate the highest peak demand it might experience. This peak might be much higher than the average level of electricity demand, which means that a power plant sized for peak demand will spend much of its time with extra generating capacity. The more customers a power plant serves and the greater the variation in their electricity usage, the more these peaks will average out, requiring less extra capacity to service the demand. In the early 20th century, this sort of scaling drove electrical utilities to expand their customer networks as much as possible.[393]

Backups, such as spare parts or emergency supplies, are another common example of statistical scaling. A small factory that produces parts using a single machine will need a certain number of spare parts on hand to keep the machine operational. A larger factory with a greater number of machines will require proportionately fewer spare parts because it's unlikely that every machine will break down simultaneously. In his memoir, *Sixty Years with Men and Machines,* machinist and journalist Fred Colvin describes the proliferation of spare parts in a railroad machine shop that resulted from there being no standard locomotive type:

> "That there stock pile," said the superintendent, "is my biggest headache in twenty years of railroading. And there's nothing I can do about it. We have over fifty different types of locomotives running on this road, and I have to carry spare parts for each type. Maybe there's only one or two engines of a certain make out of the thousand on the line, but I gotta carry a complete stock of spare parts for them just the same... Half a million dollars worth of spare parts just because we got fifty different kinds of loco-motives running."[394]

Other examples of statistical scaling include repair people (the number of maintenance staff rises more slowly than the number of machines), inventory control (the amount of safety stock needed to ensure that random variation in demand can be met rises proportionally with the square root of demand), cash reserves (the amount of cash a bank needs to keep available for withdrawals rises less than proportionally with the volume of deposits), and quality control (the number of items that need to be inspected to ensure a given level of quality rises with the square root of output).[395]

*Influence scaling*
The fourth mechanism behind economies of scale is influence scaling. Also sometimes called pecuniary effects, this type of scaling follows from the fact that the greater the production volume, the bigger and more important the company doing the producing becomes, which may allow it to reduce its costs. A bigger firm may receive bulk purchasing discounts from its suppliers, either because its importance to the supplier grants it greater bargaining power or because the supplier can profitably sell larger volumes more cheaply due to their own economies of scale. Large homebuilders, for instance, can have lower costs because they have access to better financial terms and

lower interest rates than smaller homebuilders.[396] A bigger company may also be able to force suppliers to bear the costs of any required adaptations to production, or they might require the supplier to meet quality control targets. Toyota, for example, has strict quality control standards that its suppliers must meet, and Apple requires its suppliers to adhere to a 206-page code of conduct.[397]

*Learning curve effects*
The fifth and final mechanism behind economies of scale is the learning curve. Improvements to a production process tend to accumulate at a rate proportional to cumulative output, so each doubling of production volume causes costs to fall by some constant percentage. This phenomenon—known variously as the learning curve, the experience curve, or Wright's law—means that increased production volume often results in the accumulation of cost-saving improvements as a producer or industry gains experience. The cost of wind-generated electricity in California, for example, fell by roughly 18 percent with each doubling of total installed turbine capacity between 1980 and 1994 as the design and technology of wind turbines improved.[398] We'll take a more in-depth look at learning curves in Chapter 7, so we will largely pass over the topic here.

*Network effects*
Network effects are, strictly speaking, different from economies of scale because they don't reduce production costs directly, but the two are sufficiently related that it's worth mentioning them here. Network effects occur when a product or service becomes more useful or valuable as more people use it. Social networks such as Facebook and Twitter, as well as communications technology such as telephones and fax machines, are examples of network-effect products. The more people have telephones, the more useful it is for you to have one; the more people use Facebook, the more useful it is for you to have an account.

Although network effects aren't economies of scale, they can enable economies of scale in a kind of virtuous circle. When a product has network effects, growth begets more growth. As more people use the product, it becomes more useful, which attracts more users, which makes the product even more useful, which attracts even more users, and so on. As the number of people purchasing the product increases, economies of scale—enabled by one or more of the five mechanisms outlined here—will drive down the price of the product as it becomes more valuable, reinforcing the cycle.

Airlines are a good example of this phenomenon. If an airline operates on a hub-and-spoke model, each new city introduced to an airline's network adds an increasing number of routes, since each new city creates a route to every other city already on the network. An airline that has a hub in Memphis and flies one route to Chicago is limited to serving customers who are interested in going from Memphis to Chicago and vice versa. If the airline then adds a flight from Memphis to Atlanta, the number of passengers on the Chicago-to-Memphis route will increase, because those flights will now include people flying from Chicago to Atlanta (via Memphis). Every new city introduced to the network adds more passengers to existing routes. And the more passengers on any given route, the greater the potential economies of scale, as one textbook on the economics of corporate strategy highlights:

> As its traffic volume increases, an airline can fill a larger fraction of its seats on a given type of aircraft... Because the airline's total costs increase only slightly, its cost per RPM [revenue passenger miles] falls as it spreads the flight-specific fixed costs over more traffic volume. As traffic volume on the route gets even larger, it becomes worthwhile to substitute larger aircraft (e.g., 300-seat Boeing 767s) for smaller aircraft (e.g., 150-seat Boeing 737s). A key aspect of this substitution is that the 300-seat aircraft flown a given distance at a given load factor is less than twice as costly as the 150-seat aircraft flown the same distance at the same load factor. The reason for this is that doubling the number of seats and passengers on a plane does not require doubling the sizes of flight and cabin crews or the amount of fuel used, and that the 300-seat aircraft is less than twice as costly to build as the 150-seat aircraft, owing to the cube-square rule.[399]

This sort of virtuous circle was crucial in the formation of FedEx, which was founded in 1971 as an airline offering overnight package delivery. Early in the company's formation, it shipped packages between a small number of cities and its package volumes were low; the initial network of five cities only generated between 10 and 12 packages a day. Operating costs greatly exceeded revenue, and the company required continued injections of funding to stay afloat.[400] But, as one history of the company notes, "each new city added to the network raised the [package] volume levels geometrically."[401] Only when enough cities were added to the network did package volumes reach the point where the company could operate profitably.[402]

More generally, economies of scale are often self-reinforcing. Increased output results in reduced costs, which, if the savings are passed on to the customer, leads to more sales and even greater output. This can create a powerful moat, as potential competitors will face a business with structurally low costs that can only be matched by making enormous, risky investments in large-scale production. Conversely, this virtuous circle can turn into a vicious death spiral if customer volume drops. Fewer customers means higher costs, which means raising prices, which results in even fewer customers. For example, in the early 2000s, as customers abandoned landlines in favor of mobile phones, the high fixed costs of the landline network had to be spread over a smaller number of customers, raising costs and prices for the remaining customers.[403]

### Diseconomies of scale and cost curves

In addition to economies of scale, there are also diseconomies of scale—factors that cause unit costs to *increase* as production output increases. Like economies of scale, diseconomies of scale are caused by a variety of mechanisms: administration costs, demand effects, geometric diseconomies, and statistical diseconomies.

*Administration costs*

As a company grows, it incurs more overhead costs. Assume, for example, a manager can have 10 direct reports. A division that can be managed by one person when it has 10 people will, therefore, need 11 managers when it has 100 people—one manager for every 10 employees and a manager for the 10 managers—representing a 10 percent increase in overhead cost. More generally, the more employees a company has, the more layers of management are required to oversee them. A company that doubles in size will, all else being equal, more than double the amount of management required.

Similarly, coordination costs within a company increase disproportionately to company size due to what we might think of as a reverse network effect. For an organization with $n$ employees, every new hire adds $n - 1$ potential lines of communication. The 10th employee adds nine potential lines of communication, but the 1,000th adds 999. Each new employee is someone who every existing employee might need to give information to or get information from, while each new manager is someone whose input might be needed before a decision can be made. Each layer of management slows down how quickly upper management gets on-the-ground information. As a result, all else being equal, larger companies will have more overhead than smaller companies, be slower to make decisions, and have a more difficult time implementing changes. This can result in larger companies only being able to profitably fulfill a minimum order size or complete a minimum project size: The administrative overheads manifest as large transaction costs, which need to be amortized over a large enough project for operations to be profitable. It can also cause additional downstream problems. One study of firms that increased in size through an acquisition found that these companies were 50 percent more likely to have civil lawsuit judgments against them due to these diseconomies of managing—that is, the increased complexity and loss of oversight that came with increased firm size.[404]

*Demand effects*

As production output increases, companies may find it increasingly difficult to obtain the inputs they need. A company may exhaust nearby sources of materials or other inputs (labor, for example), which will require it to source these inputs from farther and farther away. Similarly, early customers may be close by, but reaching new customers and new markets may require transporting the product over longer distances.

The American steel industry provides an example of this sort of demand diseconomy. After WWII, domestic sources of high-grade iron ore were nearly exhausted, so American steel companies were forced to make enormous investments in developing foreign sources of iron ore. A consortium of seven steel producers spent an estimated $175 million in 1950 ($2.2 billion in 2023 dollars) to develop sources of ore in Canada, including building a railroad to transport it. Republic Steel began to ship large quantities of ore from the African nation of Liberia, while US Steel began to import ore from Venezuela after its geologists found what one historian described as "literally a mountain of ore, eleven miles long and one mile wide, rising two thousand feet from the surrounding plains."[405] Imports of iron ore, which were around 2 percent of domestic production in 1920, skyrocketed to over 30 percent at the peak of American steel production in the late 1960s and early '70s.[406]

Alternatively, the increased difficulty or cost of acquiring inputs might require using less desirable or more expensive substitutes. In the steel industry, as high-grade ores containing a large percentage of iron were increasingly mined out, steelmakers were forced to use lower-grade ores that contained less iron.[407] But to use these ores in a blast furnace, they needed to undergo a process known as beneficiation, which turned low-grade ore into sinter or pellets with a higher iron content. Today, almost all American ore undergoes beneficiation, and the cost of beneficiation can be three times as high as mining the ore itself.[408]

*Geometric diseconomies*

In some cases, making something physically larger requires proportionally more, rather than less, material. This is known as a geometric diseconomy of scale. Imagine a beam spanning two supports that is supporting a load applied along the length of it. The bending force in the beam, known as the bending moment, is proportional to the beam's length squared; its deflection, or how much it sags under load, is proportional to its length to the fourth power. Therefore, doubling the length of the beam more than doubles the forces it must resist, which will require more than double the material if compensating measures aren't taken (like making the beam deeper).

A similar effect occurs as vertical structures get taller. In skyscraper construction, every new floor adds more cost than the floor below it and adds more weight to the floors below, requiring heavier structural framing to support the whole thing. Lateral forces rise nonlinearly with height—bending moments rise with the square of height, while sway and deflection rise with height to the fourth power—requiring stronger lateral systems and a stiffer structure as the building gets taller. Additionally, the taller a building, the more complex its mechanical systems become and the more elevators it needs, both of which encroach on the rentable space below. (Shorter buildings, on the other hand, might not need elevators at all.) A taller building also has more stringent fire safety requirements, necessitating the use of expensive noncombustible materials or fire protection systems. And because it takes longer to move people and materials to higher floors, every additional floor in a building involves proportionally more construction time.[409] Because of this, on a per-square-foot basis tall buildings are typically more expensive to build than short buildings, and skyscrapers have a maximum economic height beyond which additional costs exceed additional benefits.[410]

*Statistical diseconomies*

Diseconomies of scale can also arise due to statistical effects. Larger machines and equipment will have more parts than smaller machines, which can result in proportionately more downtime. All else being equal, a machine with 1,000 critical parts can be expected to break down roughly 10 times as often as one with 100 critical parts. For this reason, higher-capacity equipment might be less reliable than lower-capacity equipment.

In the 1970s, for instance, larger power plants were often found to have lower availability—times when they could be producing power—than smaller plants.[411] This was attributed in part to the greater number of components in larger plants and in part to the fact that a component failure in a large plant will be proportionally much more costly than in a smaller plant. As one study notes:

> Consider two [power] supply systems, one of which has a 500 MW [megawatt] unit and the other ten units each of 50 MW capacity. Very roughly, the large boiler has about the same amount of tubing as the ten small boilers. If the quality control in tube manufacture is such that one tube failure can be expected per year, then the single unit system will suffer a loss of availability of 500 MW for, say, three days, whilst the second system will lose 50 MW for three days, i.e., a ten-fold improvement in quality control is necessary in order to maintain an equivalent standard of availability.[412]

**Cost curve shapes**

Economies of scale typically show diminishing returns—that is, each absolute increase in production output results in proportionately less cost savings. In blast furnace construction, going from a 500,000-ton furnace to a million-ton furnace yields much greater cost reductions than going from 1 million to 1.5 million tons.[413] Similarly, in container shipping, the cost reduction per container going from a

1,000-container ship to a 3,000-container ship is much larger than going from a 3,000-container ship to a 5,000-container ship. This is both what we observe in actual costs and a natural consequence of modeling economies of scale with a scaling exponent, which results in a constant cost reduction (in percentage terms) for every doubling of output: Going from 10 to 20 units will show the same level of cost reduction as going from 10,000 to 20,0000 units.

The existence of diminishing returns and diseconomies of scale has traditionally led economists to posit a U-shaped cost curve for production operations: Costs fall as output rises to a certain point, after which they begin to increase. Empirically, however, what seems to be more commonly observed is an L-shaped cost curve: Costs fall as production output increases for a while but eventually taper off, at which point further increases in output don't result in either lower or higher costs.[414] In both cases, there's an elbow to the curve—a production volume after which further cost savings, if any, are limited.

In manufacturing, this relationship is typically expressed in the idea of minimum efficient scale—the plant size needed to capture most economies of scale and be competitive in the marketplace. The minimum efficient scale will vary from industry to industry and will be a function of both production economies and demand for the product. In some cases, the minimum efficient scale is a nominal fraction of total demand and an industry can support many small producers. Machine tool manufacturing, for instance, has historically had a very small minimum efficient scale (less than 1 percent of the market), resulting in hundreds of machine tool manufacturers competing against one another.[415] In other cases, the minimum efficient scale is a significant fraction of the entire market for a product (or even greater than the entire market), and only a limited number of players can efficiently compete. Commercial aircraft manufacturing, for instance, serves a sparse market, delivering fewer than 2,000 large jets every year. Because manufacturers must operate at a large enough scale to recoup the costs of developing new aircraft—a type of fixed cost spreading—the market has converged on just two major manufacturers, Boeing and Airbus.[416]

### Reliance on scale effects
Methods of efficiency improvement and cost reduction often either act through the mechanism of scale effects or are otherwise reliant on them.

The Origins of Efficiency

Consider, for instance, division of labor, the phenomenon famously described by Adam Smith in his observations of a pin factory. By dividing the factory's operations among numerous individual workers, the factory could produce many more pins than if each worker worked separately:

> One man draws out the wire, another straights it, a third cuts it, a fourth points it, a fifth grinds it at the top for receiving the head; to make the head requires two or three distinct operations; to put it on is a peculiar business, to whiten the pins is another; it is even a trade by itself to put them into the paper; and the important business of making a pin is, in this manner, divided into about eighteen distinct operations, which, in some manufactories, are all performed by distinct hands, though in others the same man will sometimes perform two or three of them... Ten persons, therefore, could make among them upwards of forty-eight thousand pins in a day... If they had all wrought separately and independently, and without any of them having been educated to this peculiar business, they certainly could not each of them have made twenty, perhaps not one pin in a day.[417]

There are several mechanisms by which division of labor increases efficiency, many of which take advantage of scale effects. Dividing labor allows each worker to perform a single task much more frequently, gaining experience and skill faster than they would if they were performing every step of the process. Smith also suggests that the more time workers spend doing a job, the more they figure out better ways to do it, inventing improved machines or better methods. Both of these are learning curve effects, a type of scale effect. In both cases, the benefits are predicated on increasing the frequency with which a worker performs a specific task.

Smith also notes that dividing up labor "[saves] the time commonly lost in passing from one sort of work to another." A weaver who also works on a farm "must lose a good deal of time in passing from his loom to the field, and from the field to his loom."[418] Time spent walking back and forth or changing tools can be thought of as a fixed setup cost. A worker who focuses on just one task will have fewer setups, but they'll also be able to spread the costs of the remaining setups over a larger volume of work. In other words, they'll take advantage of fixed cost spreading.

Fixed cost spreading is also at work in another benefit of division of labor described by Charles Babbage in his 1832 book *On the Economy of Machinery and Manufactures*. Division of labor means that each worker learns fewer tasks, which results in less time and materials wasted while a worker learns a task.[419] In this case, there are benefits from reducing the amount of training required overall, but there are also scale effects at work: The costs of training can be thought of as a fixed cost, which gets spread more thinly when each worker does just one task.

More generally, production process improvements typically have up-front, fixed costs that can only be justified if there's a sufficiently large production volume to spread them over. Many of the efficiency improvement methods we've touched on so far have this sort of cost structure. Manufacturers of high-volume products can afford to spend much more on design and optimization efforts, including DFMA, than manufacturers of smaller-volume products. Ford spent $2.7 billion developing a new model of the Taurus in the 1990s—more than 100,000 times the cost of the car itself. This was only feasible because Ford expected to sell a large number of Tauruses.[420] By contrast, a new apartment building will be built just once, and the developer will spend roughly one-fiftieth of the cost of the building on its design (say, $200,000 for a $10 million building). The lack of production volume means that proportionally much less can be spent on design, which affords the manufacturer much less room for design optimization and DFMA.

Similarly, in the mid-20th century, the use of interchangeable parts could dramatically reduce unit costs by eliminating the time and effort required to fit parts together. But developing a system of interchangeable parts had enormous up-front costs, since it required creating new gauges, fixtures, and other precision tooling that could only be recouped under high production volumes.[421] Likewise, GE's decision in the 1950s to train and employ 120 value analysts collectively saved the company millions of dollars, as it enabled the company to redesign products to use cheaper components and production methods, but the analysts' high salaries were only justifiable because GE was a large company with a substantial volume of output; even a small per-unit savings would greatly exceed the cost of the value analysis group when multiplied by the millions of light bulbs, refrigerators, and electric motors GE was producing.

## Mass production and mass consumption

An obvious constraint on capturing economies of scale is that they're only feasible with a sufficiently large market. In other words, all that product must have somewhere to go. A pin factory that produces 48,000 pins a day is only viable if those pins can be sold. If customers are only buying 48 pins a day, the factory can't be justified, no matter how efficient it is.

Historically, this has been a major limitation on achieving large-scale production volumes. Take the cast-iron stove, for example. In the late 1700s and early 1800s, American homes were mostly heated by woodburning fireplaces. However, these did a poor job of heating a house—much of the heat escaped up the chimney rather than heating the room itself—and consumed large volumes of firewood. The cast-iron stove, which delivered a much larger proportion of its heat to the room itself, was a far superior home heating technology, and it had often been used for home heating in parts of Europe.[422] But in America in the early 1800s, transporting a stove from where it was made to the consumer was difficult and expensive. Because stoves were cast from molten iron, they were made at blast furnaces, which tended to be built in remote locations near sources of wood, ore, and moving water. Stoves were heavy—an early stove might weigh as much as 500 pounds—and, as one historian observes, "transporting them over unimproved dirt roads or even hard-surfaced turnpikes by horse- or ox-drawn wagon was costly, difficult, and slow."[423]

But transportation technology in the early 1800s was rapidly improving. The Erie Canal, completed in 1825, reduced the cost of transporting goods from Buffalo to New York City by 90 percent and was followed by many other canal construction projects.[424] By 1840, over 3,000 miles of canal had been built in the US.[425] Canal building was accompanied—and then greatly exceeded—by railroad construction. In 1830, there were only 40 miles of railroad in the US. By the 1870s, there were nearly 50,000 miles. Between 1810 and 1860, the costs of land transport of heavy items fell by 95 percent.[426] This allowed bulky items like stoves to be shipped long distances—in some cases thousands of miles—at cost-effective rates. The result was an explosion in cast-iron stove production. The number of stoves made per year in the US rose from 25,000 in 1830 to 2.1 million in 1870.[427] By 1860, stoves made up nearly a third of the value of all cast-iron products made in the US.[428] In terms of value add (the total price of the final goods minus the cost of the inputs), stove making was as large an industry as making rails for railroads.[429]

By justifying soaring production volumes, the ability to ship stoves over long distances also meant stove production could increasingly take place in large, centralized stove foundries, which manufactured stoves by the thousands. In 1850, 33 percent of all stove production in the US occurred in just two New York cities, Albany and Troy.[430] And as production volumes surged, costs fell. Stoves became the first widely adopted consumer durable good.[431] No longer expensive luxuries, they had become affordable and reliable appliances.[432]

The effect of falling transportation costs on production volume had an enormous impact on the economy. As Chandler Jr. notes in *Scale and Scope*, it was the advent of the railroad that allowed the modern, large-scale industrial enterprise to appear in the 19th century:

> Economies of scale and scope could rarely be attained as long as the flow of goods depended on the energy provided by horse, man, wind, and current... In the half century before the coming of the railroad and the telegraph, although the total volume of goods produced, transactions handled, and number of enterprises increased enormously, the size and scale of industrial operations remained small.[433]

The railroad, along with the telegraph, made it possible to bring in huge volumes of raw materials, transform them into finished goods, and quickly send them back out for sale. This allowed for production volumes the likes of which had never been seen and was followed by a series of industrial innovations, such as high-output machinery, that were as transformative as the Industrial Revolution had been in 18th-century Britain.[434]

Similarly, cost reductions in ocean transportation following the introduction of the shipping container in the 1970s changed the face of global manufacturing. As long-distance shipping became increasingly inexpensive, it became feasible for manufacturers to produce a single item or component in the production process, achieve enormous scale, and let other companies handle the rest. As economist Marc Levinson explains in his history of the shipping container, *The Box*, by relying on other companies to supply the necessary inputs and specializing in a narrow range of products, each supplier "could take advantage of the latest technological developments in its industry and gain economies of scale in its particular product lines. Low transport costs helped make it economically sensible for a factory in China to produce Barbie dolls with Japanese hair, Taiwanese plastics, and American colorants, and ship them off to eager girls all over the world."[435]

When significant economies of scale exist in an industry, there are often a small number of large-volume producers. The auto industry is a good example of this. But it may also result in more complex arrangements. For example, although the cost of a single transistor has fallen precipitously, the up-front costs of building semiconductor fabrication facilities capable of producing ever-smaller transistors have risen enormously.[436] The skyrocketing fixed costs and subsequent urgent need to maximize production volume—to fill the fab, in industry parlance—has precipitated a shift toward fabless semiconductor companies. Whereas historically companies would design and then produce their own chips, this has become less and less tenable as the construction costs of fabs have exploded. Today, only a very small number of companies with enormous production volumes, like Intel, can hope to amortize the billions of dollars needed to build a leading-edge fab. Instead of manufacturing their own semiconductors, chip companies like AMD and Nvidia often opt to simply design the chips, which are then manufactured by another company, such as TSMC or GlobalFoundries.[437] These pure-play semiconductor foundries, which only manufacture chips designed by other companies, can achieve the large production volumes required to justify the high up-front costs by making semiconductors for a large number of smaller companies.[438]

### False economies and false diseconomies
Historically, economies of scale have been one of the major drivers of falling production costs and lower prices. But in practice, they're a double-edged sword. Achieving economies of scale requires spending enormous amounts of money on large-capacity equipment, much of which might be extremely specialized, as well as on high-volume production facilities. Not only is this a huge barrier to entry but it's also risky: If there's less demand for a product than anticipated, that large, efficient production facility can become a financial millstone.

Because of this, it is, paradoxically, often useful to develop production methods that eliminate economies of scale, allowing for low production costs without requiring enormous production volumes. In practice, many economies of scale turn out to be false, in the sense that simply rearranging the production process can achieve as low or lower production costs without requiring high production volumes or expensive high-capacity equipment. Several modern efficiency improvement systems, most notably the Toyota production system and lean methods, can be thought of in terms of avoiding these false economies of scale.

As an example, the conventional way of addressing fixed costs is to spread them out over as much production volume as possible. If a machine requires a long setup, then the costs of that setup should be amortized over a large number of goods by running the machine for as long as possible. Historically, this was how large-volume auto manufacturers operated. After WWII, US automakers' large production volumes allowed them to minimize the costs of die changes by having dedicated presses for certain parts, which they would run for months or even years without changing the dies.[439] But if the cost of a setup could be lowered—if the time it took to change the die was reduced from 12 hours to, say, 12 minutes—not only would overall costs go down (since less time and effort would be spent on the setup), but it also wouldn't be necessary to produce parts in such enormous volumes or have many separate presses dedicated to a single part. The manufacturer would be free to change the die much more often, which would allow it to use the equipment to produce many different parts. This is the goal of SMED, as discussed in Chapter 3: to make small-lot production economical by decreasing the time required for setups. In the case of die changes, spreading the fixed cost of a long setup time is a false economy of scale because it's possible to greatly reduce the fixed cost, yielding the same benefit without increasing production volume at all.

Similarly, high-volume production has historically resulted in the accumulation of large inventories. In their analysis of how lean production revolutionized car manufacturing, James Womack and Daniel Jones report that when they inspected a conventional, mass production-style auto plant, they found that "next to each workstation were piles—in some cases weeks' worth—of inventory."[440] Because the fixed costs of the plant's equipment were so high, plant operators wanted to keep the equipment running constantly. With a large stock of in-progress parts, every machine could be continually kept busy, even if a machine earlier in the process broke down. But a high production volume that merely sits in a plant as inventory doesn't reduce cost; it *adds* cost. Not only has money been spent on producing goods that aren't yet needed and haven't yet been sold, but all that additional inventory also requires more space to store it, more equipment to handle it, and more coordination to keep track of it.

Keeping equipment busy by producing something before it's actually needed is another false economy of scale. Rather than maximizing equipment utilization and accumulating a large amount of inventory in the production process, it's preferable to match the overall production rate with the sales rate, minimize the amount of inventory in the system, and address the root cause of machine failures and other delays so they don't cause slowdowns. Similarly, it's better to have cheaper lower-capacity equipment that matches the necessary rate of production than more expensive higher-capacity equipment that manufactures more parts than needed, even if the per-unit costs are lower for the latter machine when operating at full capacity. These sorts of improvements are the focus of the Toyota production system and its descendants, which we'll discuss in Chapter 6.

More generally, the high costs and high risks that come with building large high-output production facilities mean that it's useful to find production technologies that can lower the minimum efficient scale of a plant. One steel executive noted in the 1970s that "the difficulties in raising large blocks of capital, as well as the desire to establish regional steel-producing facilities, has led to a search for methods of producing iron in facilities with a smaller economic size than the modern blast furnace."[441] Today, the blast furnace has largely been replaced in the US by electric arc-based mini mills, which have lower capital costs and can profitably operate at a much lower production volume.

In addition to false economies of scale, there are also false diseconomies of scale—things that appear to add cost as scale increases but that can, in practice, be avoided. For example, in the first decades of the commercial deployment of wind turbines, many experts thought that costs would outpace gains in energy generation as the turbines got bigger, and that the optimal size for a wind turbine would be around a 40-meter rotor diameter. The argument, as articulated in a 1995 book that advocated for smaller turbines, was simple: "Wind energy increases with the square of rotor diameter, while weight and cost increase with the cube," so it would be more economical to have many small turbines rather than fewer, larger turbines.[442]

However, in the 25-plus years since that book was written, average wind turbine size has increased by a factor of 10. Average rotor diameter for land-based turbines went from 40 meters in the 1990s to over 120 meters in 2021.[443] Offshore wind turbines, which aren't constrained by the need to be transported over land, have gotten even bigger, with an average rotor size of more than 150 meters in 2021.[444] This expansion in size, and the associated economies of scale from bigger turbines—a combination of geometric scaling and fixed cost scaling—has been one of the main mechanisms through which wind-generated electricity has gotten cheaper. Between 1984 and 2022, the cost of wind power fell by roughly 90 percent.[445]

Like advocates of traditional mass production methods, small-wind advocates assumed a constraint was binding when it could actually be relaxed. For mass production, the constraint was setup times, which were often treated as fixed but could in practice be significantly reduced. With wind turbines, the perceived constraint was turbine mass, which was treated as a pure function of turbine size but could in practice be decreased. A modern wind turbine blade weighs 90 percent less than an equivalent blade from the 1990s.[446]

Wind turbines offer a good illustration of the complexities and multifaceted nature of economies of scale: Many simultaneous scale effects—learning curve effects, geometric scaling, statistical scaling, fixed cost effects—are at work, which often pull in different directions. It's not necessarily obvious, even to experts, which effects will dominate.[447]

### Bottlenecks

A production process consists of many steps, and it's often not possible to have all steps operate at the same rate or to ensure that every step is always busy. Some steps in the process may have extra capacity and are frequently idle, while others may always be occupied. When this is the case, these bottlenecks will govern the rate of the overall process, and the production rate can be increased simply by increasing the output of the bottleneck processes.[448] Conversely, if one or more bottlenecks exist, improving production rates at other steps in the process won't boost the overall production rate.

As an example, a manufacturer of truck trailer lamps was able to raise its overall production rate by maximizing the production rate of its bottleneck process: injection molding of lenses and lamp housing. The manufacturer did this by changing its scheduling system to prioritize the most urgent molding operations, reducing the time it took mechanics to collect the necessary molding equipment, and reassigning tasks from the molding mechanics to other workers, giving the mechanics more time for molding.[449] Similarly, a manufacturer of truck bumpers was able to increase its overall production rate by reassigning tasks from welders to nonwelders, boosting the production rate at the welding station, which had been a process bottleneck.[450] A car manufacturer was likewise able to increase its engine block production by 5 percent by ensuring that the bottleneck step in the process had a large enough buffer of work in process from which to draw.[451]

Identifying and exploiting bottlenecks is the prime strategy of the theory of constraints, a system of manufacturing improvement that became popular in the 1980s and '90s.[452] Bottleneck analysis also forms the basis of the critical path method, a project management tool developed at DuPont in the 1950s.[453] The critical path method analyzes the time needed to complete a project by considering the time it takes to perform each step and figuring out which steps need to be completed in which order to find the critical path, or the steps that determine how quickly (or slowly) the project will be completed. By focusing on reducing the time it takes to perform the critical path steps, producers can shorten the overall project duration:

> In essence, the critical path is the bottleneck route. Only by finding ways to shorten jobs along the critical path can the over-all project time be reduced; the time required to perform noncritical jobs is irrelevant from the viewpoint of total project time. The frequent (and costly) practice of "crashing" all jobs in a project in order to reduce total project time is thus unnecessary. Typically, only about 10% of the jobs in large projects are critical. (This figure will naturally vary from project to project.) Of course, if some way is found to shorten one or more of the critical jobs, then not only will the whole project time be shortened but the critical path itself may shift and some previously noncritical jobs may become critical.[454]

By expanding the capacity of a bottleneck, overall production rate can be increased, allowing for greater economies of scale.

### Shifting economies of scale

As discussed in Chapters 2 and 3, production processes exist within a constantly changing landscape. Input costs fall or rise as patterns of demand fluctuate. Expensive processes become cheaper and low-quality processes improve as technology advances. Consumer tastes, demographics, and government policies all change over time, and with them the relative costs and benefits of different production processes.

These sorts of changes also affect economies of scale. The range of scale effects that can be taken advantage of is constantly evolving. In some cases, variations in the landscape make larger scales possible: As transportation technology developed over the 19th and 20th centuries, it became economical to ship goods farther, enabling the creation of large-volume plants that could achieve enormous economies of scale. Technological advancements for equipment like blast furnaces and wind turbines made it possible to increase their size, with corresponding geometric scaling effects that reduced the cost of their output.

In other cases, shifts in the landscape have meant that lower-volume production becomes more economical. Advances in electric arc furnace technology, along with the widespread availability of steel scrap thanks to a century of US steelmaking, made it economical to produce steel in smaller-scale mini mills instead of in large-scale integrated steelworks. The mass production methods of early 20th-century US automakers gave way to the lean methods of Japanese automakers, which could profitably produce cars at much lower volumes. And the development of 3D printing and other additive manufacturing methods made low-volume part production much cheaper than it had been previously.

As with other production process decisions, operating in this dynamic landscape requires a certain amount of trying to predict the future. A producer must operate based on some guess of what the market will be: whether it will be big enough to justify large production volumes, and whether changing technology will benefit large-volume or small-volume production. Trying to account for this dynamic landscape and anticipate what the resulting economies of scale are likely to be is far more difficult and nuanced than simply trying to minimize unit costs by maximizing production volume.

## Summary

As we've noted over the course of this chapter, economies of scale have historically been a major driver of increased efficiency. By spreading fixed costs over larger volumes of product, building bigger plants and equipment, smoothing out variation with increased output, and taking advantage of the other scaling mechanisms discussed here, manufacturers can drive down production costs precipitously.

Not only do economies of scale make things cheaper, but their very existence is part of what makes industrial civilization possible. Greater scale not only creates wealth, by increasing the amount of goods that can be created for a given amount of input, but it also pushes technology forward, by making it possible to recoup large investments in its development. Greater scale, enabled by the emergence of transportation options like railroads and canals, is what spurred the advancement of new high-volume production technologies that could manufacture goods in enormous volumes and distribute them all over the country. Scale is what makes it possible to push the boundaries of semiconductor technology and build increasingly advanced fabrication facilities. By continuing to drive energy technologies like solar photovoltaics and batteries down the learning curve, making them ever cheaper, scale may even be what saves the planet.

# 5

# Removing a Step

Over the past two chapters, we've explored improvements in efficiency that come from using fewer, cheaper inputs. Transitioning to lower-cost materials, less expensive production processes, or larger-capacity equipment can all drive down the price of a production process. In this chapter, we'll look at a different strategy for reducing the inputs of a process: removing a production step, along with all the resources that step consumes.

Since the most efficient process step is no step at all, cutting an unnecessary step out of a process is the best possible option for making it more efficient. But why would a production process include any unnecessary steps in the first place?

Broadly, steps in a production process can be thought of as either value-adding or non-value-adding. Value-adding steps perform the transformations required to make a product. Returning to the cut-nail example from Chapter 2, value-adding steps include cutting the nail body from the nail plate and squeezing the top of the nail body to form the head. These steps transform the inputs and create the actual nail by cutting and molding the metal. Other types of value-adding operations include chemical transformations, such as iron oxide and carbon reacting in a blast furnace to form liquid iron and carbon dioxide, and assembly operations, like attaching the engine of Ford's Model T to the chassis on the assembly line. Non-value-adding steps, by contrast, don't perform a required transformation of the input materials. In the nail-making example, a worker carrying a sheet of iron from storage to the nail machine would be a non-value-adding step, since it doesn't transform the iron sheet in any meaningful way. Other non-value-adding steps include replacing the cutting blade on the machine when it gets dull and clearing the machine when it jams, because neither of these steps physically transform input materials.

Non-value-adding steps can be thought of as scaffolding for the production process. They don't create value themselves, but they support the steps in the process that do. Carrying iron sheets to the machine doesn't turn them into nails, but it does make it possible for the machine to carry out the subsequent transformation. However, because non-value-adding steps don't result in any physical change to the product, they can be removed from the process without impacting the product. If we can rearrange the nail-making process so that the iron sheets are stored right by the machine and automatically fed into it, cutting out the carrying-iron-sheets-from-storage step, this doesn't affect the nail itself. Similarly, by switching to a more reliable and durable machine, we can forgo, or at least reduce, repairs and inspections. This has no impact on the product itself, because we've simply removed non-value-adding steps.

Determining whether a step is value-adding or not is typically done by taking the customer's perspective. If a step contributes to the functionality and usefulness of a product in the eyes of the customer, it's value-adding. If it doesn't, it's non-value-adding.[455] Using the customer as the arbiter of value is one of the hallmarks of the Toyota production system and lean methods, though it has also long been recognized by economists.[456] From this perspective, there will typically be many more non-value-adding steps than value-adding steps in a process.

Consider the following process for attaching a bracket to a truck chassis, which is taken (and slightly modified) from a process described in Jeffrey Liker's *The Toyota Way*:

| 1 | Walking from the assembly line 20 feet to pick up the bracket |
|---|---|
| 2 | Removing the cardboard cover to expose the bracket |
| 3 | Reaching for the bracket |
| 4 | Picking up the bracket |
| 5 | Picking up bolts for the bracket |
| 6 | Walking back 25 feet to the chassis |
| 7 | *Positioning the bracket on the chassis* |
| 8 | Walking to the impact wrench |
| 9 | Reaching for the impact wrench |
| 10 | Walking with the wrench to the bracket on the chassis |
| 11 | Moving the impact wrench to the bracket |
| 12 | *Placing the bolts in the bracket* |
| 13 | *Tightening the bolts to the chassis with the power tool* |
| 14 | Walking back 25 feet for the next bracket[457] |

Of this 14-step process, only three of the steps—positioning the bracket on the chassis, placing the bolts in the bracket, and tightening the bolts to the chassis—are value-adding. Every other step is a non-value-adding scaffolding step. This sort of breakdown is typical. Most business processes, Liker finds, are "only 10 percent value-added work."[458] Similarly, the time spent on value-adding steps is generally a small proportion of the total amount of time the process takes, typically because a product spends much of its time waiting for the next step. As James Womack observes in *Lean Thinking*, the total time required to produce an aluminum soda can, from mining the raw materials to putting the filled can on store shelves, is around 319 days. But the actual value-adding time—the time spent on physically transforming raw bauxite, step by step, into an aluminum soda can—is just three hours, or 0.04 percent of the overall process time.[459] Liker also describes a process for manufacturing steel nuts with a proportion of value-adding to non-value-adding time of just 0.008 percent.[460]

Building on this concept, we can broadly classify strategies for removing steps in a process as either removing non-value-adding steps or removing value-adding steps.[461] Because non-value-adding steps don't affect the form or function of the product, they can be removed without requiring a change to the product itself, which means non-value-adding steps are often (though not always) simpler to remove than value-adding steps. And because non-value-adding steps make up such a large fraction of the overall process, removing them can result in significant cost savings. By contrast, removing a value-adding step from a production process typically requires modifying the design of the product and the transformations used to produce it, because cutting out a value-adding step necessitates changing another step to do the same (or similar) job. For example, in the cut-nail-making process, it's possible to remove the step in which the body of the nail is squeezed to form the head by changing the shape of the cutting tool so that a nail with a head is cut directly from the nail plate. (In fact, some cut nails were produced this way.) But this alters the design of the nail, which changes how the nail functions. It also changes the shape of the cutting blade required, which means different equipment is needed. For this reason, removing value-adding steps from a process is often more complex than removing non-value-adding steps—though, as we'll see in this chapter, in some cases the two strategies begin to overlap.

### Removing non-value-adding steps

Removing non-value-adding steps from a process is something manufacturers have been doing to reduce costs for hundreds of years, although historically, this was mainly a by-product of a more general attempt to remove *any* step in a process rather than a discrete targeting of non-value-adding steps. Early cut nails, for example, required two separate machines for their production: One would cut the nail body from the iron sheet, and another would squeeze the top of the nail to form the head. A worker tending the machines would feed nails from one machine to the next.[462] The development of two-step cut-nail machines, which could both cut the body and create the head, eliminated a step from the process as well as the worker motions required to move the nails between the two machines. Similarly, since at least the 18th century, manufacturers have used jigs, fixtures, and other shop aids to allow workers to quickly place parts on a machine tool without needing any time-consuming (and non-value-adding) adjustment operations to position the parts in the correct place.[463]

Ford's assembly line provides one of the most important early examples of the benefits of removing non-value-adding steps. By arranging workers to perform the steps in a precise sequence and having the work come to them instead of having them go to the work, Ford greatly reduced the time it took to assemble the Model T. As Henry Ford noted in his memoir *My Life and Work*, three simple principles of assembly allowed for "a reduction of the necessity for thought on the part of the worker and the reduction of his movements to a minimum":

| | |
|---|---|
| 1 | Place the tools and the men in the sequence of the operation so that each component part shall travel the least possible distance while in the process of finishing. |
| 2 | Use work slides or some other form of carrier so that when a workman completes his operation, he drops the part always in the same place—which place must always be the most convenient place to his hand—and if possible have gravity carry the part to the next workman for his operation. |
| 3 | Use sliding assembling lines by which the parts to be assembled are delivered at convenient distances.[464] |

Ford focused relentlessly on cutting out unnecessary movements and steps from the production process, but he doesn't appear to have drawn a distinction between value-adding and non-value-adding steps. Instead, this dichotomy can be traced to the rise of scientific management in the late 19th and early 20th centuries. A workplace efficiency improvement system, scientific management was first developed by American mechanical engineer Frederick Taylor through his efforts to "discover, by scientific methods, how long it should take men to perform each given piece of work."[465] Taylor believed that "among the various methods and implements used in each element of each trade, there is always one method and one implement which is quicker and better than any of the rest," with all other methods requiring wasteful or unnecessary actions.[466] Through his technique of job analysis, Taylor would "divide each job into particular operations" to determine which were essential and which were wasteful.[467] Identifying the "one best way" to do a job, he reasoned, would increase output while reducing costs.[468]

Job analysis would later be supplanted by the practice of motion study. One of its inventors, Frank Gilbreth, described it as "the science of eliminating wastefulness resulting from using unnecessary, ill-directed, and inefficient motions."[469] Gilbreth had originally been a construction contractor, and it was in this setting that he first used motion study to analyze the motions of bricklayers and develop an improved method for laying bricks that cut out all unnecessary motions.[470] As Taylor would later report:

[Gilbreth] studied the best height for the mortar box and brick pile, and then designed a scaffold, with a table on it, upon which all the materials are placed, so as to keep the bricks, the mortar, the man, and the wall in their proper relative positions. These scaffolds are adjusted as the wall grows in height... and by this means the bricklayer is saved the exertion of stooping down to the level of his feet for each brick and each trowelful of mortar and then straightening up again... As a result of further study, after the bricks are unloaded from the cars... they are carefully sorted by a laborer, and placed with their best edge up on a simple wooden frame... In this way the bricklayer avoids either having to turn the brick over or end for end to examine it before laying it, and he saves, also, the time taken in deciding which is the best edge and end to place on the outside of the wall.[471]

Removing unnecessary steps from the bricklaying process eliminated more than 70 percent of the motions needed to place a brick, and nearly tripled the rate of bricklaying.[472]

Gilbreth and his wife, Lillian, went on to be among the most famous practitioners of scientific management, largely through their work on motion studies. Together they developed chronocyclegraphs: long-exposure photographs of workers performing tasks while wearing special rings with lights attached. The lights would illuminate the path of a worker's hands on the resulting photograph, revealing potential wasteful motions. The Gilbreths also developed a high-level taxonomy of standard process steps—operation, inspection, transportation, delay, and storage—and their associated symbols, which is still in use today.[473] Of these five fundamental process steps, it's notable that only one—operation—is a value-adding step. The rest are non-value-adding scaffolding.

In 1915, the Gilbreths developed a lower-level taxonomy of 17 standard motions into which every more complex motion could be broken down, which they called therbligs.[474] This taxonomy is notable for taking into account the difference between value-adding and non-value-adding steps, though it didn't use those terms. The therbligs that were necessary to perform the transformation, such as reach, move, grasp, and assemble, were classified as effective therbligs. Others, such as search, plan, inspect, and hold, were ineffective therbligs—not necessary to perform the actual work and, therefore, prime targets for removal.[475]

Using therbligs, the micromotions of a particular task could be analyzed on simultaneous-motion (SIMO) cycle charts. These recorded a list of the step-by-step motions each hand performed during a task, as well as how long each motion took to perform, measured down to the thousandth of a minute.[476] By analyzing these micromotions, manufacturers could develop better production methods that required fewer movements. A two-man riveting process for an aircraft manufacturer, for example, originally required 34 motions to perform. By changing the process to have one worker ready the next rivet while holding the current rivet to be driven, 22 motions were eliminated (mostly holding, waiting, and signaling motions).[477]

This sort of microanalysis was further expanded with the development of predetermined time systems: enormous tables, created by observing thousands of worker motions, that listed individual motions and the time required to complete them. Using predetermined time systems, it was possible to analyze an operation and calculate the time it would take to complete it without performing an empirical study of that particular task.[478]

In Jefferson Cowie's study of RCA, he describes the use of one of these systems, called the work factor system, in the company's television assembly plant:

The system, based on extensive research designed "to eliminate

| EFFECTIVE | INEFFECTIVE |
|---|---|
| Reach | Search |
| Move | Select |
| Release | Position |
| Grasp | Inspect |
| Pre-position | Plan |
| Use | Unavoidable delay |
| Assemble | Avoidable delay |
| Disassemble | Rest |
| | Hold |

**Table 2.** List of therbligs. Data from Aft, *Work Measurement and Methods Improvement*.

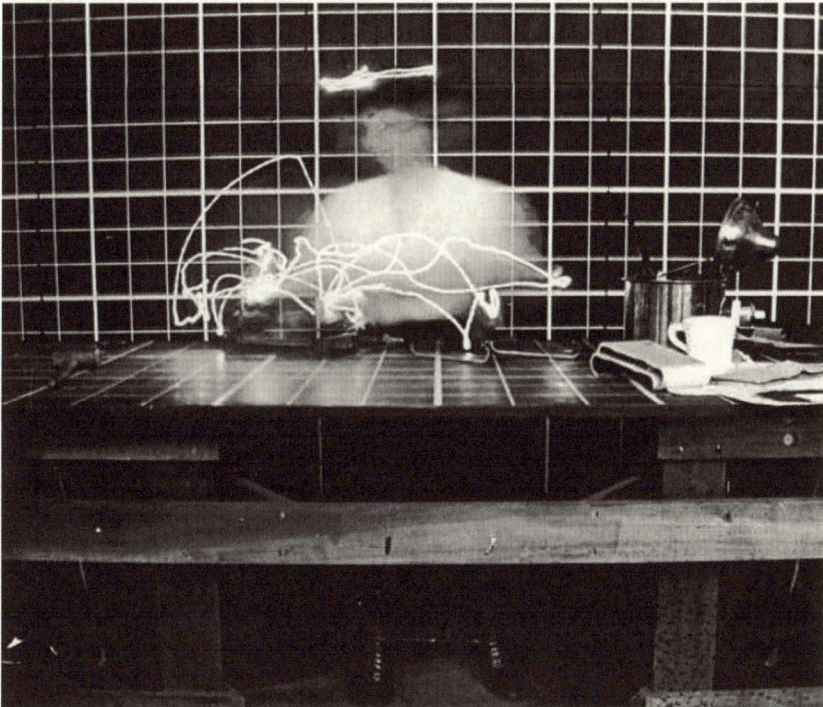

**Figure 8.** A Gilbreth chronocyclegraph. Reproduced from National Museum of American History.

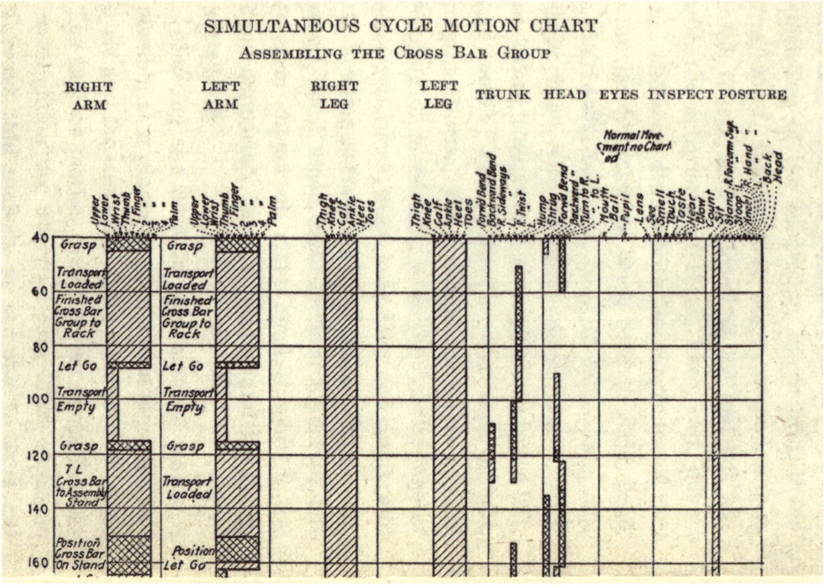

Figure 9. SIMO chart. Reproduced from Gilbreth and Gilbreth, *Applied Motion Study*.

human judgment in setting output rates," classified the distance any part of a worker's body needed to move, the body part or parts used, the type and degree of manual control involved in each motion, and the weight or resistance encountered in the operation. Each motion segment had been quantified into a "work factor unit" that equaled 1/10,000 of a minute. Using an intricate formula that compensated for the time required for a worker's body part to change directions, the time necessary to synchronize different motions, the degree of visibility of an operation to the worker, the amount of control and dexterity required, and the amount of "mental process" involved, the manager could "objectively" determine the time required to complete any task from values derived from reams of tables without recourse to a stopwatch. The time required for a given movement could vary with the obstacles or cautions involved. All of the work factor calculations for each movement in the assigned job could then be added up to a single aggregate amount of time, or "work process." The assembly of the entire television set consisted of hundreds of separate processes performed by each operative.[479]

The use of predetermined time systems was also thought to

improve work methods by helping to identify unnecessary motions, which could then be excised from the process. As Richard Otto Schmid observed in his study of such systems, by breaking down a process into individual motions, "the analyst is, in a sense, forced to examine the operation step by step, and motion by motion. A questioning attitude is raised as to which motions are really necessary, and it is indeed most difficult to find an operation that could not be improved."[480]

The ideas of motion study and eliminating unnecessary steps in a production process were further developed in the 1930s by industrial engineer Allan Mogensen, a disciple of the Gilbreths, into an offshoot of scientific management known as work simplification. While work simplification would end up encompassing many strategies for process improvement (such as redesigning parts to reduce assembly efforts), it primarily focused on analyzing each step in a process and finding ways to remove those that were unnecessary.[481] A 1963 text on work simplification emphasizes that "every operation, every transportation, every inspection, every storage is studied for the possibility of eliminating the unnecessary and simplifying the rest," and that "one must be absolutely positive that the job cannot be eliminated before attempting to work out a better way of doing it."[482]

To see how this system worked, let's look at a pea cannery process that was improved by work simplification. In this process, peas from the fields were loaded into large four-wheeled carts, called hoppers, which would then be taken by truck to a loading dock. There, the hoppers would be carried out of the truck and brought to a tenderometer, which would squeeze a sample of the peas to measure their tenderness, assigning them a grade of 1 through 4. The hopper would then be wheeled to the bin matching its grade, and the peas would be dumped in the bin. The hopper would then be washed, weighed, and wheeled back into an empty truck to return to the field.

This process consisted of 23 steps, only two of which—dumping peas from the hopper and washing the hopper—were found to be value-adding operations. Of the remaining 21 steps, 10 were movements, four were non-value-adding labor operations (lifting the hoppers on and off the trucks), five were delays or waiting steps, and one was an inspection step. A work simplification analysis of this process yielded the following suggestions: Move the trucks' unloading location to a dock area adjacent to the tenderometer; level the ground by the dock so the hoppers could be wheeled on and off the trucks instead of being lifted; and move the tenderometer off the edge of the wall so empty hoppers could pass behind it. These simple changes eliminated the lifting and weighing operations and shortened the travel path. The new process had 16 steps instead of 23 and a travel distance of 190 feet instead of 222. It also avoided considerable waiting time and confusion on the dock. Eliminating these steps resulted in a significant reduction in costs—an investment of $51 ($631 in inflation-adjusted dollars) yielded annual savings of $464 (or $5,792).[483]

By the 1950s, work simplification had become a routine feature of good management in large organizations. A 1954 survey of 207 large US companies found that 129 of them, including Sears, Honeywell, and DuPont, had implemented work simplification programs.[484] It also became popular in government agencies, with both the US Navy and the Air Force adopting work simplification programs.[485]

As a result of scientific management and work simplification, other efficiency improvement systems increasingly recognized the value of minimizing wasteful motions. A 1945 factory management textbook listed its first two principles for materials handling as "Reduce all transportation to the lowest possible limits" and "Reduce lines of travel, and wherever possible have the machines and operations in sequence."[486] Texts on value analysis made note of the potential cost savings from designing parts to be symmetrical to decrease unnecessary motions in orienting them correctly.[487] Group technology, an industrial improvement system developed in the late 1950s and early '60s as a way to transfer the benefits of mass production to small-lot production (and a predecessor of modern lean methods), advocated grouping machines common to a particular process into cells, so the output of one machine could flow immediately into the next without unnecessary handling.[488]

The importance of eliminating unnecessary motions is also evident in nonmanufacturing settings. Since the late 19th century, chefs in commercial kitchens have used mise en place to keep the most important tools and ingredients close by to minimize how much they must move when preparing dishes.[489] Likewise, the spread of typewriters in the late 19th century was followed by the creation of keyboard layouts designed to improve typing efficiency by minimizing finger movement.[490] The Blickensderfer typewriter, which first appeared in 1893, placed the most-used letters—D, H, I, A, T, E, N, S, O, and R—in one row to lessen hand movement.[491] In 1936, August Dvorak, a time and motion study expert, created the Dvorak keyboard layout with similar goals.[492]

The focus on eliminating non-value-adding steps characteristic of scientific management and motion study has been taken up in our own time by modern industrial improvement methods, particularly the Toyota production system, the manufacturing improvement method developed by Toyota that has spread around the world in the form of the lean manufacturing philosophy. The primary purpose of the Toyota production system is to achieve cost savings by eliminating waste, of which there are eight different types: overproduction, waiting, unnecessary transport, overprocessing, excess inventory, unnecessary movement, defects, and unused employee creativity (which was added later).[493]

Of the seven original types of waste, five—unnecessary transport, unnecessary movement, overprocessing, defects, and waiting—are non-value-adding steps. The other two—overproduction and excess inventory—are based on inventory reduction, which lean practitioners would likely argue *also* entails removing non-value-adding steps, because if you're producing something you don't need yet, the steps to make it aren't value-adding. By cutting these wasteful steps out of a process, costs fall and quality and responsiveness rise. According to lean practitioners, "in a lean improvement initiative, most of the progress comes because a large number of non-value-added steps are squeezed out."[494] When performed repeatedly, these improvement initiatives can reportedly cut the costs of a production process by 15 to 25 percent.[495]

The Toyota production system and lean methods offer a large collection of tools and strategies for finding and removing unnecessary steps.[496] SMED production methods, for example, find ways to reduce setup times, often by cutting out nonessential steps. Shigeo Shingo, the inventor of SMED, offers the following example of how SMED methods removed wasteful steps from a cutting operation:

At Taiho Industry, one worker performed a multiprocess handling operation with fourteen cutting machines and had to manually start the machines at each process. To improve this procedure, a one-touch button for machines 1 through 7 was installed near the eighth machine so all seven machines could be started at once;

another button for machines 8 through 14 was installed near the first machine so all those machines could also be started at once. This improvement reduced the cycle time from 35 to 4.2 seconds.[497]

Shingo also created what's known as *poka-yoke,* or mistake proofing (though, like most other improvement methods, it has deep historical roots).[498] Poka-yokes are devices that prevent an error from happening or make the error immediately noticeable if it has already occurred. This can improve quality and reduce the need for separate inspection steps. For example, so-called go-no-go gauges can be used to automatically remove parts that exceed specified tolerances. Consider one ingenious example for a machine that produces stem tighteners:

> Periodically, [the machine] produced tighteners that were too thick or thin. To detect such machine mistakes quickly, the company installed go-no-go gauges across the discharge chute from the machine. Angled across this chute are two bars mounted at different heights. They mounted the first bar at the height of the upper specification. Any tighteners that cannot fit under this bar are shunted to a side tray. They set the second bar at the lower specification limit, and any parts that fit under this bar continue to the end of the chute where they also fall into a tray. The second bar diverts all of the tighteners within specifications into a bin for accepted parts. If a tightener lands in either of the defect trays, an alarm sounds and the machine stops automatically.[499]

By installing many poka-yoke devices, numerous errors and the extra process steps they cause can be prevented. In the 1990s, Toyota averaged 12 poka-yokes per machine.[500] Toyota also uses what is known as autonomation, or automation with a human touch. Toyota machines are often designed to work automatically but stop if an error is detected. This allows one worker to tend multiple machines while preventing them from producing large numbers of bad parts if a problem arises. And to make it even easier for workers, lean production also often involves the creation of work cells, where machines are placed to be within easy reach of a worker (such as by arranging them in a U-shape) and the output of one machine feeds immediately into the next.

A final Toyota production system method is the five whys:

Whenever a problem occurs, workers must trace the problem to its root cause by asking *why* the problem occurred five times. For example: Why did the machine break? It overheated. Why did it overheat? It's placed too close to the boiler room. Why is it placed by the boiler room? And so on. By locating and eliminating the root cause of a problem, the producer can prevent scrap, rework, and potentially inspection steps.

## Removing value-adding steps

Although removing non-value-adding steps is often the easier course of action, it is also possible to remove value-adding steps from a production process. But unlike removing non-value-adding steps, removing value-adding steps will typically require changing the design of the product itself, as well as that of the other value-adding and non-value-adding operations.

Recall, for example, our discussion in Chapter 3 about DFMA, which reduces a product's cost by redesigning it to be made using cheaper materials, fewer parts, and less expensive production methods. The benefits of DFMA mostly come from reducing the number of parts in a product.[501] Redesigning a product to require fewer parts cuts out many assembly steps, since each part involves its own assembly operation. And the more value-adding assembly operations there are, the more non-value-adding scaffolding is needed to support them, because there will be a higher risk of error (requiring more inspection), more needed setups, more scrap, and more timekeeping, counting, and paperwork. By reducing part count and removing assembly steps, all these ancillary steps can be removed as well.

Similarly, as we noted in Chapter 3, many industrial improvement systems (including DFMA, but also value analysis and producibility guidelines) contain recommendations for replacing expensive production operations with cheaper ones. This often involves replacing a multistep process with a single-step process. For example, machine parts, which are produced on a series of metal-cutting machines, can often be made much more cheaply using operations such as casting or extruding, which can create parts in a single step.

More generally, advances in production technology often make it possible to remove value-adding steps in the process. Consider the manufacture of plate glass: smooth, clear glass suitable for commercial windows and automobiles. In the mid-20th century, plate glass was manufactured by pouring molten glass onto a solid surface, rolling it into a sheet, sending it through an annealing furnace, and putting it through a series of grinders and polishers to remove any imperfections. The grinding and polishing operations were value-adding steps—they produced the smooth, defect-free surface that differentiated plate glass from cheaper but lower-quality sheet glass—but they were also expensive, time-consuming, and prone to failure.[502]

In the 1950s and '60s, this production process was superseded by the superior float glass process.[503] Instead of pouring the molten glass onto a solid surface, it would be poured on top of a molten bath of tin. As the molten glass flowed over the tin, it would gradually cool until it solidified. Because the surface of the tin was perfectly flat, the resulting glass was as well. This change eliminated the grinding and polishing operations, along with several hundred feet of the production line. The new process generated 15 to 25 percent less scrap and required 50 percent less energy and 80 percent less labor.[504] Not only was glass made by the float process cheaper than any other kind of plate glass, but it also ultimately became cheaper than lower-quality sheet glass. It was so cheap, in fact, that the separate categories of low-quality sheet and high-quality plate were eliminated. Today, essentially all flat glass is high-quality plate glass.[505]

The development of interchangeable parts offers another example of how technological advancements can allow for the removal of value-adding steps. Prior to the emergence of interchangeable parts, no two products were exactly alike. Two guns of the same model made by the same gunsmith would have slightly different triggers, barrels, and hammers, and parts from one gun couldn't be used to replace parts on another. Because every part was slightly different, getting the parts of a gun to fit together required each part to be manually filed and adjusted by skilled fitters.[506] By making parts interchangeable—rendering every trigger, every barrel, and every hammer on a gun identical—not only could one part easily replace another, making repair easier, but the filing and fitting operations could also be eliminated.[507]

In the early to mid-18th century, however, producing interchangeable parts greatly *increased* the time and labor required in a production process.[508] Due to the imprecision of the manufacturing methods and machine tools of the time, making parts interchangeable required great amounts of manual working and filing to ensure each part was exactly the same. Rather than eliminating filing and fitting operations, then, early interchangeable parts added more of these operations.

As a result, interchangeable manufacture was only used when other benefits of interchangeability, such as ease of repair, were worth the extra cost. Being able to achieve interchangeability economically required the invention of new technologies like more precise and accurate machine tools and cutting tools capable of cutting hardened steel (because hardening parts after they were machined would often distort them).[509] These efforts took many decades to refine. As late as the 1930s, interchangeability still had high up-front costs that could only be justified by large-volume production.[510]

## Value-adding versus non-value-adding

The distinction between value-adding and non-value-adding steps is a useful framework for understanding the kinds of changes that will be required to eliminate steps in a production process. But the border between value-adding and non-value-adding steps can be fuzzy. In some cases, an operation may appear to be value-adding because a physical transformation is taking place, but the operation may not be necessary. It might be duplicating effort that is taking place elsewhere in the process or providing a feature that doesn't add functionality or that the consumer of the product—the ultimate arbiter of value—doesn't care about. In lean terminology, these non-value-added steps are known as overprocessing.

A 1944 factory operations manual provides several examples of this type of operation. In one instance, a manufacturing process for meat dishes made of nickel silver involved polishing the underside of the dish. But it turned out that consumers didn't care whether the undersides of their dishes were polished, so the step could be eliminated.[511] In another example, a multistep washing operation in which knives were washed on a wheel, racked, cleaned, and dried in sawdust was eliminated, and the knives were simply wiped with cloths to a similar effect.[512]

Tesla provides a more modern example. In one of its factories, a robotic system for installing fiberglass mats on top of the battery pack frequently failed to install the mats correctly. But further testing showed that the mats, whose purpose was supposedly to reduce cabin vibration and noise, had no impact on noise levels within the car. The mats could, therefore, be eliminated, along with the robots used to install them.[513]

Because non-value-adding operations can masquerade as value-adding ones, improvement methodologies such as work simplification often emphasize that each operation should be "put on trial" to make sure it's necessary.[514] Indeed, Elon Musk has said that after making sure there are no unnecessary requirements in a manufacturing process, the way to try to improve a manufacturing process is to "try and delete part of the process."[515]

In some cases, removing a non-value-adding step may require the kinds of changes usually reserved for removing value-adding steps. While doing DFMA on the Ford Taurus in the 1980s, Ford altered the design to use uniform screw sizes throughout. This wouldn't eliminate any value-adding operations, because the same number of screws would need to be tightened either way, but it did remove the non-value-adding steps of changing torque heads on the screwdriver.[516] Still, removing these non-value-adding steps required a change to the design of the product, making it akin to removing a value-adding operation.

Similarly, in his discussion of SMED, Shingo notes that when using bolts as fasteners, only the final turn of the bolt—the one that actually tightens it—is a value-adding step. None of the other turns actually tighten the bolt, so they are non-value-adding. Yet removing these non-value-adding bolt turns might require redesigning the attachment mechanism.[517] Just as a value-adding step may be supported by a non-value-adding scaffolding, a non-value-adding step might be embedded in a matrix of other process steps and require significant changes to the product or process to eliminate.

### Summary
Cutting a step out of the process is the ultimate efficiency improvement. Not only does it remove 100 percent of the inputs that step requires but it can also remove an entire scaffolding of support operations. What's more, cutting a step out of a process will, all else being equal, make that process more reliable, because an operation that doesn't exist is an operation that can't fail. This is especially true given how few of the steps in a production process are of actual value in the eyes of the customer. A comparatively small amount of time and effort is typically spent on value-adding operations like cutting or joining metal, assembling parts, mixing chemicals, or other steps that actually make the product. Often, 90 percent or more of the time spent in a process is on scaffolding steps like waiting, inspection, moving things back and forth when they don't strictly need to be moved, and

so on. And even steps that are value-adding can often be eliminated by redesigning a product to have fewer parts or use production methods that require fewer operations.

It is perhaps unsurprising, then, that cutting steps out of a process was arguably the defining characteristic of industrial improvement methodologies during the 20th century. It forms the core of Ford's assembly line, the basis of Taylor's scientific management system and its many descendants, the primary driver of improvement in DFMA, and the foundation of the Toyota production system and its offshoots. Antoine de Saint-Exupéry famously quipped that "perfection is finally attained not when there is no longer anything to add, but when there is no longer anything to take away."[518] Production processes, which can be steadily whittled down to a small number of critical operations, are no exception.

# 6

# Variability, Knowledge, and Control

A production process is a series of transformations that turns a given set of inputs into a finished product, step by step. In the production process for light bulb blanks we discussed in Chapter 1, various chemicals (sand, lead, potassium carbonate) were melted together in a furnace. The molten mixture was gathered on a blowpipe, placed in a bulb mold, and blown into the shape of a light bulb. The bulb blank would then be cut off from the blowpipe, and the process would repeat.

This process did not, however, produce the exact same bulb each time. Some bulbs were slightly different shapes; some had slightly different chemical compositions; others displayed minor differences in the thickness of the glass. In some cases, the process didn't work at all: the bulb might have shattered when trying to remove it, or the blob of molten glass might have fallen off the blowpipe, or the chemicals comprising the glass might have been mixed in the wrong amounts. No matter how hard we try, a production process will inevitably have some amount of variation. Not even the most carefully controlled process can produce a perfectly uniform output.

Why does this variation occur? For one, a production process will be at least partly performed by people, and people are fallible. Highly skilled glassblowers will still sometimes drop the molten glass when trying to gather it or fail to blow hard enough to fill the mold. But even setting aside clear-cut mistakes, people can't perform an action in a completely uniform manner. Glassblowers will gather slightly different amounts of glass each time and blow slightly different amounts of air into the pipe. The time it takes to gather the glass, fill the furnace, or blow the bulb will likewise vary slightly, if for no other reason than the natural variation in human movement.

This lack of uniformity also applies to machinery and equipment. Like people, machinery will sometimes fail. Parts will break unexpectedly or wear down over time. And even when a machine is working properly, it will inevitably have some amount of variation in its output. There's a limit to how accurate it can be. Injection molding, for instance, uses steel molds that are typically machined to within 0.005 inches accuracy, so an injection-molded plastic cylinder with a diameter of 3 inches might actually be anywhere from 2.995 inches to 3.005 inches when the equipment is working properly.[519] Other production methods may have higher or lower tolerances, but they will likewise have limits to their accuracy.

There will also be variation in input materials. Raw materials mined directly from the earth will have naturally occurring variation, while processed inputs will have variation due to the differences in the upstream processes that produce them. Structural steel, for example, is typically made to conform to ASTM A6 standards, which specify how much the chemical composition and dimensions of steel products are allowed to vary. A steel plate specified to be a quarter-inch thick might actually be anywhere from 0.22 to 0.28 inches thick.[520]

More broadly, in addition to the natural effects the production process harnesses to invoke a given transformation—the heat that turns solids into liquids, or the air pressure that forms the shape of a bulb—there will be innumerable other natural effects that influence the output of a process over which the producer has little or no control. On a small scale, changing environmental conditions can affect how a process behaves. Fluctuations in temperature can cause machines and equipment to expand or contract, slightly altering their behavior. Variations in ambient humidity levels might impact how liquids behave or cause components to dry out. High-energy cosmic rays might flip bits in electronic equipment and modify the behavior of software. On a larger scale, a tornado might strike the factory, the supplier of a critical component might declare bankruptcy, or a competitor with an improved product might cause demand to collapse.

Even the effects that a process purposely harnesses might not be completely understood. Steel was produced for hundreds of years before steelmakers understood that it was an alloy of carbon and iron.[521] Historic production of steel goods, such as Japanese swords, often relied on the smith choosing acceptable lumps of steel out of what the furnace happened to produce.[522] And while more modern processes are better understood, they still often behave in unexpected ways. As one mid-20th-century factory supervisor noted, "All our production troubles can be divided into two classes: the obvious and the mysterious. The obvious we understand and we can correct them easily. The mysterious we do not understand, and they are a continual headache!"[523]

So, due to uncontrolled or uncontrollable factors, any production process will inevitably have some amount of variation in it. In some cases, this variation may not matter. The behavior of a light bulb is unlikely to be affected if it's made of glass that's 0.0001 inches thicker than average. Tiny differences in the chemical composition of steel might not matter, so long as the steel is strong enough for its intended use. But in many cases, variation will have negative effects. Parts, components, and materials all have design tolerances, or differences they're designed to accommodate. If the variation exceeds those tolerances, the output isn't acceptable and must be thrown away or reworked. We'll refer to this as destructive variation. If the opening on a particular bulb blank is too big or too small, it won't fit in the base and will have to be thrown away or remelted, wasting the time and effort it took to make it. In fact, destructive variation is possible even when a part is within tolerance. The Japanese engineer and statistician Genichi Taguchi recognized that any deviation from a part's ideal value often results in that part working less well, even if it is within tolerance.[524] In car production, for example, Ford observed that transaxles with more variation in their dimensions produced much higher warranty claims, even if they were within the allowable tolerance.[525] Sony noticed something similar in the production of television sets.[526]

A second distinct cost of variation concerns the misalignment between demand and production. Whenever two adjacent steps of a production process operate at different rates—when the rate at which one step requires an input is different from the rate at which the previous step is producing it—there's a misalignment between demand and production. This misalignment will ultimately require some sort of buffer that will require resources to provide and maintain, which will add cost to the process. In some bulb blank factories, glass was made in large batches once a week. But in the actual blowing process, a team of glassblowers would form the blanks one bulb at a time. So the process involved a large buffer of molten glass waiting to be turned into bulbs.

To illustrate how variation in the time production steps take to complete can cause this sort of misalignment, consider a simplified five-step production process for pins. Raw wire comes into the factory and goes through a series of machines, where it gets cut, straightened, and sharpened, has the head added, and is painted. Each step in the process takes exactly 10 seconds, after which the pin is passed to the next step.

In this process, there are no buffers. After a pin is straightened, it is immediately passed to the sharpening step with no waiting, because the sharpening step finished with the previous pin at the exact same moment the straightening step finished with the current input.

No material accumulates between the machines, and each machine is occupied 100 percent of the time.

However, if we change this process so that each step takes *on average* 10 seconds but has some variation—say, a standard deviation of three seconds—we now have a misalignment between demand and production. Previously, every step was perfectly aligned with every other step. One step finished and was ready for another pin right when the previous step passed it along. But with the variation in process times, the steps are no longer aligned. Sometimes a step will complete and the next step won't be ready for it yet, so the pin will have to wait to be processed. Likewise, sometimes a step will complete and the previous step won't have finished yet, so the step will be forced to wait idly.

If we run a simulation of this process, we find that after 100,000 steps, significant amounts of material, or work in process, has accumulated in front of each machine. This work in process acts as a buffer, decoupling a machine from the variation of its predecessor. A step can now draw from its buffer whenever it needs more material without having to be perfectly aligned with the previous step. The buffer, of course, represents an additional cost, because it needs to be stored, managed, and accounted for.

There are several ways a process might be buffered. In the pin factory example, the buffering took the form of extra inventory between steps in the process. Another way of buffering would be through extra capacity. If we run the same simulation but cap the amount of work in process to just five units, the machinery will simply sit idle much of the time. As a result, the overall production rate will be much lower.

These two costs—destructive variation and misalignment—are closely related, since destructive variation will generally cause misalignment. For example, if each step in the pin-making factory takes exactly 10 seconds but has a 1 percent chance of failure, then 1 percent of the time a step won't pass a pin to the next step and the downstream machinery will sit idle while it waits for a pin. So not only do we have waste from destroyed pins, but the process steps are also no longer aligned. Even though each step produces at the same rate *on average*, the variation in the process results in wasted capacity, or time the machines aren't making pins.

There are two strategies for dealing with variation in a production process. The first is to reduce or eliminate the variation by achieving greater control over the factors that cause it. This requires both an

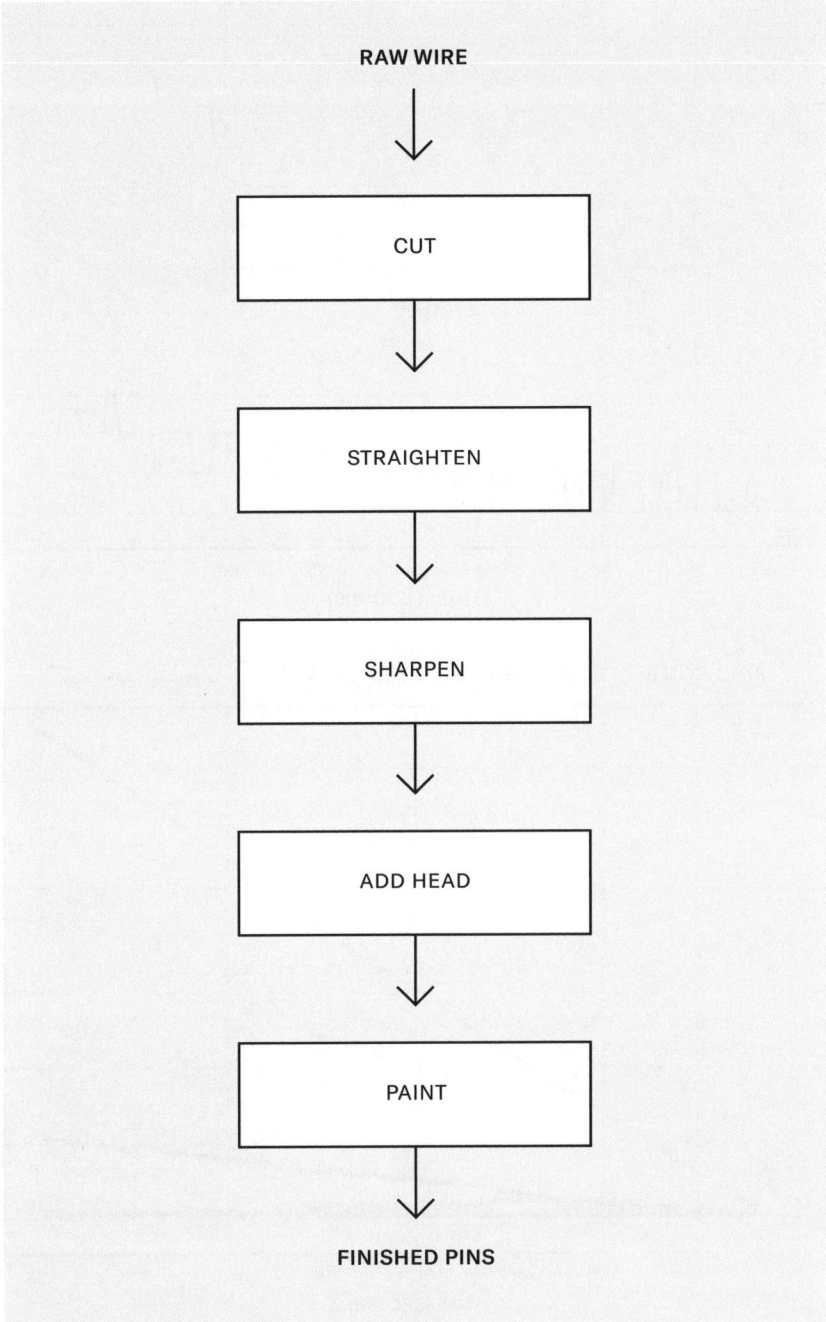

**Figure 10.** Process flow diagram of a pin factory.

PINS COMPLETED AND BUFFER SIZES OVER TIME (2OO SECONDS)

PINS COMPLETED AND BUFFER SIZES OVER TIME (1,600 SECONDS)

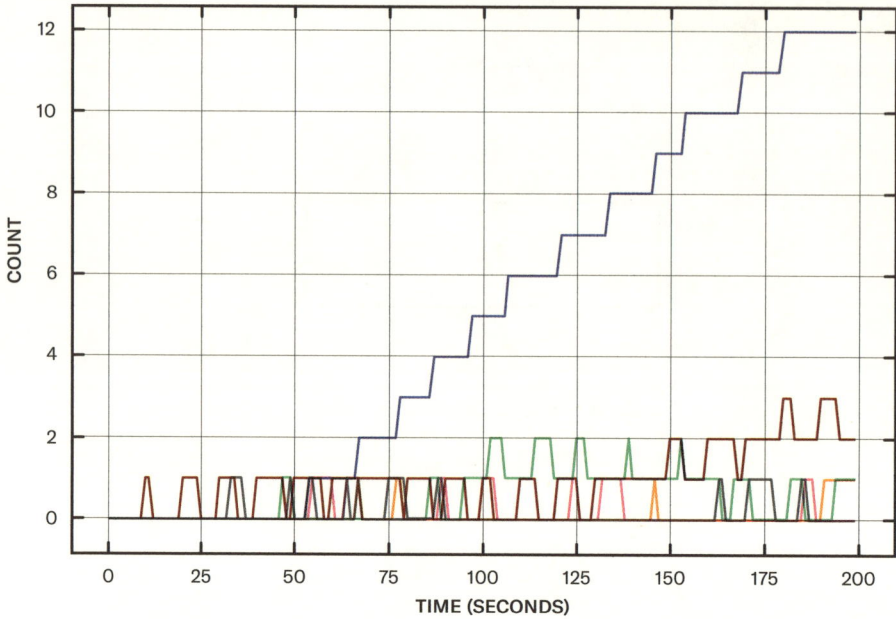

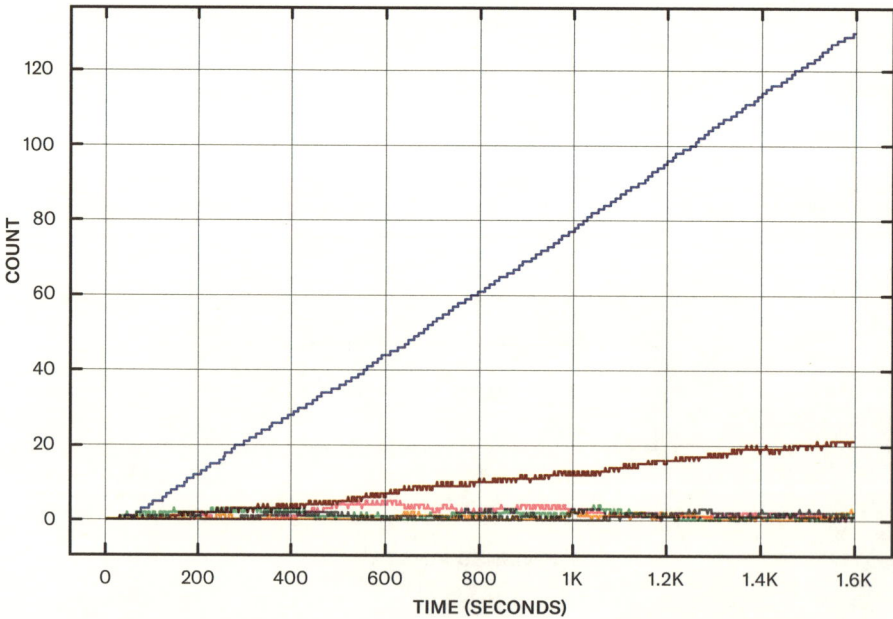

● COMPLETED PINS    ● STRAIGHTEN BUFFER    ● ADD HEAD BUFFER

● CUT BUFFER    ● SHARPEN BUFFER    ● PAINT BUFFER

**PINS COMPLETED AND BUFFER SIZES OVER TIME (400 SECONDS)**

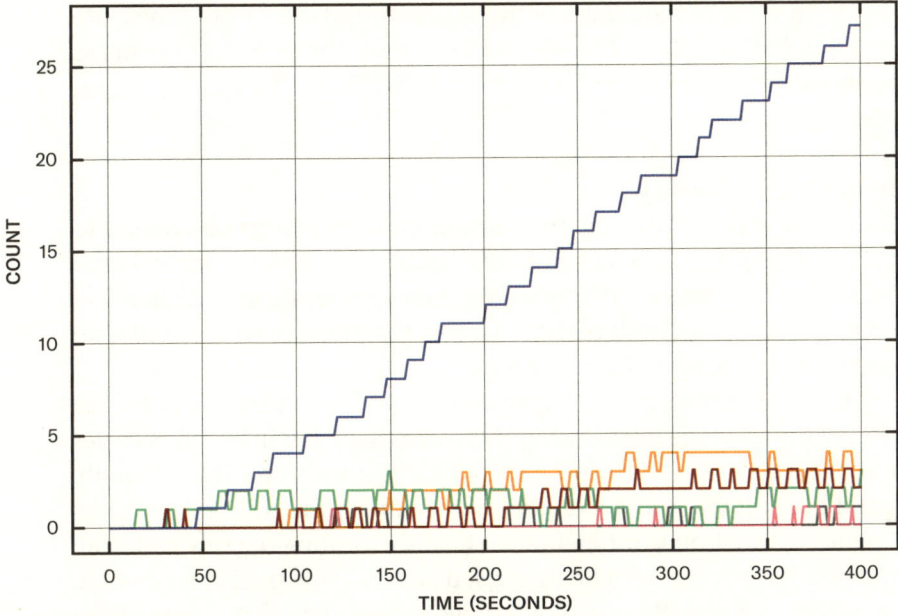

**PINS COMPLETED AND BUFFER SIZES OVER TIME (10,000 SECONDS)**

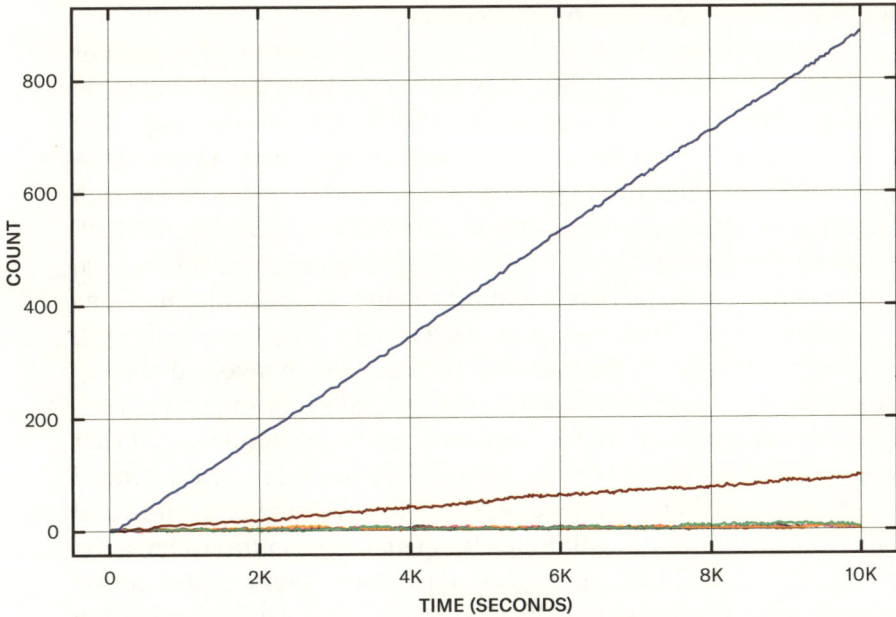

**Figure 11.** Simulation of a pin factory with an average of 10 seconds and standard deviation of three seconds for each step in the process to run. Over time, buffers of material accumulate in front of the machines.

understanding of the causes of the variation and an ability to measure and manipulate them. The second is to make the product or process more robust to variation that does occur so it doesn't affect the output of a process.

### Reducing variability

Reducing variability in the production process requires an ability to control the factors that affect it. If the quality of glass produced in the bulb blank process is affected by furnace temperature and chemical composition, then both of these factors will need to be controlled to ensure the process behaves reliably.

Control, in turn, is aided by knowledge of the factors that influence the process and the relationships between them. It will be much easier to control the temperature of the glass furnace to produce reliably good glass if we know exactly how temperature affects the glass quality—if we know, for instance, that after adding the raw material, the temperature should be increased slowly. This makes it possible to predict what will happen when we manipulate some part of the process, which enables us to achieve the output we want. Reducing variability is, therefore, a question of having increased knowledge and increased control of the production process.

How do we acquire knowledge about a production process? For most of human history, production processes were largely based on craft production, performed by hand by artisans and craftspeople and heavily reliant on the individual skill of the worker. In craft production, knowledge of the process was primarily acquired through experience, with masters passing on their skills to apprentices. Much of the knowledge would exist in tacit, difficult-to-articulate forms, such as performing specific movements or relying on physical sensations or a worker's intuition. For example, wrought iron in the early 19th century was made via the puddling process, which involved heating cast iron from the blast furnace in a special puddling furnace.[527] Making wrought iron in this fashion was a highly skilled craft that required ironworkers to constantly stir the iron and carefully control the atmosphere in the furnace. W.K.V. Gale, a scholar of the iron and steel industry, notes that although he spent many hours studying the puddling process, had handled all of the tools, and could describe exactly how the process was done, he could "not earn [his] living as a puddler," as it was an empirical trade that required years of practice and couldn't be learned from books.[528] Even something as seemingly simple as feeding a furnace with coal was a complex skill that required extensive practice to master.[529]

With their basis in human skill, along with a lack of any sort of scientific theory underlying them, most historical production processes functioned "despite nobody's having much of a clue as to the principles at work."[530] Against this backdrop, workers were often reluctant to adjust their methods, instead relying on traditional practices, which might vary greatly from practitioner to practitioner. Trades would often jealously guard their methods, making the spread of process knowledge difficult. Historian Barbara Keyser notes, for instance, that the dye industry in the late 18th century had little basis in science and was steeped in tradition, with one manufacturer describing nearly all practices "ow[ing] their origin to remote ages." Dyers would prevent people from observing their work to protect their secrets, and the practice of applying colors to fibers differed in nearly every dye shop.[531] Similarly, in the making of clay crucibles for metallurgy, some crucible makers were reluctant to replace the traditional method of working clay with their feet for hours with the much less labor-intensive pugging mill, believing that "human toil produced a better result."[532] This widespread conservatism meant that industrial processes often continued to be used for centuries, even millennia, after better processes had come along.[533]

Conservatism or a lack of concrete scientific knowledge don't preclude a production process from evolving toward greater control and less variability. Steel was probably originally produced in western civilization as an accidental by-product of the bloomery process for making iron, in which lumps of iron ore were heated along with charcoal in a large furnace, producing a spongy bloom of partially melted iron. In some cases, the process would produce a few bits of steel if the furnace happened to get hot enough and have the right chemical conditions.[534] But by the fourth century CE, steel could be produced reliably via the cementation process, in which iron from the bloomery was placed along with charcoal in clay chests and then heated for several days.[535] Similarly, an early technique for producing charcoal in pits was eventually superseded by a method that used large covered mounds, because charcoal makers found that the latter process could produce a purer charcoal with fewer impurities. And the production of brass, an alloy of copper and zinc, evolved from brass makers relying on copper ores that happened to have high zinc content to using closed crucibles that allowed them to closely control the alloy composition.[536]

More generally, producing *anything* requires figuring out how to sequence a series of transformations to achieve a particular result with some reliability. So the very existence of historical production processes implies some degree of understanding and control, which has been refined over time. But that process of improvement was often slow and haphazard, relying on undirected experiments and random chance. It could take decades, or even centuries, for improvements to occur under such conditions.

Over time, of course, knowledge of production processes, as well as the methods used to acquire such knowledge, became much more robust. Around the 18th and 19th centuries, producers gradually began to deliberately experiment with their processes to understand the relationships between different variables and how they could be manipulated.[537] As economic historian Joel Mokyr notes, experiments "created situations that did not occur 'naturally' and thus vastly expanded the realm of phenomena that could be cataloged and then harnessed."[538] This made it possible to learn much more about the process than could be gleaned simply by operating it. Toward the end of the 19th century, these industrial experiments often became quite involved. Frederick Taylor, of scientific management fame, spent more than 20 years performing more than 30,000 experiments in machine shops to determine the factors that affected how quickly metal could be machined, publishing this work in 1906 as *On the Art of Cutting Metals*.[539] In addition to making various discoveries about the properties of metal cutting (for example, that a heavy stream of water aimed at the cutting area increased cutting speed by up to 40 percent), Taylor established a series of mathematical formulas for the relationship between factors such as tool diameter, cut depth, tool life, and cutting speed that could be used to establish the optimal machining speed.[540] Similarly, in the 1890s at the German chemical firm Bayer, any new potential dye color—of which there were thousands every year—was subject to "tedious, meticulous" experimentation to determine whether and under what conditions it would tint any of the common fibers or other such items as wood, paper, leather, fur, or straw.[541] Through deliberate experimentation and controlled manipulation of production variables, the relevant causes and effects in a production process could be uncovered and understood.

The 18th and 19th centuries were also characterized by important changes in the dissemination of knowledge, as the relatively closed practices of traditional craftspeople gave way to more open sharing and distribution of knowledge about production processes. Knowledge that had previously been tacit was written down and shared in magazines, books, and technical publications, such as Diderot's famed *Encyclopédie*. Books on topics ranging from windmill construction to furniture design to chemical recipes to different types of mechanical movements became widespread. Engineering drawing practices became more finely developed, and industrial practitioners made greater use of mathematics, making it possible to more precisely describe and convey information about quantities and physical relationships. Standardization of measurement systems, such as the metric system, made it easier to share information. All these developments enabled knowledge about production processes to accumulate and spread on a massive scale.[542]

The development of measuring instruments that could more precisely quantify factors such as temperature, pressure, or flow rate also aided the acquisition of knowledge about production processes. Historically, determining the state of a production process had been limited to the capacities of the human senses. For example, medieval production of saltpeter, a key component of gunpowder, relied on the worker tasting a liquid mix of water, earth, and lime; when it tasted "biting" enough, the liquid was ready for boiling.[543] Similarly, heat-treating steel historically relied on a worker scrutinizing the color of the metal to estimate the steel's temperature.[544] In the late 18th century, however, instruments began to be developed that could measure not only existing production variables with greater precision but also variables that were entirely beyond the realm of human senses. The English potter Josiah Wedgwood developed a pyrometer for measuring the temperature of ceramics within a kiln, and American engineer Clemens Herschel invented the Venturi meter for measuring the flow of liquid.[545] By the early 20th century, large factories often had thousands of instruments to display and record production process variables.[546] Precisely quantifying factors that had previously only been understood qualitatively, if at all, created a clearer picture of what, exactly, was happening in a production process and how it could be manipulated.

Producers also made increasing use of scientific knowledge to understand natural effects, particularly those that couldn't be observed directly, such as chemical composition. Fertilizer, for instance, was greatly improved by the development of organic chemistry and systematic agricultural experimentation by scientists like Justus von Liebig, John Lawes, and J.H. Gilbert.[547] By the end of the 19th century, chemical fertilizers designed to provide specific nutrients, such as phosphorus, were in widespread use.[548]

Similarly, in the second half of the 19th century, iron and steel production began to increasingly draw on science and scientific methods. Carnegie Steel hired its first chemist after chemical testing showed that the flue cinder being thrown away as waste was in fact chemically identical to another type of cinder used earlier in the process, and was thus useful. By testing ores to determine their iron percentage, "nine-tenths of all the uncertainties of pig iron making were dispelled under the burning sun of chemical knowledge."[549] By the early 20th century, chemists were common in the iron industry, and the chemical reactions of the blast furnace had been mapped in great detail.[550]

Scientific theories that explained natural phenomena greatly aided the acquisition of knowledge about production processes by narrowing the scope of investigation. As Mokyr notes, "When no one knows why things work, potential inventors do not know what will *not* work... The range of experimentation possibilities that needs to be searched over is far larger if the searcher knows nothing about the natural principles at work."[551] Nevertheless, science provided only a rough guide to the production process landscape, and much of the work of developing and improving processes demanded empirical investigation. For example, despite using well-understood materials and technology, the development of flush riveting processes for aircraft manufacturing required extensive experimentation and testing to determine what factors influenced the process and how it could be made to work reliably.

By the early 20th century, experimentation, science, and measurement had made it possible to acquire a detailed, accurate picture of what was happening within a production process and what factors influenced it. One study from the 1930s describes how these factors came together to improve understanding of a coal-burning process:

It was discovered in early studies of fuel combustion that the $CO_2$ content of flue gases provided an index of combustion efficiency.

The Origins of Efficiency

If the supply of air were insufficient, there would not be enough oxygen in the mixture; then CO would be formed instead of $CO_2$, and less heat would be developed. On the other hand, if too much air were present, this excess air would be heated, and considerable heat would be wasted. By measurement and experimentation the most economical fuel-to-air ratio was determined, resulting in an appreciable increase of combustion efficiency.[552]

Heat-treating of steel provides another example. In heat-treating, steel needed to be heated above the point where its crystal structure changes; this is known as the decalescence point or critical point.[553] This point, which varies depending on the composition of the steel, had historically been judged by skilled heat treaters who relied on experience and visual observation.[554] But in the early 20th century, steelworkers observed that when steel was undergoing a phase change (that is, moving from one type of crystal structure to another), the rate of heat increase in the steel slowed down, which made it possible to precisely determine when the steel should be removed from the heat treatment furnace using specially designed thermometers.

Knowledge is only one side of the coin, however. It can tell us what's happening in a process and what factors affect it, but reducing variability ultimately requires the ability to control those factors. There are, broadly, three possible options for doing this: removing the source of the variation, shielding the process from the variation, or trying to compensate for the variation.

### Eliminating the variation's source

Eliminating the source of variation is in some ways the simplest method to increase control over a production process. But it's impossible to eliminate all variation within a production process, no matter how hard we try. In what cases, then, does it make sense to try to eliminate the source of variation?

The idea of deliberately trying to understand and control the sources of variation in a production process can be traced to the work of American physicist and engineer Walter Shewhart in the 1920s. According to Shewhart, there were two types of variation at work in any process: assignable cause variation and chance cause variation.[555] Chance cause variation is the result of numerous tiny factors: slight fluctuations in material composition, slight differences in air temperature or pressure, the natural variation in machinery or human actions. This type of variation has no direct assignable cause; it consists of many small fluctuations in conditions that are beyond the scope of our knowledge, either because they're fundamentally impossible to completely predict or because the effort required to try to predict them would vastly exceed the potential benefits.[556] Because chance variation has no specific cause, it can't be removed. However, numerous small factors will form a statistical distribution of variation, and if the underlying factors causing the variation don't change, this distribution will be stable and, therefore, predictable over time.

Assignable cause variation, by contrast, can be traced to some specific source: a machine setting, a work method, an environmental factor. Because it comes from a specific causal factor, this type of variation alters the conditions of the process, making it no longer statistically stable. However, because such causes are traceable (at least in theory), they can be found and eliminated. Removing all assignable cause variation from a process will leave only chance variation, so the process will act in a statistically well-defined manner. This type of work is known as statistical process control or statistical quality control.

The primary tool of statistical process control is the control chart, which graphs a measurement of the output of a process over time. The two primary types of control chart are the X-bar chart and the R chart. The X-bar chart measures the average of a sample of some quantity, such as length or diameter over time. The R chart measures the range of a sample: the maximum minus the minimum over time. The upper and lower bounds of the control chart would typically be defined as three standard deviations on either side of the average. A process that is in control will have control charts in which nearly all data points sit safely within these boundaries. Data points outside these boundaries or measurements with noticeable patterns indicate the presence of assignable sources of variation that should be identified and eliminated.

Using control charts, assignable cause variation can be hunted down and eliminated. One example comes from a company that was manufacturing electrical switches.[557] Initially, variation in the switch's magnetic intensity, a measure of switch performance, was incredibly high, with many data points outside the acceptable range of tolerance and a very low process yield. Plant researchers performed a study to identify sources of variation and eliminate them, one by one. First,

The Origins of Efficiency

the researchers found that operators were making adjustments to the machine to try and correct for the variation. Eliminating that practice uncovered new patterns, which ultimately revealed that the fixtures holding the switch assembly in place during one operation were loose. So the fixtures were changed and a magnetic material holding system was added. This modification brought to light another issue: In many cases, switches were being removed from a machining fixture without being given sufficient time to cool. So an automatic timer was added to prevent this from happening. This fix revealed yet another issue: The way the machine tool moved while in operation caused parts to occasionally become bound in a fixture. So the machine's motors were relocated, which solved the problem. At this point, the process was under control. All data points were within the expected variation, which was within the allowable tolerance of parts. Any future deviations could now be noticed and addressed quickly.

Statistical process control can help identify and remove assignable cause variation, but it's ultimately limited by the capabilities of the process and the nonassignable cause variation within it. (In fact, trying to reduce variation beyond the level of consistency the process is capable of only serves to *increase* variation.)[558] This frontier can only be pushed back by introducing new technology that operates with fundamentally less variation than existing technology. For example, in the 17th and early 18th centuries, producing cannons typically involved casting a solid piece of bronze or iron and then drilling out the barrel using large vertical boring machines. This process had limited accuracy, and cannons were often scrapped due to inaccurate boring. But in 1715, Swiss inventor Jean Maritz developed a horizontal boring engine mounted to a heavy stone foundation that, unlike existing vertical boring engines, remained stationary while the cannon was rotated around it. This new technique greatly improved the accuracy of cannon boring.[559] English industrialist John Wilkinson likewise used a horizontal boring engine to produce the highly accurate cylinders that made the Watts steam engine possible. Similarly, the replacement of natural dyes with synthetic dyes in the mid-19th century reduced color variation, because the synthetic dyes were "perfectly standardized so the exact color could be reproduced repeatedly with absolute certainty."[560]

In general, the introduction of automatic machinery tends to reduce the amount of variation within a process. Although machines don't behave in a perfectly uniform fashion, they're capable of much

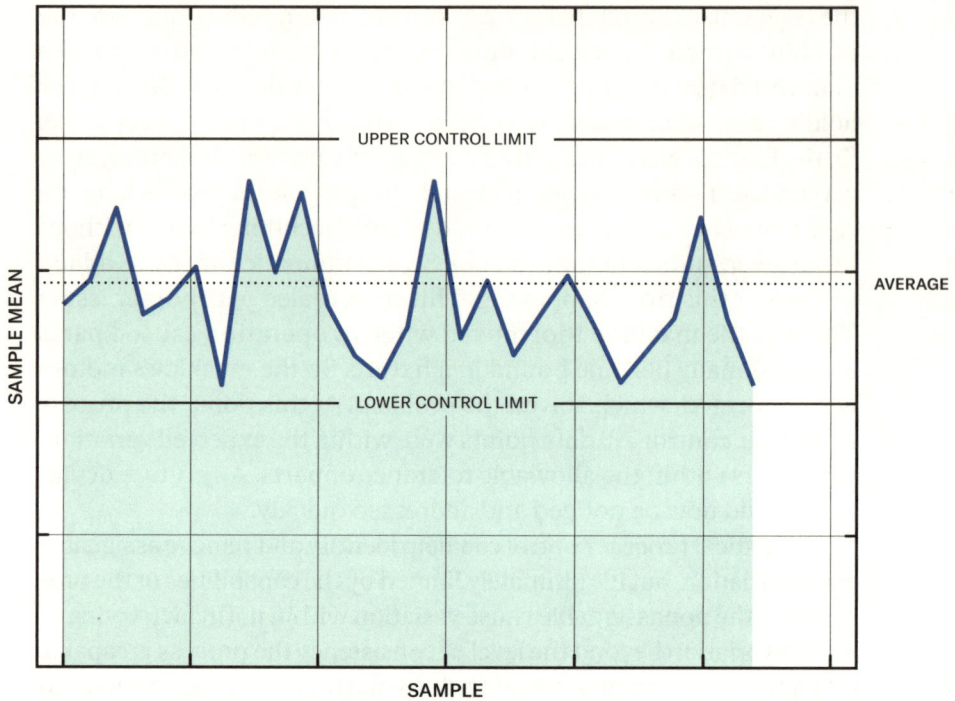

**Figure 12.** Example of an X-bar chart.

greater uniformity than people. Automatically controlled open-hearth furnaces were much more regular than human-operated ones, resulting in higher production volumes and less fuel used per ton of steel produced. Similarly, automatic heat-treating furnaces produced a more uniform, reliable output than the skilled heat treaters they replaced.[561]

The development of new technology can also extend the consistency frontier by affecting upstream processes. If the quality of input materials rises, variation in the output of the process will fall. Rayon manufacturing, for example, replaced expensive cotton with cheaper wood pulp as a source of cellulose as the quality of wood pulp improved.[562]

### Shielding from variation

In some cases, it will be impossible to eliminate a source of variation. But in such cases it's often possible to shield the process so the variation can't affect it. In essence, this strategy seeks to break the causal path between the source of the variation and the production process.

The Origins of Efficiency

The simplest example of this is moving a production process into a factory building. Enclosing the production process within a building shields it from variations in weather conditions, preventing the process from being affected by wind, rain, or snow. On a smaller scale, encasing production equipment in a protective cover can prevent interference from water, dust, and people touching it. And shielding the production process from one source of variation will often have the ancillary effect of protecting it from additional, unobserved, or poorly understood sources of variation. An enclosed factory protects against known sources of variation, like garden-variety bad weather, as well as ones we might not anticipate, like freak hailstorms or clouds of locusts.

In practice, most processes are shielded from variation to some extent. But completely shielding a process from sources of variation is often impractical. It would be prohibitively expensive to try to make most factories fully airtight to shield the interior from any variation in outside air, for instance. And it's infeasible to shield a production process from things that are integral to the process itself, like the behavior of workers and machinery or the properties of input materials. And, as with relying on buffers to damp variation between steps, shielding a process from variation adds expense. For example, in the early 20th century, chemical plants were often located in cold northern locations, which required burying pipes beneath the frost line and putting equipment within enclosed, heated buildings. When plants began to be built in the much warmer Texas climate, shielding operations from the cold became unnecessary and plants adopted open construction techniques that didn't require enclosed structures. This reduced the cost of building a chemical plant by between one-third to one-half.[563]

### Compensating for variation: control systems
If it's not feasible to shield the production process from a source of variation, it may be possible to adjust the process to cancel out the effect of the variation. The most common way to do this is with a feed-back-based control system, which measures the output of a process, compares it to some desired value, and then adjusts the process to try to bring the two values into alignment.

A thermostat is a classic example of a feedback-based control system. Indeed, temperature fluctuations are often an important source of variation that producers seek to eliminate. It would be impractical to try to shield the interior of a factory from changes in temperature—no matter how thick the insulation, heat would eventually leak in or out. Nor is it possible to eliminate the sources of variation in heat, unless one has control over the sun and the clouds. But with a thermostat hooked up to an HVAC system, a factory can compensate for temperature variations, turning on a heater when the temperature falls too low or an air conditioner when the temperature gets too high, keeping the interior temperature close to constant. In fact, air conditioning was originally developed as an industrial control technology for maintaining a precise level of humidity within a factory. Operations such as textile mills, bakeries, and chocolate makers were highly sensitive to the amount of humidity in the air, and using air conditioners to keep humidity within a narrow range significantly improved the consistency of their production processes.[564]

One advantage of a control system is that it can eliminate variability even without being able to predict everything that might affect the system. With a thermostat and an HVAC system, we don't need to understand, anticipate, and control every single factor that might influence the temperature inside a factory—we just need to be able to measure, raise, and lower the temperature. Control systems thus make it possible to compensate for unexpected disturbances and uncertainty.

Historically, human operators acted as manual control systems, observing the state of a production process and adjusting it to bring it in line with a desired value. For example, when windmills were used to grind grain, the gap between millstones needed to be kept constant to grind the grain to the proper coarseness. But as the wind blew faster, the upper millstone had a tendency to rise. So the miller would watch the grindstone and adjust its height to keep the gap between the stones constant, an adjustment known as tentering. Similarly, if the direction of the wind changed, the miller would manually rotate the body of the windmill to keep it facing into the wind.[565]

We can also think of craft skills in control-system terms. Consider a traditional skill like making a hand ax by knocking off flakes of stone from a rock, a process known as flint knapping. Rather than performing mindless repetitions in this process, a human making a hand-ax will adjust each step based on how the last step went. If one flake is unexpectedly large or unexpectedly small, the ax maker will adjust their subsequent strikes to compensate. By adjusting

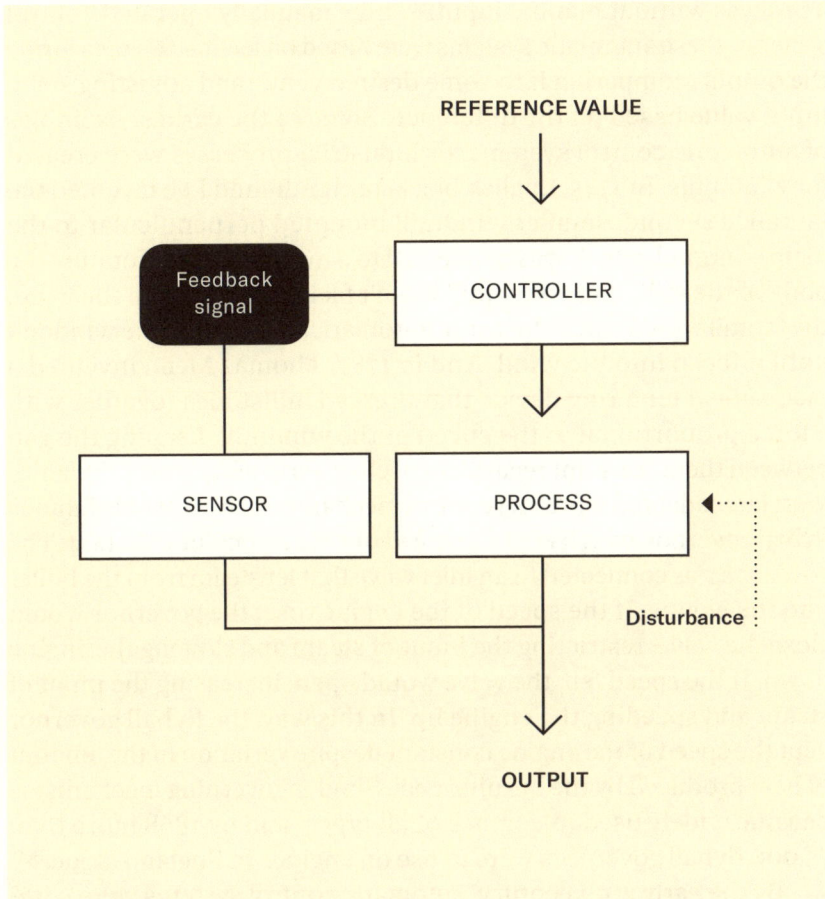

**Figure 13.** Basic structure of a control system.

the process based on feedback, it's possible to offset the natural variation in materials and human movement, making it possible to achieve a desired output (say, a hand ax of the proper shape) without being able to perfectly predict or control the effects of each action. This sort of control-based craftsmanship enabled the production of precision artifacts like firearms or clocks even though parts weren't uniform or interchangeable; craftspeople would adjust the size and shape of the parts to fit with the contours and imperfections of previous parts.[566]

In the 17th century, inventors, engineers, and tinkerers began to develop automatic control systems that could make adjustments to a process without manual input.[567] Like manually operated control systems, these automatic systems were based on feedback—measuring the output, comparing it to some desired value, and adjusting some input value based on the difference. Some of the earliest examples of automatic control systems for industrial processes were created for windmills. In 1745, English blacksmith Edmund Lee invented the fantail, a second, smaller windmill mounted perpendicular to the main windmill, which was connected to a mechanism for rotating the body of the mill. If the windmill wasn't facing directly into the wind, the fantail would catch the wind, automatically rotating the windmill until it faced into the wind. And in 1787, Thomas Mead invented a mechanical tentering device that pressed millstones together with a force proportional to the speed of the windmill, keeping the gap between them constant regardless of changes in wind speed.[568] James Watt later adapted tentering control mechanisms to create his famous flyball governor, which kept the speed of a steam engine constant. The governor was connected to an inlet valve that let steam from the boiler into the engine. If the speed of the engine rose, the governor would close the valve, restricting the input of steam and slowing the engine down. If the speed fell, the valve would open, increasing the input of steam and speeding the engine up. In this way, the flyball governor kept the speed of the engine constant despite variation in the amount of heat produced by the burning coal. Similar governing mechanisms became widely used in engines of all types, and by 1868 more than 75,000 flyball governors were in use on engines in England alone.[569]

By the early 20th century, automatic control systems were common in industrial processes, often replacing less reliable processes that depended on operator judgment. As with other types of mechanization, mechanical control systems could exceed human physical limits, measuring quantities a human couldn't perceive and responding much faster than a human could. These advantages, combined with the greater uniformity of mechanical operation, significantly reduced production costs. Automatic temperature control in clay kilns, for example, could significantly decrease wasted product, which was as high as 20 percent in manual kilns.[570] In gasoline production, prior to the introduction of automatic control, yields were often low because the conditions in gasoline stills were difficult to control manually. One report summarizes the challenges:

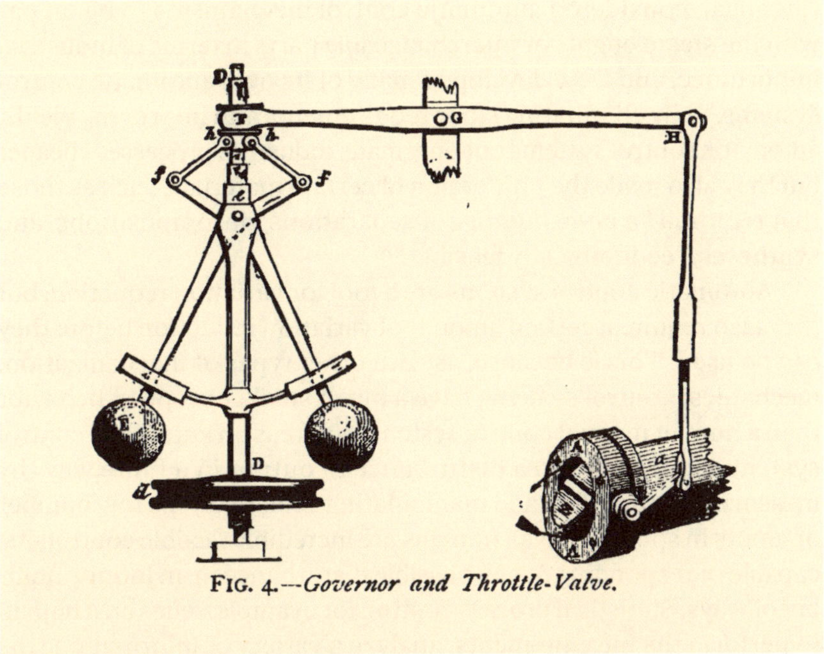

FIG. 4.—*Governor and Throttle-Valve.*

Figure 14. Watt's flyball governor. Wikimedia Commons, CC-PD-US-expired.

Maybe [the still operator] lets the temperature of a still run up, with the result that material is carried over that shouldn't go into that cut. Perhaps it is caught in time and diverted to a slop tank from which, later, it has to be rerun at more expense of time and fuel and equipment... Or if the stillman does not keep up his fire and lets the temperature drop too low, he fails to get off all the gasoline in the first still, later stills have to do some of the first still's work, and the whole mess perhaps has to be rerun and the gasoline yield in the end is less than it should have been.[571]

Automatic temperature control systems allowed gasoline plants to run for longer and have higher yields while reducing fuel consumption, coke formation, and required repairs.[572] Similarly, in cement manufacturing, automatic control systems introduced in the 1950s could react far faster than human operators to changing kiln conditions. Being able to instantly adjust a myriad of kiln variables enabled the construction "of kilns far larger than had previously been thought possible."[573] Herbert Henry Dow, the founder of Dow

Chemical, considered automatic control mechanisms to be on par with the steam engine or interchangeable parts in terms of industrial importance, and Dow developed many of its own automatic control systems.[574] By eliminating labor requirements and improving yields, automatic control systems not only made industrial processes cheaper but they also made the production of certain materials, such as those that required "a large number of separations, decompositions, and syntheses," economically feasible.[575]

Automatic control systems are a tool for variation reduction, but they also require a certain amount of variation reduction before they can be used. This is because, as with other types of mechanization, mechanical control systems have a much smaller scope of behavior than a human manual control system. Whereas an automatic control system can respond to a disturbance in output in a finite way (by measuring limited data and manipulating a relatively narrow number of inputs in specific ways), humans are incredibly flexible controllers, capable of responding to any possible disturbance in an infinite number of ways. Statistical process control, for example, relies on a human to perform the measurements, analyze a variety of information (the pattern in the observations, the behavior of workers and equipment, the properties of various input materials), find the source of the variation, and make any number of possible adjustments (repairing equipment, changing operator procedure, modifying material requirements) to eliminate them.

In the second half of the 20th century, reducing process variation became a prime focus for manufacturers. Many popular industrial improvement systems, notably Six Sigma, the Toyota production system, lean, and total quality management, are heavily focused on it.[576] The Toyota production system provides an especially interesting case study of how different methods of variation reduction can be applied in a production setting, so let's examine it in more detail.

**Variation reduction in practice: the Toyota production system**
The Toyota production system (TPS) was based on adapting Ford's mass production methods to the Japanese market (which was both smaller in volume and demanded greater product variety) and combining them with quality control ideas adapted from W. Edwards Deming and Joseph Juran.[577] The central goal of TPS is to reduce the costs of production while simultaneously increasing quality. The two

primary strategies for lowering costs are reducing the amount of labor and decreasing the amount of inventory and work in process. As we discussed earlier in this chapter, inventory functions as a buffer that can damp variation between two steps in a process, and TPS considers buffers waste that should be eliminated. Unlike conventional mass production, in which parts are produced in large batches, the goal in TPS is one-piece flow—producing one part at a time, at exactly the rate it is needed, continuously. Minimizing inventory and achieving a state of just-in-time production eliminates all of the costs of that inventory, including handling, storage, and investment.

But producing parts just in time requires an extremely predictable and reliable process. Without buffers to damp variation, any disruption will be immediately felt. As Taiichi Ohno, one of the creators of TPS, notes, "An upset in prediction, a mistake in the paperwork, defective products and rework, trouble with the equipment, absenteeism—the problems are countless. A problem early in the process always results in a defective product later in the process. This will stop the production line or change a plan whether you like it or not."[578] So to achieve just-in-time production, factors that might disrupt or inject variability into the production process must be found and eliminated. TPS has a variety of tools for doing this.

One source of variation is fluctuating demand, which requires either extra inventory, to cover capacity shortfalls when demand rises, or extra capacity, which will sit idle when demand falls. To ensure a steady daily production rate that doesn't require inventory buffering, TPS uses a strategy called *heijunka*, or production leveling: The company determines how many cars of each type will be needed over a certain period (say, a month), then produces them at a steady, continuous rate, shielding the process from variation in demand. To keep the production rate as close to constant as possible, Toyota and other lean Japanese carmakers might cut prices during downturns to keep demand from falling.[579]

True one-piece flow is often difficult to achieve because different machines and processes often run at different rates. To accommodate situations where true one-piece flow can't be implemented, TPS uses the *kanban* production control system. The system works roughly as follows: Between any two steps in the process, there will be a collection of physical kanban cards, with each card representing an order that a process must fulfill. When a worker completes a step in the process,

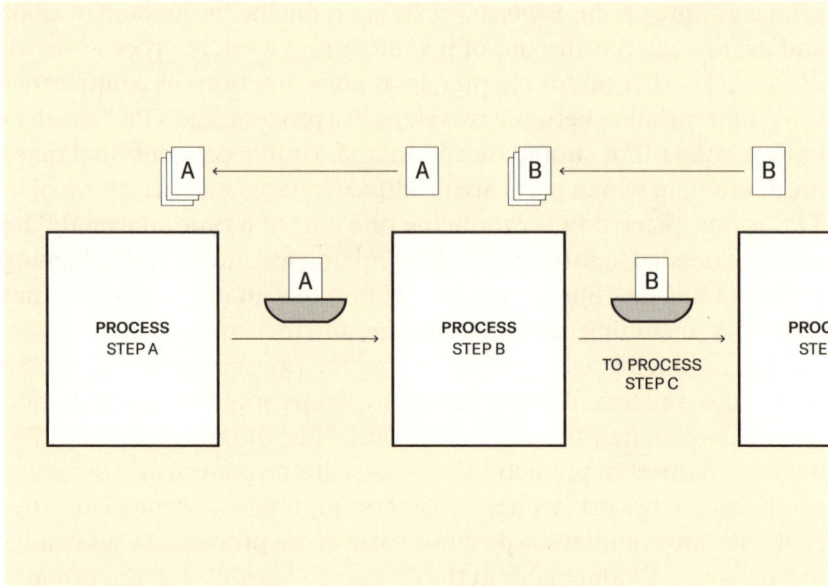

**Figure 15.** A kanban system. When process step A completes a part, it sends the part along with a kanban card to process step B. When step B starts its work on that part, it sends the card back to A; when step B finishes, it sends the part with a card to step C, and so on. Each step can only proceed if it has kanban cards available.

they send that part, along with a kanban card, to the next step in the process. When the next step completes an operation on its buffer of parts, it sends a kanban card back to the previous step. The first step can only produce a part if it has a kanban card from the next step—if it doesn't, it stops production. This prevents steps in the process from producing parts that subsequent steps don't need yet and limits how much inventory can accumulate in the system, since buffers between two steps can't exceed the number of kanban cards. Kanban acts as a feedback-based control system that compensates for variation in production times by starting and stopping production operations. While this doesn't eliminate variation, it does limit the size of the inventory buffers. And, as we'll see later, kanban and similar systems that cap the amount of work in process also add robustness to a process.

With one-piece flow and kanban, there are either no buffers or very small buffers of extra inventory in the process. This means that any disruption to the process—a machine failure, a bad part—immediately shuts down the production line. But in TPS this is considered a feature rather than a bug, because the disruption reveals problems in the process that the inventory buffers were masking. This approach—reducing the amount of work in process to uncover problems—is often referred to as "lowering the water level to reveal the rocks."[580] The problems are then diagnosed and fixed with the aid of techniques such as the five whys. As Taiichi Ohno puts it, "By repeating why five times, the nature of the problem as well as its solution becomes clear."[581] In his book on TPS, Ohno gives an example of applying the five whys to a machine that has broken down:

| | |
|---|---|
| 1 | Why did the machine stop? |
| | There was an overload and the fuse blew. |
| 2 | Why was there an overload? |
| | The bearing was not sufficiently lubricated. |
| 3 | Why was it not lubricated sufficiently? |
| | The lubrication pump was not pumping sufficiently. |
| 4 | Why was it not pumping sufficiently? |
| | The shaft of the pump was worn and rattling. |
| 5 | Why was the shaft worn out? |
| | There was no strainer attached and metal scrap got in.[582] |

When a breakdown or other issue occurs, a worker will pull an *andon* cable, which signals to other workers that there is a problem at a workstation. Like statistical process control, andon signals act as part of a higher-level, flexible control system that allows workers to quickly notice and take action to correct problems.[583]

Features of TPS discussed in prior chapters, such as poka-yoke (mistake proofing) and autonomation (automation with a human touch), can also be seen as ways of reducing variation, both by preventing errors from happening and by making it obvious when they do happen so that they can be quickly corrected. For example, in one Toyota factory, an assembly operation required installing a cotter pin on a ball joint. To ensure the pin was installed, the tray of pins was separated from the worker by a light sensor. If the worker didn't reach for the pin and trigger the sensor, an alarm would sound. Similarly, to prevent machine breakdowns from disrupting the production process, TPS emphasizes preventive maintenance to ensure machines will behave reliably. This often takes the form of total productive maintenance, where the whole organization focuses on keeping production equipment in proper condition.[584] In this way, TPS extends the statistical process control idea of identifying sources of assignable cause variation. Rather than simply noticing when an error occurs, TPS aims to prevent errors from happening in the future, with the ultimate goal of having zero defects.

Achieving the level of control TPS requires to function demands deep knowledge of the production process. TPS, therefore, encourages managers to "go and see" (*genchi genbutsu*) the shop floor to understand how work is actually done.[585] Because individual workers are invariably the most knowledgeable about how processes work, Toyota uses quality control circles where small groups of workers study problems and try to generate solutions. Workers can submit suggestions for improvements, which are taken seriously by management. Indeed, in 2005, Toyota implemented nine suggestions per worker per year on average.[586] Similarly, TPS cultivates worker insight by rotating workers through many jobs, yielding many different perspectives and ideas for solving problems. The job rotation system also enables TPS to flexibly adapt to variation in market demand. As demand for different car models rises and falls, workers can easily be reassigned to different production lines to meet changing production targets, in what is known as *shojinka* (flexible staffing).[587]

Like any car manufacturer, Toyota does not produce all its own parts but purchases parts and materials from suppliers. Minimizing variability in supply requires close coordination and relationships with suppliers, which might include stationing employees at a supplier and having the supplier station an engineer in the manufacturing plant.[588] Because of its close supplier relationships (often known in Japan as *keiretsu*, or a network of businesses with close ties, including cross-ownership), Toyota can reliably get the parts it needs exactly when it needs them, minimizing the amount of inventory it must hold.[589]

**Making a product or process more robust to variation**
The other way to reduce the effects of variation is to make the product or process more resilient to it. With this approach, the variation still impacts the process—it is neither eliminated, shielded, nor counterbalanced—but it doesn't affect the output of the process or the functionality of the product.

As an example, it may be possible to relax the tolerances on a product without affecting its functionality. Just as there may be production steps that don't add value and can be eliminated, there may be specified tolerances that are stricter than necessary. Because achieving strict tolerances is often expensive, relaxing them can significantly reduce production cost. One manufacturer, for example, was producing a small steel valve part that consisted of a head and shaft. The dimensions were specified to within half a thousandth of an inch—a very tight tolerance—resulting in high manufacturing costs. Upon investigation, it was found that while the head of the valve needed to be this precise, the shaft did not. Relaxing the tolerance on the shaft to two-thousandths of an inch reduced manufacturing costs for the part by about 98 percent.[590]

It may also be possible to redesign the product so that wider tolerances are acceptable. For example, if mating surfaces—the places where one part is pressed against another part—can be eliminated, a much higher level of surface roughness is acceptable. Similarly, since each part produced will have its own tolerances, layering one part on top of another will cause these tolerances to build up, in what is known as tolerance stacking. If a product can be redesigned to have fewer parts, there will be less tolerance stacking, which may reduce the tightness of the tolerances required.

Consider a rod made up of three parts that fits between two connectors. If the entire rod must have a length tolerance of ±0.1 inches, then each individual part of the rod needs a tolerance of just ±0.0333 inches; any larger than this and, in some cases, the total length might exceed the allowed tolerance. But if the rod can be redesigned to be made from just one part, it can have a much looser tolerance of just ±0.1 inches, since tolerance stacking has been eliminated.

Because overly tight tolerances add significant manufacturing costs, production manuals and design guides often stress the importance of avoiding them. One manual notes that "it is only in exceptional cases that a mechanism cannot be modified so as to retain all functional advantages and yet allow liberal tolerances on the majority of its dimensions."[591] In the rules for design given in Roger Bolz's handbook on production process productivity, the third rule is to "use the widest possible tolerances."[592]

However, liberal tolerances aren't always possible. Some products, by their very nature, are susceptible to minute amounts of variation and require extremely tight tolerances. Semiconductors must be dosed with extremely precise amounts of certain chemicals, on the order of one part per million, and even tiny amounts of contaminants can dramatically change their properties.[593] Similarly, lenses and mirrors for telescopes are affected by minuscule changes in shape.[594]

But even in such cases, it's often possible to make the process more robust to variation without that variation being passed along to the final product. One method is known as selective assembly, in which parts are grouped together based on assembly-relevant dimensions such as diameter and length after they've been produced. Take a piston head and an engine block, for example. To work properly, the gap between the piston and the cylinder must be as small as possible while still allowing the piston to move smoothly. One way of achieving this without requiring overly burdensome fabricating tolerances is through selective assembly: Pistons made slightly too big are matched with engine cylinders that are also slightly too big. In this way, close tolerances—a piston that fits snugly into a cylinder—can be achieved and low yields avoided, even with high-variation production methods. Historically, selective assembly has been an important technique for achieving close tolerances; Ford factories used it to produce the Liberty airplane engine during WWI.[595]

Another way to make the process more robust to variation is through variability pooling. If different sources of variability can be added together, they will tend to average out, reducing variation overall. In the Bessemer steelmaking process, slight variations in the chemical composition of the iron in the blast furnace could have a large effect on the process. As historian Thomas Misa notes, "If there was a high level of silicon or carbon, say, the converters overheated; if the level was low the converter would chill solid."[596] The invention of the Jones mixer, which pooled the output of many blast furnaces into one large vat of pig iron, solved this problem. Pooling the output of multiple blast furnaces tended to even out the variation of each; high silicon content in one would be balanced by lower silicon content in another.[597]

Variability pooling can also be used to reduce variation in process times. If many stations perform an operation in parallel, process time will be more uniform if they all share the same queue. If one machine takes unusually long to complete a task, the queue can simply redirect incoming material to the other machines. This sort of queue sharing is commonly used in banks and other service industries to limit congestion in the face of highly variable process times.

Similarly, variability pooling can be used to damp variation in demand. The apparel company Benetton would produce undyed

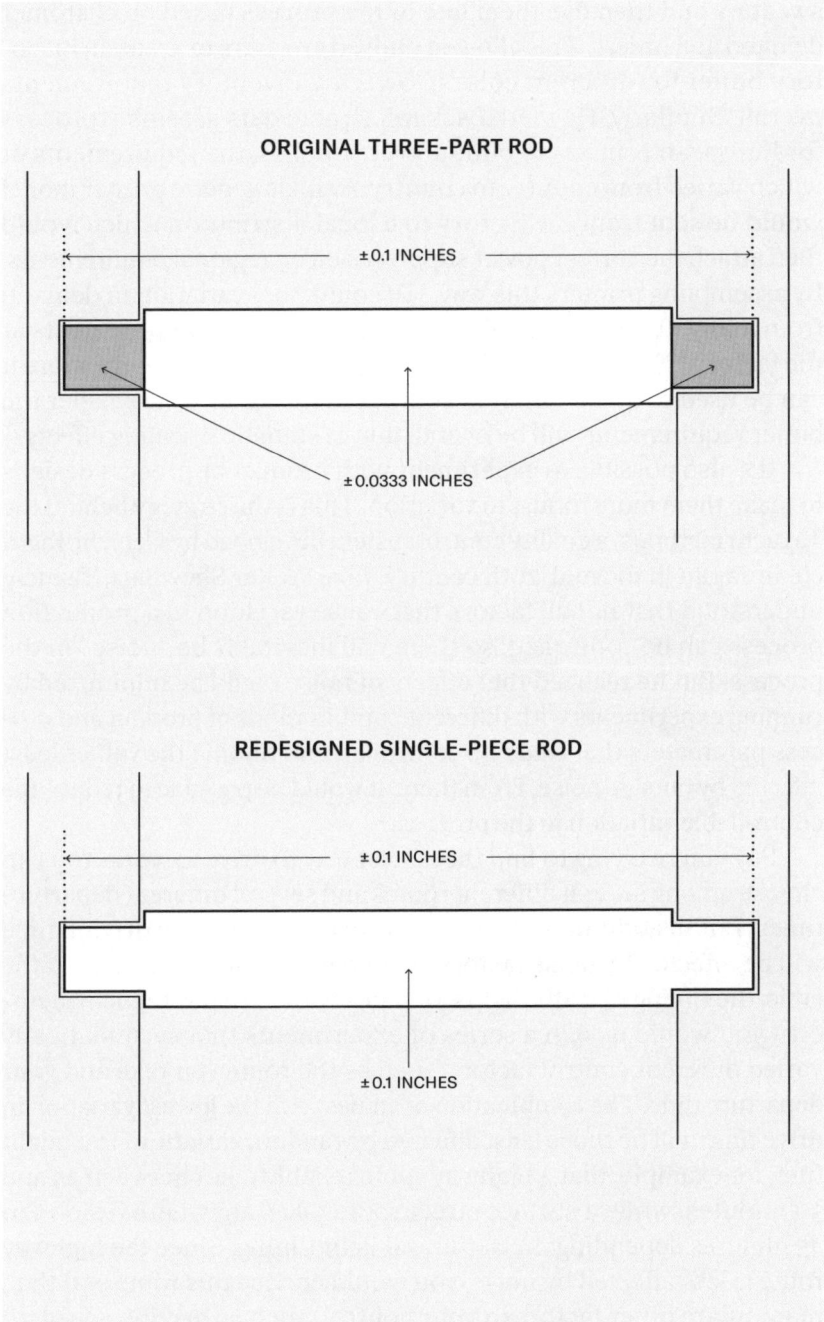

**ORIGINAL THREE-PART ROD**

± 0.1 INCHES

± 0.0333 INCHES

**REDESIGNED SINGLE-PIECE ROD**

± 0.1 INCHES

± 0.1 INCHES

**Figure 16.** Tolerance stack-up in single-piece and multipiece rods.

sweaters and then dye them late in the process based on customer demand and orders. This allowed undyed sweaters to act as an inventory buffer for different colors, lowering inventory requirements overall. Similarly, Hewlett-Packard arranged its assembly process for European printers to omit power supplies, the requirements of which varied from country to country. A single generic printer model would be sent from the factory to a local distributor, which would then attach the correct power supply based on regional requirements. By assembling printers this way, HP could pool variation in demand from many different countries, reducing inventory requirements at the factory.[598] More generally, the more flexible a buffer—the more it can be used to damp different sources of variation—the smaller the buffer requirements will be overall due to statistical scaling effects.

It's also possible to experiment with product or process designs to make them more robust to variation. This is the strategy behind the Taguchi methods, a quality control system developed by Genichi Taguchi in Japan in the mid-20th century. Like Walter Shewhart, Taguchi understood that not all factors that cause variation in a production process can be controlled, so there will inevitably be "noise" in the process. But he realized that effects of noise could be minimized by running experiments with different combinations of product and process parameters that *could* be controlled and finding the values least affected by outside noise. From there, it would be possible to reduce the controllable variation in the process.

Say you're trying to find the fastest way to drive to work. You can choose among several different routes and several different departure times. But in addition to these controllable factors, the driving time will be affected by noise factors beyond your control, such as traffic jams, the timing of traffic lights, or bad weather. Using Taguchi methods, you would design a series of experiments that systematically varied different control factors, such as the route you take and your departure time. The combination of values with the lowest variation in drive time will be those least affected by random variation. You might find, for example, that a highway route reliably takes between 25 and 30 minutes, while a surface street route takes anywhere from 15 to 45 minutes depending on traffic and light timing. Since the highway route is less affected by noise, you would choose this route and then manipulate other factors in your control, such as driving speed, to bring the average driving time closer to your goal.

An example of the Taguchi methods in practice comes from a company that was using a large furnace to bake ceramic tiles. The kiln operators found that tiles on the edges of the kiln weren't baking uniformly due to uneven temperature levels in the kiln; as a result, the tiles were coming out in a wide variety of sizes. Redesigning the kiln was too expensive, so the manufacturing team identified seven other controllable factors they thought were likely to affect the tile shrinkage. They then ran experiments in which they systematically varied these controllable factors, along with tile positions in the kiln. They discovered that one factor, the limestone content of the tiles, had the largest effect on tile shrinkage. By raising the limestone content of the tiles from 1 percent to 5 percent, along with other changes in material composition, the tile defect rate fell from 30 percent to less than 1 percent.[599] Although the temperature variation couldn't be reduced, the kiln operators were able to make the tiles more robust to this source of variation by changing their material properties.

One final way to make a process more robust to variation is to use a pull system rather than a push system for inventory control. In a push system, material is released into the production process on a set schedule. In a pull system, inventory is pulled into the production process only when the process is ready for it. A push system can, therefore, continue to accumulate work in process even if subsequent stages of the process aren't ready for the material, but a pull system will cap the amount of work in process. The kanban system discussed earlier in this chapter is perhaps the most famous example of a pull system.

Capping the amount of work in process gives pull systems much less variable cycle times than push systems, since in a push system inventory, and thus buffer size, can simply continue to rise.[600] In addition to ensuring lower, more predictable cycle times, mathematical analysis shows that a pull system is more robust to small deviations from the ideal level of inventory. In a pull system, relatively large differences in the work-in-process cap have a small impact on the production rate, and there's comparatively little risk from getting the cap slightly too high or slightly too low. But in a push system, small errors in the release rate have much larger effects. Set the release rate too high, and work in process will continue to build up. Set the release rate too low, and the production rate will be lower than what the factory is capable of achieving, adding cost by spreading overheads over a smaller amount of output.[601]

### Variability trade-offs

These two strategies for dealing with variability in a production process—reducing it or making the process more robust to it—point to a fundamental tension: Reducing variability has the potential to both increase and decrease production costs.

On the one hand, reducing variability often lowers costs. The less variable a process, the less it fails, improving yields and decreasing waste and rework. Reliable and consistent processes also require smaller buffers between process steps, need fewer inspections, and will have fewer quality problems in the field (and, therefore, have lower warranty costs). Quality control advocates note that "every penny you don't spend on doing things wrong, over, or instead becomes half a penny right on the bottom line."[602]

But it's equally possible for variability reduction to increase cost. Shielding, eliminating, or compensating for variation requires additional equipment and labor, which add expense to a process. Semiconductor fabrication facilities are expensive to build in part because they require costly environmental controls to keep conditions inside the fabs as uniform as possible. Similarly, making parts more uniform requires tighter tolerances and costly machining operations. Tighter tolerances might also mean lower yields, more expensive and higher-quality inputs, or additional inspection steps. This is why DFMA manuals and manufacturing recommendations emphasize having the widest possible tolerances on parts and omitting the unnecessary precision of making parts more accurate than they need to be. One manufacturing manual from the mid-20th century went so far as to state that if you never produce a part out of tolerance, you're spending too much on machining operations.[603]

Reducing variability, then, comes with a trade-off: Making a process work more reliably and uniformly reduces cost, but the time and effort to achieve these desiderata is itself expensive. How can we determine if this trade-off is worth it?

While the specifics will inevitably vary depending on the production process, one good heuristic for whether reducing variability will decrease costs is whether it adds or subtracts operations from the process on balance. If it adds operations overall, there's a good chance reducing variability will add cost. If it subtracts operations, it will likely lower costs.

To see how this calculus works, consider interchangeable parts.

Using interchangeable parts is an obvious way to reduce variability in a process, since every part produced is the same. However, as we've discussed, when interchangeable parts were first introduced, they increased rather than decreased production costs. Machine tools at the time weren't accurate enough to produce interchangeable parts on their own, so making parts uniform enough for interchangeability required extensive manual work—lots of manual hand filing, lots of careful checking against gauges. In this case, reducing variation added many operations to the process. But by the early 20th century, machine tools were accurate enough to produce interchangeable parts without additional manual work. Absent these extra manual operations, interchangeability reduced the number of operations in the production process overall by eliminating the filing and fitting operations needed to assemble noninterchangeable parts, and greatly simplified repair operations.

**Variability over time**

Despite the trade-offs, the historical tendency has been for production processes to embrace greater control with less variability.

Agriculture, for instance, was historically a highly variable process. Crop yields were influenced by uncontrollable factors such as the weather or the presence of insects or pests. But over time the process has tended toward increased control. Fertilizers were developed to provide the exact nutrients that plants require, and mechanical irrigation systems were built to provide a reliable source of water. Insecticides and other pest control technologies were created to prevent the effects of pests. Harvesting and planting became mechanized and highly uniform. Today, plants are manipulated at the genetic level to possess specific traits, such as harvestability or pest resistance, and precision farming techniques can vary the amount of fertilizer each portion of a field receives.

Or consider firearm production at Beretta. Historically, guns were made by craft production. Even though they were made with the aid of patterns, models, or simple drawings, each was a unique product of the artisans who had made it.[604] In the early 1800s, Beretta adopted English-made machine tools, which could operate with higher precision, and replaced crude sketches with dimensioned engineering drawings that clearly specified the result. The general-purpose English machines were then replaced with special-purpose American machines in 1860, at which point Beretta began to produce guns with interchangeable parts.[605] As time went on, production tolerances became even tighter. When Beretta began to manufacture the M1 Garand after WWII, it required an order-of-magnitude increase in precision, and Beretta built its own machine tools to achieve it.[606] To keep the machines from wandering, or deviating from their desired cutting path, Beretta adopted statistical process control. Later, manually operated machines were replaced with numerically controlled ones, which automated the machining operations and achieved even greater uniformity.

This trend toward greater uniformity is due in part to advancing technology, which makes it less expensive to reduce variability. As technology improves, manufacturers squeeze more and more variability out of their processes to reduce their production costs. But new technology also often requires greater precision and control to produce. Over time, technology tends to get smaller and smaller, which means greater and greater precision is needed to manufacture it. Transistors, for example, have gone from 200 per square millimeter in 1971—already orders of magnitude smaller than the first transistors in the 1950s—to 5 million per square millimeter in 2011.[607] At this size, even tiny effects can impact process yield, and factors that would be considered random noise in other processes must be carefully controlled. As one discussion of Intel's semiconductor fabrication methods notes, "Items that were once considered second-order effects, such as barometric pressure and ultrapure water temperature, are now important variables affecting process results," and even factors like changing the lengths of cooling hoses can have "catastrophic effects."[608]

New technology also often depends on effects that are difficult to generate or maintain, requiring greater control to achieve them. A modern jet engine, for example, operates at temperatures high enough to melt steel. To survive this temperature, turbine blades are made of carefully formulated alloys and have extremely precise machined holes to allow air to flow through them to keep them cool. Even minute defects in jet engine parts can cause disastrous failures.[609] Returning to semiconductor manufacturing, even minuscule fluctuations in chemical composition will interfere with semiconducting behavior. When semiconductors were being developed at Bell Labs, technicians discovered that contamination from a tiny number of copper atoms caused by touching copper doorknobs was enough to cause the semiconductors to fail.[610]

This tendency toward greater control is, in some ways, a manifestation of the technological S curves we discussed in Chapter 2. In the early days of a technology, when the major relevant effects are still being mastered, the technology often works intermittently or unreliably. Effects that have a large influence on a production process are necessarily brought under control in these early stages, or the process will never be viable. Then, as time goes on, further experimentation and development affords producers control over effects of lower and lower significance that occur less and less frequently, which slowly but steadily reduces the level of variability in the production process.

### Good variability

While reducing variability makes a production process more predictable, less wasteful, and more efficient, increasing variability can also have benefits. For instance, it can enhance producers' knowledge of the process and reveal ways it might work even better. In his 1964 study of learning curves, Winfred Hirschmann describes a forging operation where hot steel was molded into wrenches by large metal dies. Over time, the dies expanded and eventually needed to be replaced. One evening, a new worker began fiddling with the furnace temperature, causing the forging dimensions to vary erratically. Upon investigation, the company discovered that steel could be successfully forged at much higher temperatures than previously thought. "New dies were subsequently made smaller, and the metal first forged at the minimum practical temperature," Hirschmann noted. "As the die wore (got larger), the forging temperatures were increased. The hotter forgings shrank more on cooling, and in this way continued to fall within tolerance. This practice permitted a die to produce triple the normal output."[611]

As we've seen in the industrial experimentation of the 18th and 19th centuries, practices like Taguchi methods, exploratory tinkering, trial and error, and deliberate experimentation can yield new and better ways of doing things. The choice between increasing and decreasing variation can be thought of as a specific case of the more general phenomenon of exploration versus exploitation: We can choose whether to explore—to tinker with our process to find new and potentially better production methods, thereby increasing variability—or exploit—to try to get our existing process to work as reliably as possible, thereby decreasing variability.

The cultivation of variability has historically been especially important in cases where biological elements play a key role in a process, such as in agriculture or in certain types of pharmaceutical production. Because our present understanding of and ability to manipulate biological machinery is quite limited, processes that rely on biological mechanisms often take advantage of random variation to improve these mechanisms over time. For example, scientists vastly increased penicillin yields by deliberately introducing mutations in penicillin-producing mold strains and screening the resulting mutants to find ones that produced more penicillin. As a result of this process, modern mold strains produce 4,000 times as much penicillin as Fleming's original.[612]

Similarly, in agricultural production, most modern crops were improved over hundreds or thousands of years by deliberately selecting for certain random variations. Seeds of the best-performing plants were retained and planted, while seeds from plants with less desirable traits were not. Over time, this selection process resulted in crops that were able to produce much more fruit and were much easier to harvest. Modern corn is the result of 9,000 years of breeding a plant called teosinte, which has tiny cobs of just six to 12 kernels that shatter when ripe, scattering the seeds. Over millennia of selective breeding, teosinte was transformed into the high-yield corn plant we know today, and further improvements in output per acre have come about thanks to the tremendous natural variation of the teosinte plant.[613]

The relative benefits of reducing versus increasing variability likely depend on where a production process sits on the technological S curve. In the early days, at the bottom of the curve, the process doesn't yet work well, so there's lots of low-hanging fruit in the form of improvements that will yield substantial benefits. During this period, the potential value from exploring the process is high. But at the top of the curve, when the process is mature, most of the obvious improvements will have already been found and most of the knowledge landscape will have been mapped. In this case, it's more beneficial to exploit the existing process and get it to work as reliably as possible.

### Buffers and the challenge of predicting demand
Over the past several decades, manufacturers have become adept at eliminating buffers, and modern supply chains typically run extremely lean. However, lean operations have downsides. Reducing buffers can make the process fragile, as any effects that haven't been accounted for will now disrupt the process. Reducing the buffer size—cutting

slack out of the process—leaves the process vulnerable to unexpected sources of variation.

Take the supply of semiconductors during the Covid-19 pandemic. Covid was a large, unpredicted disruption to production processes worldwide. People couldn't go to work, factories halted production, and many supplies became scarce or unavailable. The car manufacturing industry in particular felt the brunt of these disruptions. In the early days of Covid, automakers canceled many of their orders for automotive semiconductors, while at the same time widespread work-from-home policies increased demand for semiconductors in other electronic products, such as laptops, headphones, and wireless routers. When demand for cars recovered, automakers found that semiconductor fabricators had reallocated their production to electronics, and automotive chips weren't immediately available. Because car manufacturers had little to no stock of chips on hand, many of them had to shut down their production lines, costing the industry an estimated $110 billion.[614]

Interestingly, Toyota, the pioneer of just-in-time manufacturing, was initially least affected by the chip shortage. In 2011, when the Fukushima nuclear accident disrupted Japanese supply chains, Toyota realized that the long lead time for semiconductors put them at risk of significant production stoppages. To prevent this from happening, Toyota ordered its suppliers to stockpile up to six months' worth of chips. As a result, when General Motors, Volkswagen, and Ford were forced to scale back production in 2021, Toyota was able to increase production with the stock of chips it already had on hand.[615]

A Toyota spokesperson at the time described this as a classic lean solution. Indeed, despite being known for minimizing inventories and getting parts just in time, many aspects of the Toyota production system emphasize being able to flexibly adapt to variation that can't be controlled. For example, SMED and cross-trained workers make it possible for Toyota to reallocate production to different models as demand changes. More generally, the lack of buffers means that the system has very short cycle times, so Toyota can produce to actual orders and is less reliant on trying to predict demand.[616]

There will always be factors that affect production beyond the control of even the most careful and well-planned process. So despite the trend toward increasing control over time, making processes robust to inevitable variation will always be important.

## Summary

The great, ongoing project of civilization is to explore, understand, and ultimately master our world, and to use that mastery and understanding to allow people to lead better, longer, and richer lives. One pillar of this effort is science, which, bit by bit, turns mysterious natural phenomena—why the sun rises and sets, what keeps the planets in orbit, what makes fire burn—into events that can be predicted, allowing us to build an increasingly accurate model of how the world works.

Technology is another pillar of this effort. While science seeks to tease apart and explain natural phenomena, technology seeks to harness those phenomena, chaining them together to produce useful outcomes. Coupling electricity and magnetism, electrical resistance, incandescence, the properties of noble gases, and innumerable other phenomena gave rise to incandescent light bulbs; combining combustion with the expansive powers of steam resulted in the steam engine; reacting carbon with iron oxide in the presence of high heat created the blast furnace and cast iron. Together, science and technology are constantly enhancing our understanding and control of the world around us, making it possible to enrich the material conditions of humankind.

Efficiency improvements and control over production processes form a small but important part of this grand project. Production technologies consist of an enormous number of different phenomena chained together in unfathomably complex ways to produce the goods and services that make civilization possible. As we refine our understanding and control over production processes—as we tease out the complex chains of cause and effect within them—we can boost their performance. With understanding and control, we can eliminate causes of failure, reduce waste, and increase yield. We can also make processes work more smoothly and consistently, preventing the expensive buffering required when processes operate unpredictably. And, of course, it's only by making a process work with some degree of reliability that it's able to be used at all. To return to our discussion of S curves, it's only when we advance beyond the flat portion of the S, when the technology is understood well enough to be useful, that it begins to have a major impact.

This great effort to understand and control, both in the world overall and within the confines of a single factory, is never complete. Even if we could understand and control everything in the natural world, endless variation would still emanate from the man-made world as cultures change and nations rise and fall. Human-initiated events have a way of affecting our surroundings in unpredictable ways. The Russian Revolution in 1917 ultimately precipitated the Cold War between NATO and the USSR, which in turn drove a flurry of nuclear weapons testing, releasing unprecedented amounts of radioactive material into the atmosphere. This radioactive material then made its way into steel during the smelting process—not an enormous amount, but enough that, for a time, very sensitive instruments needed to be made from low-background steel that had been smelted before the war.[617] A critique of capitalism written in the 1860s thus slightly changed the chemical makeup of steel more than 100 years later, necessitating changes in how certain goods were manufactured. More generally, the forces of chaos, entropy, and decay are constantly threatening to disrupt our carefully arranged production processes, requiring constant vigilance to keep them operational.

Improvement methods like the Toyota production system emphasize the importance of continuous improvement—of constantly trying to find ways to do things better, increase control over production processes, and fight the forces of decay. Here, again, we can see efficiency improvements as one small part of the grand project of civilization, where the work of billions of people over the course of centuries has, bit by bit, pushed the world forward.

# 7

# Learning Curves

Our investigation into the origins of efficiency has so far examined various ways to improve the efficiency of a process: changing the production method, altering the design of the product, reducing the costs of our inputs, increasing the scale of production, making the process work more reliably and predictably, and cutting out steps or buffers. Any cost-reducing improvement to a production process will fall into one or more of these categories. But simply knowing the different types of process improvements that exist doesn't tell us much about *how* processes improve. How do these different types of improvements accumulate, and how does efficiency improve over time?

At first glance, it might seem like it would be difficult to predict how a production process will improve with any degree of accuracy, given that improvement often depends on chance discoveries, individual decisions, and a constantly changing political and social environment. Indeed, according to one of the scientists who led the effort to scale up penicillin manufacturing, developing mass production techniques for the drug was so difficult that without the urgency of WWII it might not have happened at all.[618]

Without denying that production process improvements are contingent on events that are hard, if not impossible, to predict—technological breakthroughs, scientific discoveries, government policies—when we zoom out and take a broader perspective, we see that the cost of producing a good displays surprisingly regular behavior. In general, production costs fall in proportion to the cumulative volume of production. More specifically, every doubling of output tends to produce the same percentage reduction in cost. If cost falls by 20 percent when total production goes from two to four units, then it will fall another 20 percent when eight total units have been produced, another 20 percent when 16 total units have been produced, and so on.

This relationship between production volume and cost can be described by the equation $y = ax^{-b}$ where $y$ is the cost for the $n$th unit, $x$ is the total number of units produced, $a$ is the cost for the first unit, and $b$ is a constant known as the progress rate, which determines the slope of the curve, or how much costs fall for each doubling of cumulative production. When graphing cumulative production against price, this sort of relationship will yield a straight line on a log-log plot. (See Figure 17.)

This relationship goes by a variety of names, such as the learning curve, the progress curve, or the experience curve. It's also sometimes known as Wright's law, named after Theodore Wright, who described the relationship between labor hours and production volume in airplane manufacturing in 1936.[619] We'll use the term *learning curve* to describe this phenomenon, although, as we'll see, this is something of a misnomer since the improvements it describes are only partially due to what can plausibly be characterized as learning.

As an example of the learning curve, consider solar photovoltaics (PVs). The cost per watt of solar PV panels has fallen precipitously in the past four decades, from $100 in 1975 to just $0.26 in 2021 in inflation-adjusted dollars.[620] This reduction in cost is due to a variety of improvements to the manufacturing process. Solar cells are made from high-purity silicon, which has become cheaper over time, partly due to demand from semiconductor manufacturing. PV manufacturers have also found ways to make solar cells that require less of it, cutting out material and further reducing input costs. Assembling the cells, once a manual process, is now done using high-speed automated machinery. Manufacturers have also developed new process techniques, such as screen printing the electrical contacts and using wire saws to cut the silicon wafers. Other operations, such as polishing the cells, have been eliminated, and factories have become much bigger, enabling economies of scale. Factory yields improved as the processes became more reliable, producing more cells for a given amount of input, and new cells were developed that could convert a greater fraction of sunlight to electricity. In essence, the history of solar PV production demonstrates every strategy for cost reduction and efficiency improvement we've discussed so far in this book.[621]

Many of these developments were highly contingent on specific events, including the emergence of the satellite industry, which provided the first big market for solar PV; the energy crises of the 1970s, which drove more funding into solar PV research; and the German national feed-in tariff passed in 2000, which made it very lucrative for energy providers to install solar PV capacity. And yet, the progression in solar PV cost reduction appears surprisingly regular: Every doubling of production volume has reduced prices by about 23 percent.[622]

The learning curve for solar PVs illustrates the progression of an entire industry with many producers around the world. But similar learning curves occur at the level of individual factories. Take the B-17 bomber, which Boeing produced by the thousands in its Seattle plant during WWII. Boeing's first B-17s required more than 140,000 labor hours to produce. But by plane number 6,000, that had fallen to less

The Origins of Efficiency

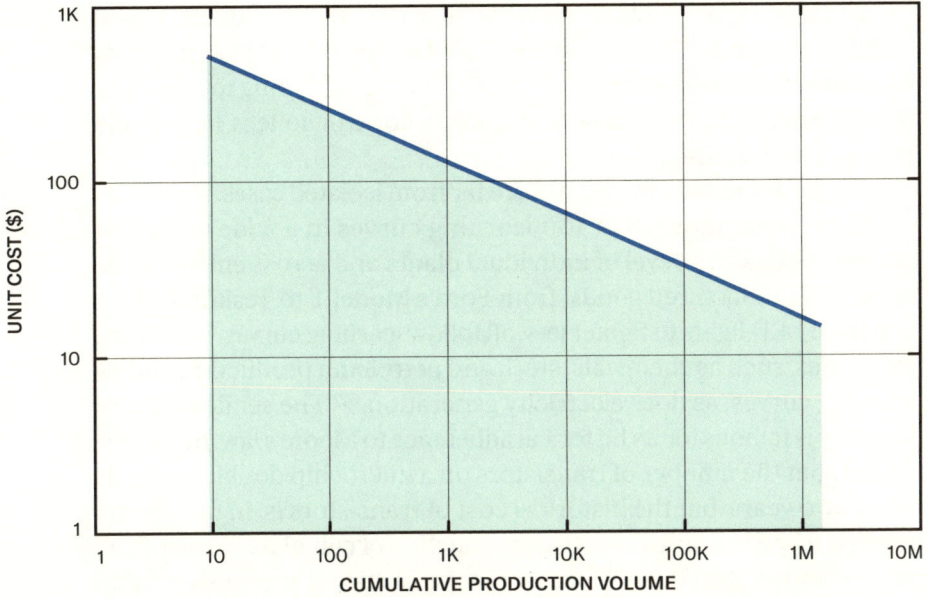

**Figure 17.** Learning curve on a log-log plot.

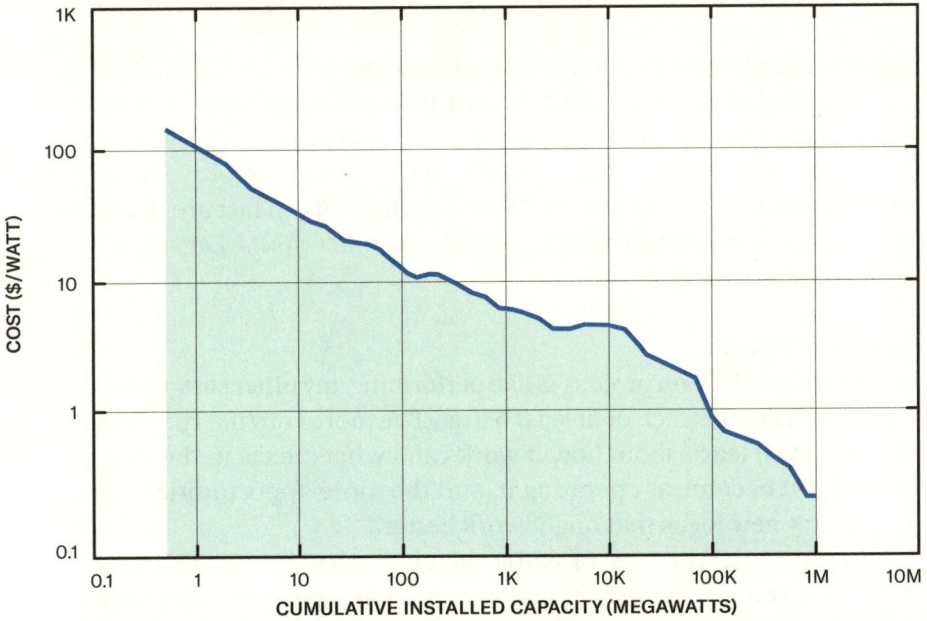

**Figure 18.** The solar PV learning curve. Data from Our World in Data, "Solar (Photovoltaic) Panel Prices."

than 20,000 labor hours.[623] As with solar PVs, this was due to a host of different process improvements, including introducing labor-saving jigs, templates, and fixtures, redesigning material routing to minimize travel time, and improving part quality control so less fitting and rework was required.[624]

Solar PVs and Boeing's B-17s are far from isolated cases. Researchers have found more than 100 learning curves in a wide variety of sectors, both at the level of individual plants and across entire industries.[625] Manufactured goods, from Ford's Model T to Tesla's Model 3 and from LED lights to fighter jets, all follow learning curves.[626] Process industries, such as chemicals, steel, and petroleum production, follow learning curves, as does electricity generation.[627] The semiconductor industry is famous for its historical adherence to Moore's law, the observation that the number of transistors on a microchip doubles roughly every two years, but the historical cost of transistors is, in fact, better predicted by a learning curve that is a function of cumulative production volume rather than time.[628] Because learning curves appear so reliably, manufacturers often use them to predict future production costs.

Why do production costs fall in proportion to cumulative production volume? There are, broadly speaking, three types of effects at work. In some cases, increased production volume affords producers opportunities to figure out how to make the process run better. We'll call these *learning effects*. In other cases, increased production volume can trigger reduced costs through mechanisms other than learning. We'll call these *indirect effects*. Finally, in some cases, improvements may appear to be related to production volume but in fact are due to entirely unrelated factors. We'll call these *artifact effects*. Let's take a look at each in turn.

### Learning effects
Running a production process is like performing any other sort of task: Practice makes perfect, or at least better. The more you run a process, the more you learn about how it works and what affects it, the more skilled you become at operating it, and the more opportunities you have to try new ideas that might work better.

In some cases, this sort of skill accumulation happens at the level of individual workers. As workers gain production experience, they learn to work faster, make fewer mistakes, and more quickly recognize and diagnose problems. For example, the textile machinery first introduced

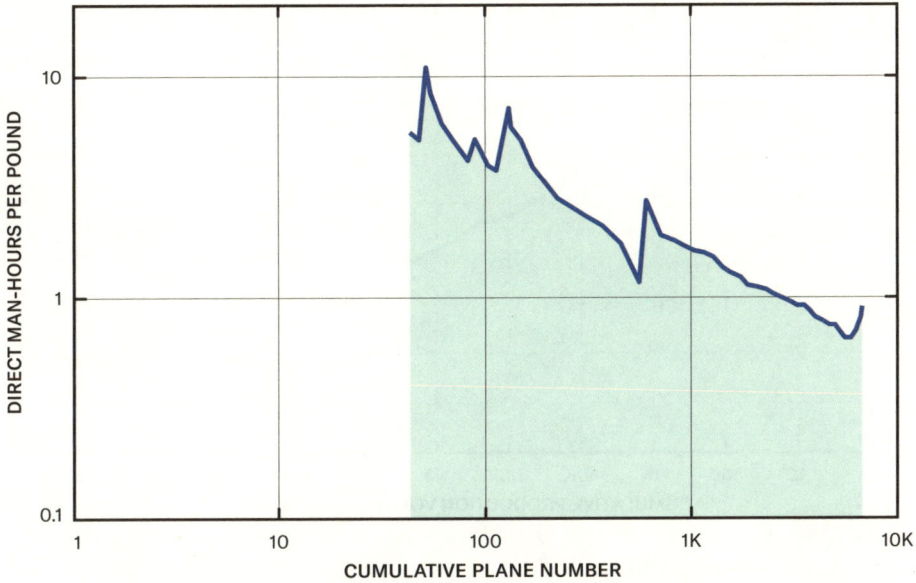

**Figure 19.** Learning curve for the B-17 at Boeing's Seattle plant. Reproduced from Reguero, "Military Airframe Industry," chart 15.

in the late 18th century took skill to operate, and that skill took time to acquire. At the famous Lowell textile mills in Massachusetts in the 1850s, a worker with 12 months of experience was four times as productive as one with one month of experience.[629] And in the early 20th century, highly skilled American textile workers were nearly six times as productive as workers in other parts of the world using the same equipment.[630]

But rather than existing in the minds and muscles of individual workers, process skill is often embedded in the process itself. As an organization gains experience, new processes are implemented or rearranged and work patterns are changed to operate more effectively, independent of the skill of the individual worker. In other words, the organization improves the production recipe, independent of the skill of the individual cooks. In B-17 manufacturing, for example, overall production efficiency increased even though the skill of the average worker actually decreased over time. While Boeing's early planes were built by experienced aircraft workers, these workers were soon drafted or promoted to supervisory positions, and their replacements on the shop floor were less skilled. And as production output continued to increase, Boeing was forced to relax its hiring standards and take whatever labor

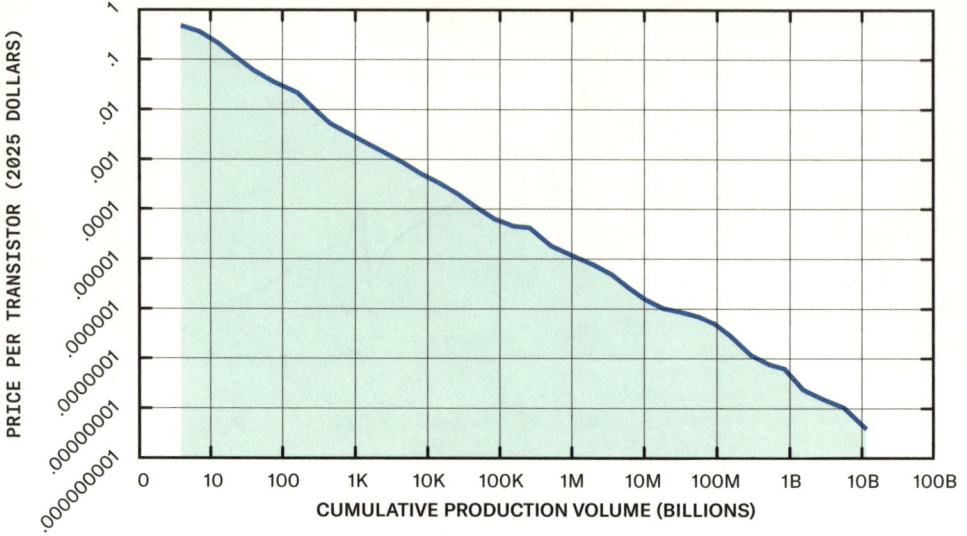

**TRANSISTOR LEARNING CURVE**

PRICE PER TRANSISTOR (2025 DOLLARS)

CUMULATIVE PRODUCTION VOLUME (BILLIONS)

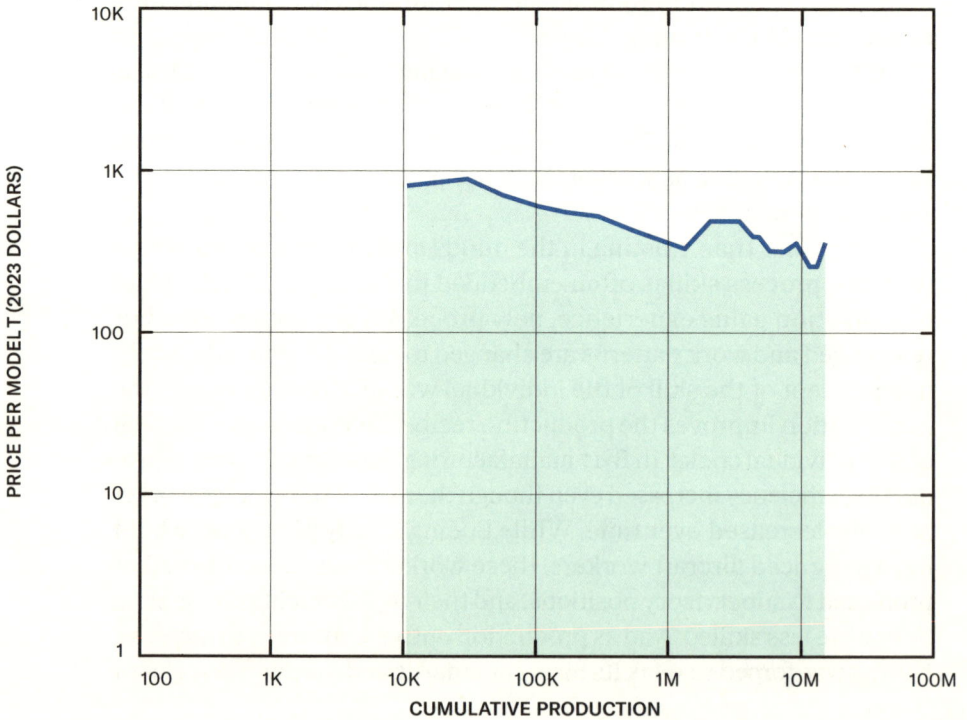

**MODEL T LEARNING CURVE**

PRICE PER MODEL T (2023 DOLLARS)

CUMULATIVE PRODUCTION

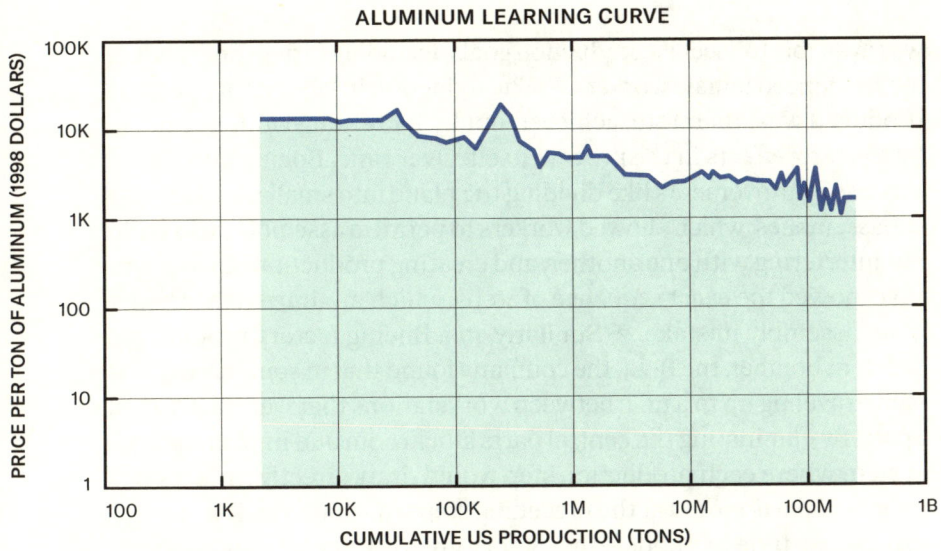

**ALUMINUM LEARNING CURVE**

Y-axis: PRICE PER TON OF ALUMINUM (1998 DOLLARS) — 1, 10, 100, 1K, 10K, 100K
X-axis: CUMULATIVE US PRODUCTION (TONS) — 100, 1K, 10K, 100K, 1M, 10M, 100M, 1B

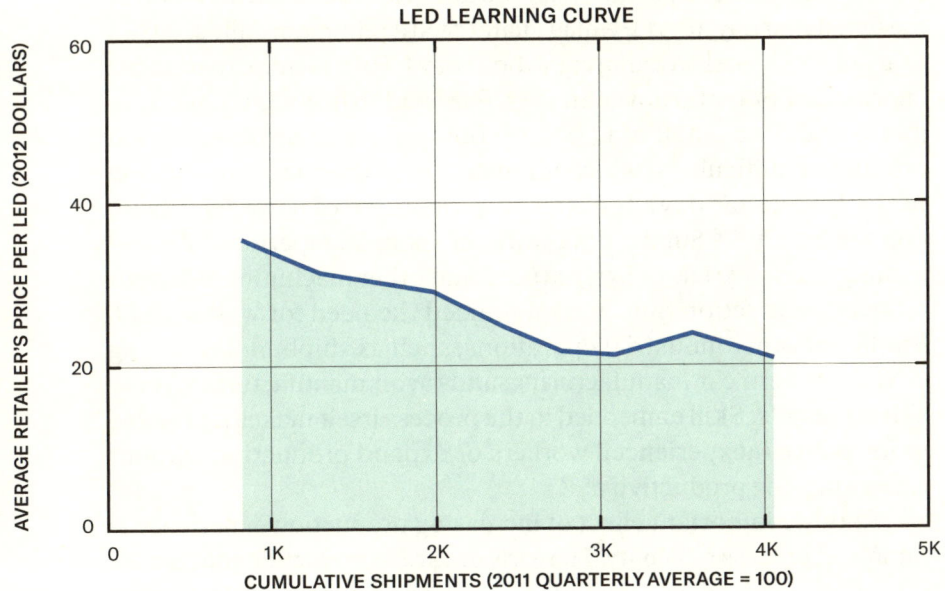

**LED LEARNING CURVE**

Y-axis: AVERAGE RETAILER'S PRICE PER LED (2012 DOLLARS) — 0, 20, 40, 60
X-axis: CUMULATIVE SHIPMENTS (2011 QUARTERLY AVERAGE = 100) — 0, 1K, 2K, 3K, 4K, 5K

**Figure 20.** Learning curves for transistors, aluminum, the Model T, and LEDs. Data from Performance Curve Database, "Transistor"; National Minerals Information Center, "Aluminum—Historical Statistics"; Wikipedia, "Ford Model T"; Gerke et al., "Household LED Lamps."

was available to meet its production goals, including hiring thousands of inexperienced female workers.[631] The reduction in labor hours per B-17 produced was, therefore, achieved not by increasing worker skill but by learning effects in the process itself. Over time, Boeing introduced process improvements like dividing the plane into smaller and smaller subassemblies, which allowed workers to perform assembly tasks without interfering with one another, and creating production illustrations that showed the exact sequence of steps, which minimized the chance of an assembly mistake.[632] Similarly, in a Boeing factory producing a different bomber, the B-24, the company found that in some cases parts were traveling up to a mile between workstations that were just 50 feet apart. By eliminating the central parts stockroom and implementing a system where each production step would draw directly from a small supply of parts stored at the preceding step—an early version of modern just-in-time methods—the factory greatly reduced handling time and effort.[633] As production experience accumulated, more and more of these types of improvements were identified and implemented.[634]

Another example of skill accumulating in the process itself comes from Ford's Model T production process. Prior to the introduction of the five-dollar day in 1914, which guaranteed a minimum daily salary of $5 to all employees working eight-hour days, Ford plants experienced enormous worker turnover. In 1913, Ford had to hire 52,000 workers just to maintain a staff of 14,000.[635] But while high turnover caused production difficulties (which ultimately prompted the introduction of the five-dollar day), it didn't stop the price of a Model T from falling steadily.[636] Subdividing work into simple, repetitive jobs and making extensive use of jigs, fixtures, and other machining aids both reduced production time and minimized the need for worker skill. Studies of other industrial operations, such as shipbuilding during WWII, modern car manufacturing, and rayon manufacturing, show similar results: Skill embedded in the process itself makes it possible to introduce inexperienced workers or expand production without compromising productivity.[637]

Another important effect of increasing production experience is improved process reliability. The early days of a production process are often characterized by numerous production problems and low yields. There's often an extensive shakeout or ramp-up period as workers learn the equipment, bugs are ironed out, and the process is adapted to its particular circumstances.[638] Economist Eric von Hippel describes the breaking-in period required for an automatic component placement machine used by an electronics manufacturer. While the machine eventually worked as intended, numerous bugs had to be resolved before it functioned properly. For example, early on the machine couldn't install components on light-yellow circuit boards. The problem was eventually traced to the vision system the machine used to position components: The machine had been developed using green circuit boards and couldn't "see" the light-yellow boards. The machine was also initially unable to place components that had tilted heat sinks because the machine designers hadn't realized these components existed, so they hadn't designed the machine to accommodate them.[639]

These sorts of bugs are common when getting a production process off the ground. An infinite number of factors can potentially affect a production process, which makes it almost impossible to get it right the first time. There will always be elements process designers haven't considered, factors that differ between the research lab and the factory, or problems that first appear on the shop floor. Only by running a process in a real-world production environment can these issues be found and fixed.

For some processes, particularly complex ones, this ramp-up period can take months or even years. In car manufacturing, the factory may not achieve its desired level of quality for the first several months.[640] In semiconductor manufacturing, it can likewise take months for certain processes to attain yields above zero, while many fabs take years to reach yields above 50 percent.[641]

But finding and correcting production problems doesn't stop once early production troubles have been overcome. Unless process yields reach 100 percent (and they never do), there are always production problems to find and fix, always errors to address. Some production problems might appear only intermittently or after long periods of operation. They might have complex cause-and-effect relationships that are difficult to unravel or hard to articulate. A research scientist at an aluminum manufacturer noted that for many plant problems "you have to spend several months trying to find out [from the plant operators] what the exact problem is."[642] In other cases, the existence of one problem may mask others, requiring problems to be peeled back one by one. The endless parade of factory floor production problems is why industrial improvement systems, such as lean and Six Sigma, focus on continuous improvement: The search for process improvements that can be made and production problems that can be fixed never ends. And it's why Toyota makes extensive use of poka-yokes. Each poka-yoke is the result of operating experience, an attempt to ensure that an error that happened in the past won't happen again in the future.

As workers and organizations gain process experience, they may also learn to improve their production equipment. Just as unanticipated factors on the shop floor can initially cause equipment to not work as designed, adapting equipment to the specifics of a production environment can make it work *better* than designed. As machines and equipment are better understood, workers learn to push them beyond their limits. Through trial and error, a steel fabricator was able to increase the output of one of its electric arc furnaces to more than double its design capacity and roll 14-inch-wide slabs of steel on equipment originally designed for 8 inches.[643] Similarly, in the mid-20th century, operators of petroleum cracking units were able to gradually increase output by taking advantage of equipment safety margins and replacing components that artificially restricted output.[644]

In addition to helping to adapt existing technology to particular circumstances, real-world operating experience is often required to design better production technology. In his examination of technological progress across a variety of fields, ranging from farm equipment to aircraft to electricity generation, economist Devendra Sahal argues that "advances in technology generally depend on the gradual acquisition of skills through actual participation in the production activity," and that "technical progress is largely a matter of learning by direct experience."[645] In the production of electrical generators, for instance,

> the new ideas had first to be patiently tried out before they became feasible. Virtually every innovation in this area has had a long history of experimentation. The earliest hydrogen-cooled generators date back to 1928. Nevertheless, the use of hydrogen became fully feasible only after the war. Similarly, a number of attempts were made to design a rotor with direct water-cooled windings as early as 1914 and 1947. However, liquid-cooled rotor windings became a real possibility only in the 1960s. That is, the fruition of many of the important advances had to await the accumulation of relevant experience.[646]

Experience is particularly useful in the creation of physically larger units. Bigger equipment is often more efficient due to geometric scaling effects, but because performance characteristics change as size increases, equipment must be scaled up gradually. As we saw in Chapter 4, modern wind turbines and blast furnaces were scaled up slowly over time so that the designs of larger units were based on workers' experience operating slightly smaller units. A similar process of gradual, experience-based scaling took place with steam turbines.[647]

More generally, as production experience increases, the causes and effects at work in a process become better understood, which makes it possible to address problems and improve performance. In some cases, understanding how different parameters affect a process can take many years of operation. For example, one important advancement in semiconductor manufacturing was the development of the diffusion process, which made it possible to create thinner, higher-quality semiconductor components by heating silicon wafers in specially designed furnaces and exposing them to different types of gas. In the heat of the furnace, the gas atoms diffused into the silicon, creating very thin, uniform semiconductor elements. But getting the diffusion process to work reliably required heating the silicon to a precise temperature and exposing it to the gas for a precise amount of time. It took several years of operation before manufacturers understood this, at which point they introduced furnaces that could be precisely controlled.[648]

In some cases, the exploration of the causes and effects in a process is accidental, as workers inadvertently discover ways to make improvements. In other cases, this exploration is done by way of deliberate experimentation, such as carefully testing machines' limits or varying the quality of the inputs. But regardless of how this understanding is achieved, data from actual operational experience is always a driver. The world is its own best model, and while much can be learned from computer simulations or small-scale prototypes before production starts, there's often no substitute for direct experience.

**Indirect effects**
Increased production volume can also trigger improvements to a process in ways that have nothing to do with learning or experience but instead happen through indirect forces. An increase in demand, for example, might force manufacturers to adopt more efficient production processes. In fiberglass manufacturing, many of the most important process improvements came about because producers

needed to fulfill larger orders. Continuous pultrusion machines, which pull fiberglass continuously through a heated die, were developed in response to a large order for fiberglass handrails.[649] Similarly, in B-17 manufacturing, process improvements were a response to increasing production demands. Since the planes were held by overhead cranes during assembly, the production rate was limited by the number of cranes. The only way to increase output was to reduce the time a plane spent hanging from a crane, which required streamlining and improving the assembly process.[650]

Higher production volume also means greater rewards for solving production problems or developing better technology. A larger market will naturally attract more research investment, entice more inventors and entrepreneurs, and inspire more companies to develop new products. A study of chemical manufacturing found that production experience on its own did not generally yield cost reductions, yet high-demand products often underwent specific R&D efforts to improve their processes, which did yield cost reductions.[651] Similarly, the enormous growth of the electric power industry in the early 20th century encouraged the industrial manufacturer Alcoa to develop aluminum transmission cable, which replaced higher-cost copper cable.[652]

Finally, greater production volume also makes it possible to take advantage of economies of scale. As output increases, it becomes profitable to adopt production technology that is only efficient at high production volumes. As we saw in Chapter 4, larger wind turbines produced electricity more cheaply than smaller ones, and larger blast furnaces produced iron more cheaply than smaller ones, but they were only economical if demand for electricity and steel was sufficiently high. Similarly, the greater the production volume, the easier it is to justify intensive DFMA efforts, replacing manual operations with automatic machinery or implementing a large division of labor.

So, beyond simply generating more knowledge and experience, increased production volume also changes production incentives, making it more valuable to develop new technology or adopt production methods that take advantage of scale effects.

### Artifact effects
In some cases, increasing production volume may only *appear* to cause production costs to fall, when in fact something else is happening instead. For instance, the direction of causality might be reversed:

Rather than increased production volume causing cost reductions, cost reductions might cause increased volume. If something becomes cheaper, then, all else being equal, we'd expect more demand for it, which would cause production volume to rise.

Similarly, production technology might simply improve over time, independent of production volume. In this situation, it might still look like cost reductions are the result of increased production volume, if production volume is also continuously rising. In one study, researchers fit the cost progression of 62 different technologies to several different predictive models, including Wright's law (in which cost falls in proportion to cumulative production volume) and Moore's law (in which cost falls steadily over time). They found that while Wright's law was the most accurate, Moore's law was close behind, largely because production volume tended to increase exponentially over time.[653] A similar study of 51 different products found comparable results.[654]

This sort of effect was likely at work in WWII-era manufacturing. For both the B-17 and the Liberty ship, production started before tooling and production facilities were complete. As these were finished, production costs fell, but this was simply due to the time the build-out took rather than the result of rising production volume.[655]

### Untangling effects

A variety of different mechanisms can be behind the correlation between cost and production volume. In some cases, it's due to learning effects; in others, to the incentives that greater production volume creates. The correlation may also be spurious, and the actual cause of the cost reductions may be due to factors unrelated to production volume.

Do we have any idea how important the various effects are? There's some suggestive evidence. During WWII, we would expect production to be mostly independent of price, as production was driven by military requirements rather than consumer demand. This would screen off reverse causality, where cost decreases result in increased output. And because production volume varied over time, it's also possible to separate the effects of time from those of cumulative production volume. In one study looking at the production of 152 different vehicles and 523 different products during WWII, experience effects (or learning effects) were found to be responsible for 67 percent of labor hour reductions and 40 percent of cost reductions, suggesting that cumulative experience plays a significant role in driving improvements.[656] But this doesn't allow us to differentiate between actual learning effects, where experience causes the cost reductions, and changed production incentives, where increased demand inspired producers to find better methods.

Other studies reveal that productivity also steadily increases in cases where no new capital is added, which screens off improvements caused by new equipment, leaving mostly learning effects (although this wouldn't differentiate between learning as a function of production experience and learning as a function of time). Between 1935 and 1950, the Horndal ironworks in Sweden increased its labor productivity by 2 percent each year despite no added capital. Similar capital-free improvements, often known as the Horndal effect, were observed at a 19th-century textile mill (2 percent improvement per year) and on an Israeli kibbutz (0.4 to 0.6 percent improvement per year).[657] The small size of these values compared to the large improvements observed in industries like aircraft and solar PV manufacturing suggests that the effect of actual learning and process organization is relatively minor compared to the development of better production equipment.

Other studies have tried to isolate the effects of scale economies. A study of DuPont's rayon plants in the mid-20th century found that the majority of cost reductions—up to 90 percent or more—came from improvements to equipment, input materials, or organizational structure, with only a small fraction due to increased scale.[658] Studies of chemical manufacturing found similar results (though in both cases, this likely doesn't include technical improvements that rely on scale effects, such as increasing a machine's production rate).[659]

While this evidence can shed some light on the question, teasing out the importance of different effects in the learning curve remains challenging. Even individual changes are likely to involve bundles of many different effects that can't be untangled.

Take the development of aluminum cable for high-voltage electrical transmission. Aluminum cable was cheaper than copper, making it one of the many improvements responsible for the electric power learning curve. Alcoa originally developed its aluminum cable in response to enormous growth in the electric power industry; in other words, the increased production volume of electric power incentivized the company to develop a lower-cost transmission cable. But it was only by accumulating operating experience that Alcoa was able to successfully evolve the technology. Its early transmission wires were made entirely of aluminum, but these all-aluminum cables vibrated excessively and tended to break under high loads. After a long period of laboratory and factory floor research—a high fixed cost made possible by the large potential market—Alcoa built an aluminum transmission wire with a steel cable core, known as aluminum conductor steel-reinforced (ACSR) cable. ACSR cables also required extensive operating experience before they could be widely adopted because utilities wanted evidence that the steel core wouldn't corrode over time. As adoption increased, ACSR cables became one of Alcoa's most important products—at one point, it made up 20 percent of Alcoa's aluminum sales—which reduced costs even further due to scale effects. This one improvement to electricity production—replacing copper cable with aluminum cable—was the result of learning by doing, deliberate research and development, incentive effects due to increased production volume, and economies of scale.[660]

This bundle of effects makes taking advantage of learning curves complicated. It's not enough to simply increase production volumes and watch costs fall in response. An electric power company resolving to double its output wouldn't, on its own, result in the invention of aluminum transmission cable, wind turbines, supercritical steam boilers, or any of the other innumerable technological improvements that have made electric power costs fall. Boeing could have increased B-17 production by building a second or third factory instead of investing in jigs, fixtures, and other process improvements, but that alone wouldn't have yielded efficiency improvements. Improvements are the result of deliberate efforts to find better ways of doing things; the learning curve tracks how these efforts bear fruit as output increases, but the increased output is not, in itself, responsible for them.

The difficulty of taking advantage of learning curves can also be observed by contrasting the production of solar PVs with that of titanium. Titanium is as strong as steel but weighs 40 percent less, making it an excellent choice in applications that require a high strength-to-weight ratio, such as aircraft manufacturing. Titanium is also highly corrosion-resistant and keeps its strength at temperatures up to hundreds of degrees. Titanium first began to be produced on an industrial scale in the late 1940s, and by the '60s it was widely used in the aerospace industry to make strong, lightweight airframes. As production volume increased, titanium fell down the learning curve, with an average progress rate of 23 percent, akin to that of solar PVs.[661] But while titanium and solar PVs had similar rates of progress, cumulative production of solar PVs far outpaced titanium, as did its price reduction. Between 1975 and today, cumulative production of raw titanium metal has increased between fivefold and tenfold. Over that same period, cumulative solar PV production has increased by a factor of 300,000. Solar PV production has had many more doublings than titanium, and, correspondingly, its costs have fallen much further. While solar PVs have become so cheap that the technology is poised to become civilization's primary method of generating electric power, titanium remains a niche metal because of its high cost.[662]

This difference in total production volume can be traced back to process improvements. The solar PV industry has continued to find process improvements and material cost reductions that allow producers to lower costs overall, so PV production continues to scale up as it is used in new applications. Solar PVs have gone from being a niche supplier of power in communications satellites and remote telephone repeaters to being used in utility-scale power installations that are millions of times larger. Titanium, on the other hand, is still produced using the same process discovered in the 1930s: a complex, energy-intensive batch process called the Kroll process. Because it's bottlenecked by its lack of process improvements, titanium remains expensive and is, therefore, relegated to specialized uses, which limits its total production volume. Consequently, it is advancing along its learning curve much more slowly than solar PVs. And trying to jump-start this process by massively increasing titanium production would fail—all the experience in the world can't magic a new chemical process into existence or whisk away titanium's high reactivity, which makes it complicated and expensive to process into components.

More generally, the relationship between cost and production volume is just one indicator of a complex ecosystem of technological and organizational improvement, so manipulating one variable in isolation is unlikely to be an effective way to increase efficiency. What can be said is that production experience and cumulative production volume create *opportunities* for improvements. As electric power output increased and time passed, plant operators had the opportunity to learn what worked and what didn't and adopt more efficient, larger-scale production technology. As power consumption increased, the large market created opportunities for entrepreneurs and inventors to create new, better electric power technology. But it's only when these opportunities are exploited that cost reductions occur. And an opportunity isn't a guarantee—production volume on its own may not result in the scientific or technical breakthroughs that continually falling costs require.

### Learning curves and S curves

One important application of the learning curve concept is that it can help us pin down, with some quantitative rigor, the S-shaped curve that characterizes technological progression. As we've discussed, a technology's performance increases slowly over time. Early in its development, fixing one problem might simply reveal more problems. It might not even be clear how to fix the problems if the phenomena at work are poorly understood. But as time goes on, these problems are addressed and the mechanics of the technology are better understood, allowing it to be pushed to higher levels of performance—increased speeds, greater scales—at a quicker pace. Eventually, the rate of the performance increase slows again as it approaches some natural limit inherent in the technology.

There's an obvious parallel between the learning curve, in which cost reductions slow down as production volume increases—that is, each unit of production experience yields ever smaller cost decreases—and the technological S curve, in which performance improvement slows down as a technology approaches its performance ceiling. If we take unit cost as the relevant performance metric, the learning curve can be understood as the second half of the S curve, when the rate of performance improvement begins to slow down.

There are some minor differences between technological S curves and learning curves. One is that the direction of the curves is typically reversed: S curves rise over time as performance increases, whereas learning curves fall over time as cost decreases. But this is easily addressed by a simple change in units. For electric power, a learning curve might have dollars per kilowatt-hour on the vertical axis; changing this to kilowatt-hours per dollar flips its orientation to match the S curve. (See Figure 21.)

Another minor difference is that the basic learning curve model gradually approaches zero (or infinity, if we reverse the units), whereas S curves approach some finite performance ceiling. But this difference is also easy to accommodate. Many learning curve studies have noted that costs eventually stop decreasing and modify the basic formula to include a steady-state phase where costs are constant.[663] Even in the traditional learning curve formulation, diminishing returns mean that, eventually, further cost improvements require enormous production volumes that will never be reached in reality. If you've already produced 100 million units, getting costs to fall another 90 percent will require producing 100 billion units, which most products will never reach. So even in the traditional learning curve formulation, costs will effectively hit a steady state.

One other potential difference is that S curves describe the performance of a single technology whereas learning curves describe the cost of an entire production process, which encompasses many different technologies. But the difference between an entire production process and a single piece of technology is fairly fuzzy. Technologies typically consist of many individual components—many subtechnologies—arranged in a specific way to achieve a particular purpose. As a technology advances along its S curve, these individual components will all be improving. The falling cost per lumen of incandescent light was due to many individual improvements operating in parallel: better filaments, better bulb atmosphere, cheaper materials, better design of the glass bulb itself, and more. Similarly, production processes consist of many different technologies arranged in a particular way. In some cases, these might be tightly bound to an individual machine, such as the cut-nail machines discussed in Chapter 2 or the ribbon machines for bulb blank production discussed in Chapter 1. In other cases, they might be split into separate machines or located in separate factories. But in both cases, the process contains a collection of different technologies arranged in a specific way for a certain purpose. Both S curves and learning curves, then, describe the cumulative improvement of a bundle of technologies that have been chained together.

So, with some minor tweaks, the learning curve can be thought of as the back half of a technological S curve. But what about the front half?

In fact, it's not hard to extend the learning curve model to encompass this. Many studies of the learning curve have noted that early performance tends to be flatter than the standard formulation would imply, suggesting that early on, experience accumulates without performance increasing much, as seen on the S curve.[664] The standard learning curve formulation, where cost is purely a function of cumulative production volume without any embellishments or additional factors,

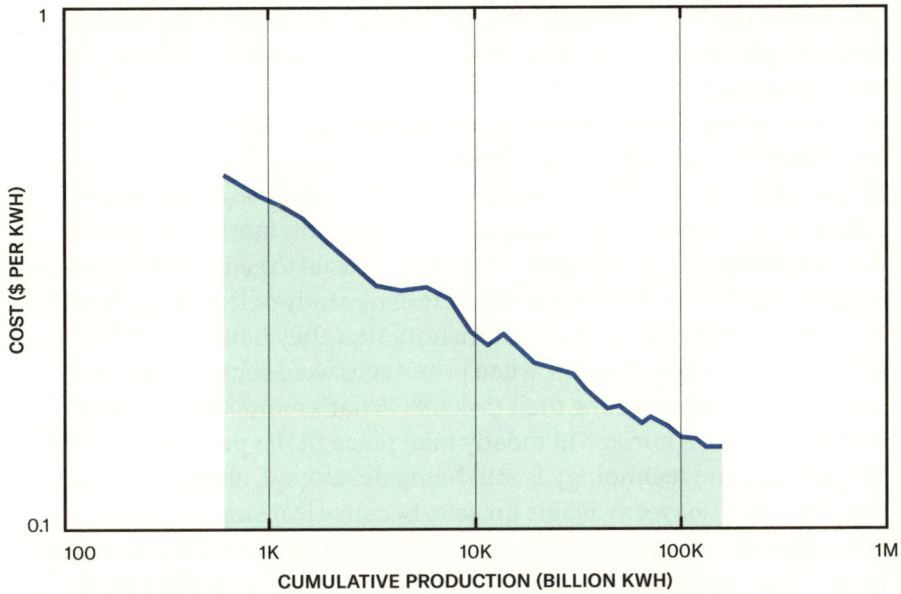

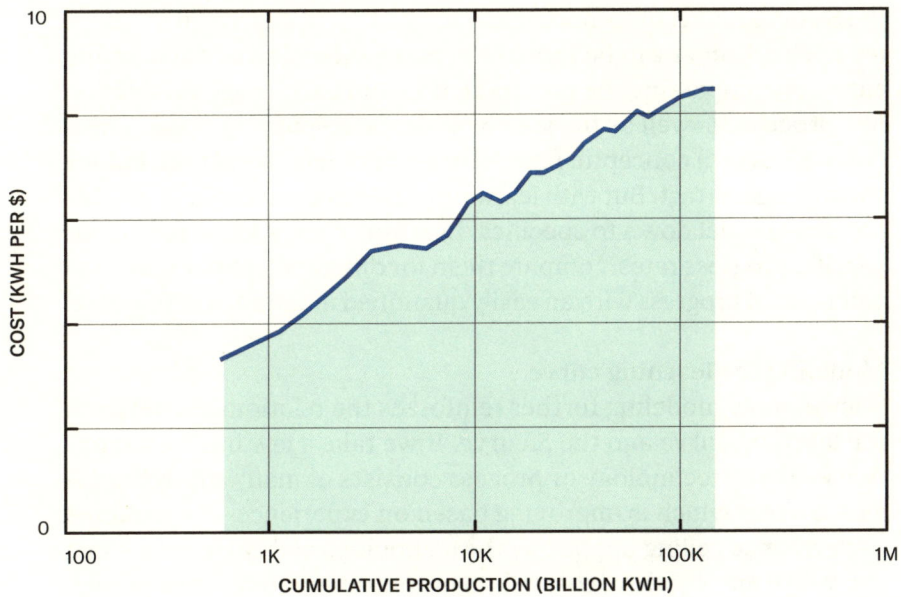

**Figure 21.** The electric power learning curve between 1958 and 1998, with dollars per kilowatt-hour and kilowatt-hours per dollar on the vertical axis. Data from Performance Curve Database, "Electric Power."

is used not necessarily because it's incredibly accurate but because it's conceptually simple, and on a log-log plot, deviations from the curve will appear very small.[665]

Curves that don't show this flat early portion have often simply omitted early production data. Solar PV learning curves, for instance, often start in the 1970s, 20 years after the technology was invented.[666] Likewise, learning curves for wind power typically start in the 1980s, despite the fact that wind turbines have been used to generate electricity since the late 1800s. In one characteristic study of learning curves in an automotive plant, the authors note that they have omitted the first few weeks of production, when prototypes were being constructed and workers were learning their tasks.[667] What's more, the flat part of a technological S curve will mostly take place in the precommercial stage, when the technology is still being developed, absorbing time and effort but not yet available for sale. Because learning curves track cumulative production volume, they will, by their very nature, screen off most of the early development work that constitutes the bottom part of the S.

So, with a few tweaks, the learning curve model largely aligns with the S curve model of technological advancement described in Chapter 2. The S curve can be thought of as an extension of the learning curve, encompassing the preproduction and early-stage production of a process, as well as the steady-state, post-learning stage. The S curve gives us a conceptual model of technological progress, but it's somewhat abstract. But with learning curves, we can begin to pin this progress model down to specifics. Learning curves let us talk about specific progress rates, compare them for different technologies, and pair rates of progress with an easily quantified measure of experience.

### Modeling the learning curve
Some simple modeling further reinforces the relationship between the learning curve and the S curve. If we take a few basic assumptions—that a technology or process consists of many subtechnologies, each of which is improving based on experience—and add the performance ceiling implied by the technological S curve, a learning curve-like model, where cost falls in proportion to cumulative output, naturally emerges.

Consider, for instance, a simple model of technology or process improvement based on random discovery. In this model, a production technology has a lowest possible cost of one and starts out costing a random amount between one and 10. Every time a unit gets produced, there's a chance of discovering a better production method, represented by drawing a random number between one and 10. If the number is below the current cost, that becomes the new cost; if not, the cost stays the same. Simulating this process for a single technology yields the curve in Figure 22.

If we model a technology or process as consisting of a large number of individual subtechnologies, with a cost equal to the sum of their individual costs, then the steps in this curve flatten out and it starts to resemble a learning curve. Figure 23 is a simulation of the cost of a process that consists of 50 individual technologies, each of which is slowly improving. Over time, it becomes less and less likely that a given innovation or improvement is superior to the current method, and because of the performance ceiling, there's less and less room for large absolute performance increases.

In this simple model, every technological component is independent. But with real technologies, each component is constrained by and constrains other components: Our choice of technology in one part of the process affects our options in other parts of the process. In light bulb production, the design of the filament affected which atmospheres could be used as well as the size and shape of the bulb. So we can extend this model by coupling each component to some number of other components, such that a potential improvement not only affects a single component but also every component it's connected to (modeled by the same random drawing process). Under these assumptions, a technology will only be replaced if the total collection of technologies it affects is cheaper.

This version of the model still presents a learning curve-like phenomenon, but the slope of the curve varies depending on how connected components are. The higher the level of connection between components—the more complex and coupled the design—the shallower the learning curve and the slower the rate of improvement.[668] (See Figure 24.)

Studies of the improvement rate of different technologies seem to bear out the model's predictions. For clean energy technologies, simple technologies with fewer components that can be mass-produced, such as LEDs and solar PV cells, have been found to have steeper learning curves—faster rates of improvement—than complex, highly

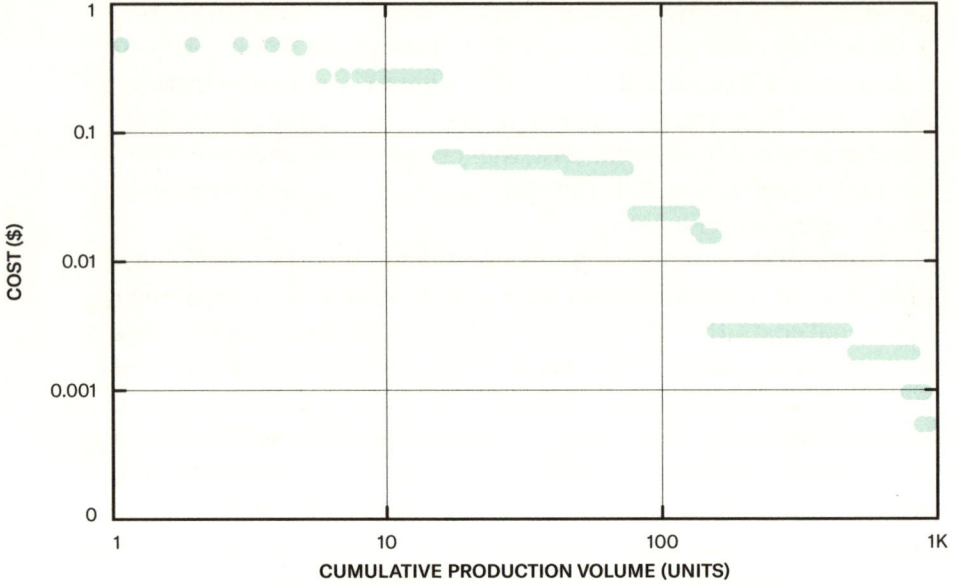

**Figure 22.** Simulated learning curve for a single technology.

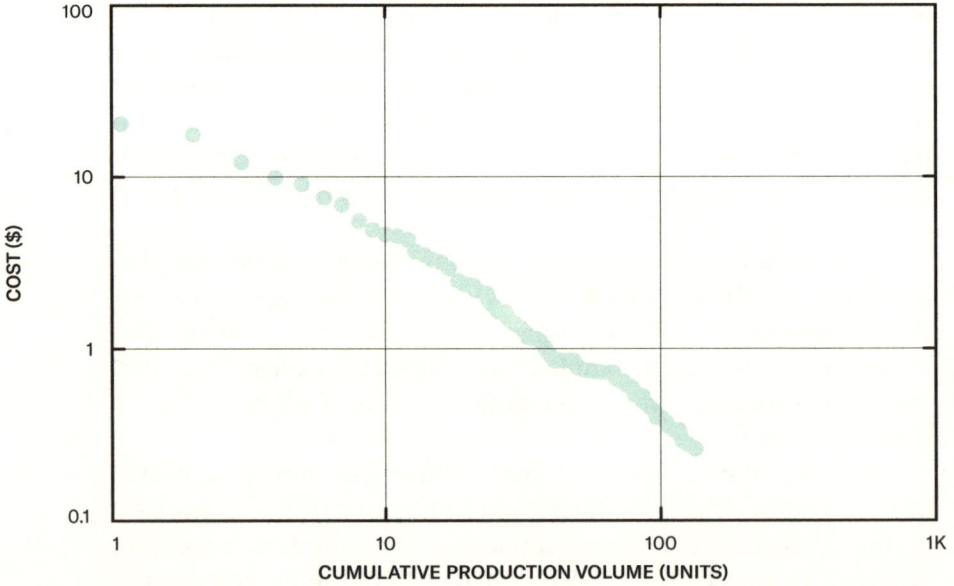

**Figure 23.** Simulated learning curve for a product containing 50 separate technologies.

interconnected products that require lots of customization, such as nuclear power plants.[669] (See Figure 25.)

When components are significantly connected with other components, there are often bottlenecks in the improvement process, such that highly connected components become difficult to improve and tend to dominate the cost of the overall technology. In this model, the more connected a component is, the less likely it is that any given change will be an improvement, since most components it's connected to will also need to be improved by the change for there to be an effect. For example, single-family homes in the US are built using timber framing; walls are made of 2 × 4 studs spaced every 16 or 24 inches. This building method is labor- and material-intensive, and there have been many attempts to develop more efficient framing systems. However, a different wall framing system would dramatically change the way other trades do their work, including how insulation is installed and how plumbing and electrical systems are routed. As a result, framing systems that are attractive on paper, such as SIPs (wall panels made from sheets of plywood separated by a layer of insulation), have had limited success.[670] When technological progress is held back by the lack of improvement in a key component, it's known as a reverse salient in a technological front.[671]

In addition to being influenced by process complexity and interconnectedness, the slope of the learning curve is affected by the type of process. Processes that involve a lot of manual labor tend to show steep improvement curves, whereas machine-intensive processes tend to show much shallower curves. That's because it's easier to rearrange the way a person operates (for example, by adding jigs and fixtures or rearranging steps to require less transportation) than it is to redesign the way a machine operates.[672] For this reason, machining operations tend to have shallower learning curves than assembly operations.[673]

This suggests that one way to take advantage of the learning curve is to manufacture products with a high degree of repetition and with little coupling in the product design—in other words, designing products or processes that are highly modular, with standardized interfaces.

Interestingly, this somewhat contradicts the impulse to adopt continuous processes, which require tightly coupled, highly integrated production systems. But, as we noted at the outset of this book and will discuss more fully in Chapter 9, continuous processes are the

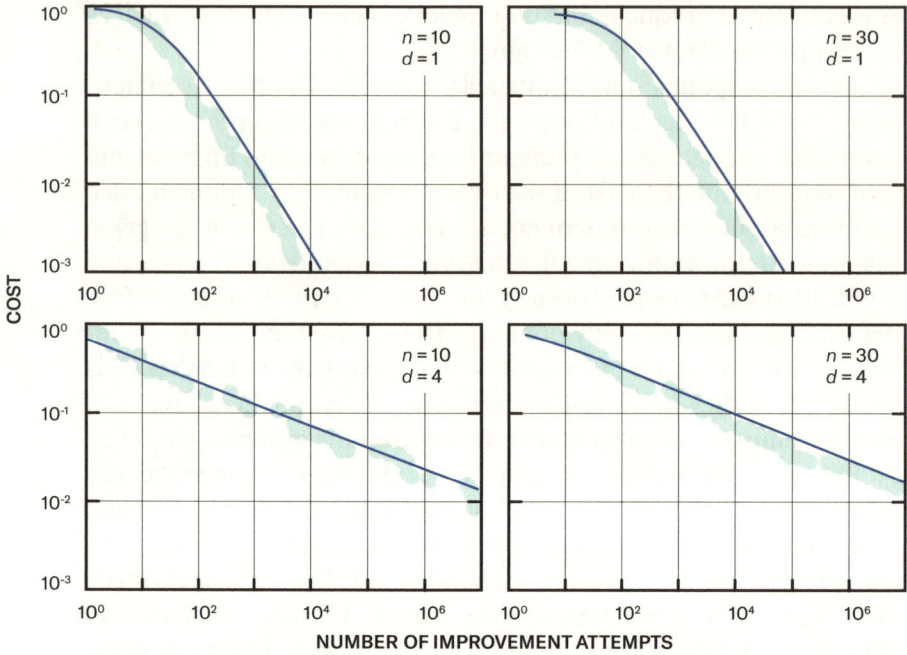

**Figure 24.** Simulated learning curve for different types of technologies. $n$ is the number of components; $d$ is the degree of interconnectedness. The greater the connectedness of components, the more difficult the technology is to improve and the shallower the learning curve will be. Reproduced from McNerney et al., "Technology Improvement."

most efficient way to arrange a production process. So how do we resolve this tension?

One possibility is by eschewing vertical integration. Each step in the process would then be performed as a tightly coupled, highly integrated process, but different companies would perform different steps, each specializing in some small aspect of the process. As an example, modern laptop computers are assembled from individual components—screens, processors, storage, batteries, power supplies, cameras—made by separate specialized manufacturers.[674] This seems to be the shape the world economy is taking as transportation and coordination costs fall and outsourcing aspects of the production process has become increasingly feasible.

### Small improvements

In our previous discussion of S curves, we noted that the performance improvement of a technology or process often results from

DEGREE OF DESIGN COMPLEXITY

STANDARDIZED     MASS-CUSTOMIZED     CUSTOMIZED

| | | |
|---|---|---|
| **STANDARDIZED COMPLEX PRODUCT SYSTEMS**<br><br>• Combined cycle gas turbine power plants | **PLATFORM-BASED COMPLEX PRODUCT SYSTEMS**<br><br>• Small modular reactor nuclear power plants<br>• Carbon capture and storage | **COMPLEX PRODUCT SYSTEMS**<br><br>• Nuclear power plants<br>• Bioenergy with carbon capture and storage | **COMPLEX** |
| **MASS-PRODUCED COMPLEX PRODUCTS**<br><br>• Electric vehicles | **PLATFORM-BASED COMPLEX PRODUCTS**<br><br>• Wind turbines<br>• Concentrating solar power | **STANDARDIZED COMPLEX PRODUCT SYSTEMS**<br><br>• Biomass power plants<br>• Geothermal power | **DESIGN-INTENSIVE** |
| **MASS-PRODUCED PRODUCTS**<br><br>• Solar PV modules<br>• LEDs | **MASS-CUSTOMIZED PRODUCTS**<br><br>• Rooftop solar PVs | **SMALL-BATCH PRODUCTS**<br><br>• Building envelope retrofits | **SIMPLE** |

NEED FOR CUSTOMIZATION

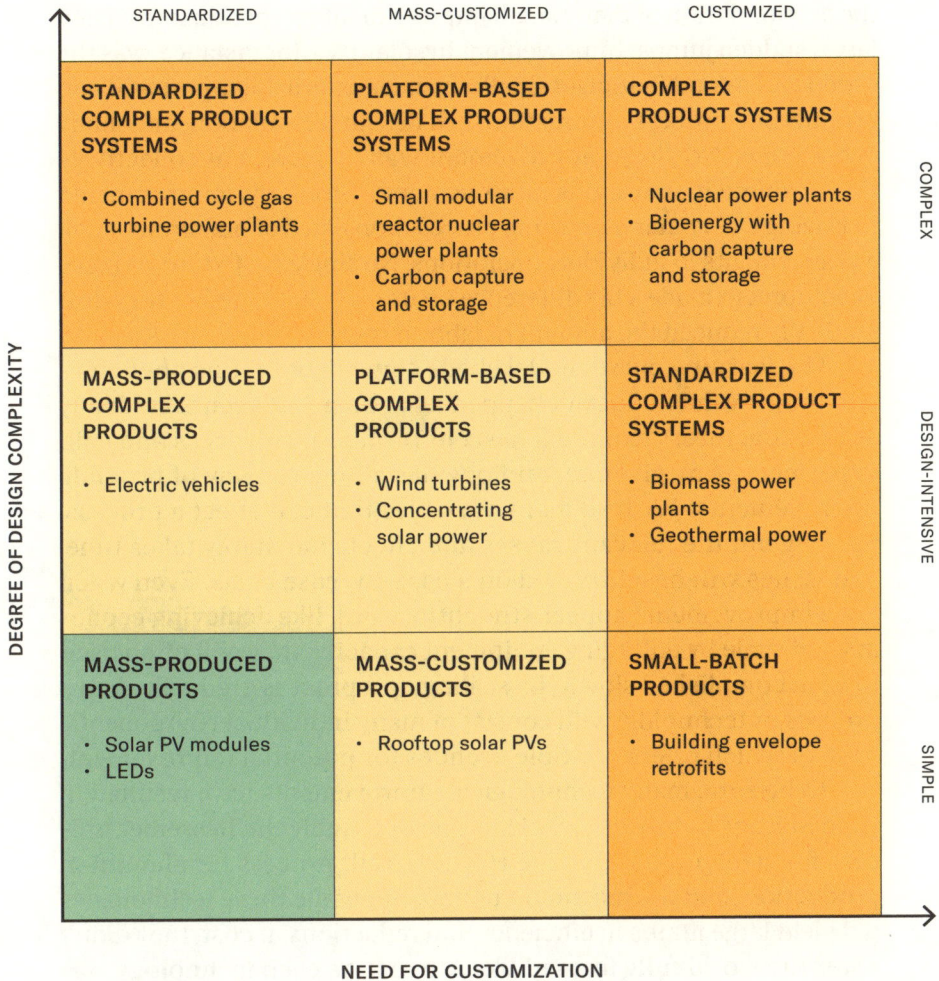

**Figure 25.** Complexity and customization versus learning rate. The more complex the product and the more it must be customized, the shallower the learning curve. Reproduced from *Joule* 4, no. 11, Malhotra and Schmidt, "Low-Carbon Innovation," fig. 1, © 2020, by permission from Elsevier.

the accumulation of many small improvements rather than from any large, sudden jumps. Improvement in solar PVs, for instance, was the product of a combination of small advancements in high-purity silicon production, process automation, cell design, installation, and many other areas. No single improvement was decisive, but collectively they resulted in enormous improvements. Similarly, improvements to Boeing's B-17 manufacturing process consisted of numerous small enhancements to individual operations. No single jig, fixture, or piece of equipment made a big difference on its own, but together they dramatically reduced the amount of labor required.

The learning curve model sheds some light on this phenomenon. Improvement doesn't happen all at once but occurs gradually as manufacturers explore a nascent production process and gain experience. Any given technology has a long runway of possible improvements: With so many variables that can affect a process, figuring out the relevant causes and effects inevitably takes time, and issues will be addressed on a case-by-case basis. Even when huge improvements appear straightforward, like achieving economies of scale by enlarging equipment capacity, they will often need to be accomplished slowly, by scaling up a process gradually. And a process or technology will consist of many individual components, each of which offers multiple avenues for potential improvement.

To be sure, many technological improvements have resulted in sudden leaps in process efficiency, most notably the Bessemer process for steel production, the Hall–Héroult process for aluminum production, and the machine-cut nail. But while these technologies did yield large jumps in efficiency and reductions in cost, they didn't enter the world fully formed like Athena. As each technology was slowly scaled up, automated, and adjusted, further cost decreases became possible. Each advanced down the learning curve by way of many small improvements.

### Disruption of the learning curve

One important finding in studies of the learning curve is that its effects are most prominent in stable production conditions with a high degree of continuity. Stable environments are what make learning from experience possible (since lessons learned will apply in the future), make it worthwhile to invest in finding solutions to a problem, and make it possible to create machinery that performs some repetitive action. When this stability is disrupted, the learning curve is often disrupted as well.

In one apparel factory, for example, production initially showed the standard learning curve, with cost falling in constant proportion to cumulative production volume. But when the factory introduced a new style of clothing, there was a reset in the curve. Costs jumped higher, at which point the learning curve started over at the new, higher level.[675]

Similar disruptions have been observed in a variety of production operations. When a steel manufacturer started making a new product, its production costs jumped, then started declining again.[676] When a car factory began making a new model, the new model had more defects than the existing model, which were corrected over time.[677] When Boeing was manufacturing the 707, production hours per plane would jump whenever starting production for a new customer because each had unique requirements.[678]

If there's too much variation in a production environment, the learning curve might not appear at all. In one study of food processing, for example, "the extreme variance in the raw produce attributable to seasonal, geographical, and other variables affect[ed] productivity to such a degree that one cannot define a true learning curve."[679]

Avoiding these sorts of resets is what inspired Intel's Copy Exactly semiconductor manufacturing policy. Prior to Copy Exactly, new semiconductor fabs achieved much lower yields than existing fabs while plant operators mastered the new, slightly different operations.[680] By making new fabs identical in every way to existing fabs, this reset was eliminated, the shakeout period was greatly reduced, and new fabs quickly matched existing fabs' yields.

Why do changes to a process result in these disruptions? A changed process means that, to some extent, prior experience and skill will not apply, portions of the process will need to be relearned, skill will need to be acquired, and new bugs will need to be worked out. As we've seen, production skill can be embedded within organizations, but unlike a physical machine, an organization's skill is a fragile, ephemeral thing, a web of habits and behaviors of many individual workers that can easily fray. And a changed process means the introduction of new causal factors that might affect the existing process in any number of ways.

Because of this, it often takes surprisingly little to disrupt a learning curve. Disruptions can occur merely by breaking the continuity of operations even if no substantive changes are made to the process.

If production only occurs intermittently, there might be significant forgetting of process methods between operations. One TV component manufacturer that altered its production schedule so production only occurred for one to three weeks during a multiweek period experienced a significant loss of efficiency and an increase in production costs.[681] Another study found a similar disruption in the construction of a concrete building that was interrupted.[682] This forgetting can be because of actual forgetting by the individual workers, or because of a more general process of decay: best practices that get ignored, tooling that gets rearranged, parts that get lost, or maintenance that isn't performed while a process is interrupted. One advantage of a consistently managed production operation is that it will avoid this sort of deterioration.

In some cases, this sort of forgetting might occur even if production slows rather than completely stopping. Learning curves often exhibit a toe-up phenomenon, where costs increase as production tapers off at the end of a run.[683] Declining production of the B-17 was followed by an increase in labor hours per airframe. Similarly, the lack of a learning curve in the manufacture of Lockheed's TriStar jet airliner, which resulted in enormous losses, has been blamed on highly variable production rates.[684]

In extreme cases, lack of stability in a process can cause a phenomenon known as negative learning, wherein costs increase over time due to constantly changing production conditions. This occurred in the construction of US nuclear power plants, where, after a period of cost decline in the 1960s, costs rose continuously in the 1970s and '80s due to the ever-evolving social and political landscape plants found themselves operating in. Increasingly strict environmental regulations meant designs had to be updated frequently, even for plants under construction, and environmental lawsuits significantly slowed projects. This was followed by a long period of no new plant construction, resulting in a considerable loss of plant construction experience and a withering of the nuclear supply chain. When US nuclear plant construction started up again in the 21st century, these plants went substantially over budget. One plant was canceled prior to completion after spending billions of dollars, and another was completed years late and billions of dollars over budget.[685]

The obvious implication is to try to keep as much stability in a production process as possible. Without continuity, previous experience

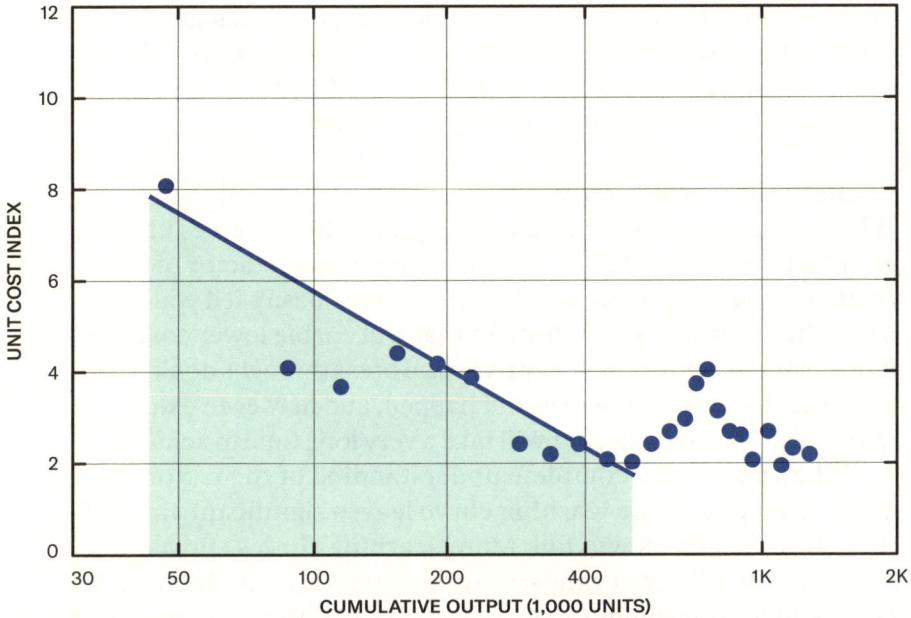

**Figure 26.** Learning curve disruption in an apparel factory. Blue dots represent the unit cost measured at different points in the production process. Reproduced by permission from *IEEE Transactions on Engineering Management*, Baloff, "Startup Management," fig. 2.

won't apply, learning won't occur, and process improvements can't accumulate. Industrial improvement systems, such as lean manufacturing, often explicitly note the importance of maintaining a stable process before embarking on any improvement efforts. As Jeffrey Liker, author of *The Toyota Way,* writes, "If the process is shifting from here to there, then any improvement will just be one more variation that is occasionally used and mostly ignored. One must standardize, and thus stabilize, the process before continuous improvements can be made."[686]

Long breaks in production should similarly be avoided to prevent process decay. The US Navy, for instance, deliberately schedules construction of nuclear subs in such a way to provide continuity of production and prevent loss of knowledge of how to build them.[687] By contrast, when the UK attempted to construct a new line of nuclear submarines after a long fallow period, the nation faced many difficulties, as many of the key skills had atrophied or been lost.[688]

The risks of disruption also provide another explanation for why progress tends to occur as the result of many small changes: While small changes will often be absorbed easily by a process or organization, large, sudden changes are much more difficult to absorb and carry with them the risk of a learning curve reset.

### Limits to the learning curve

While learning curves shed light on how and why processes improve at the rate they do, it's difficult to use them as a predictor of future costs. The mere existence of a learning curve doesn't tell you when, or if, a key technical breakthrough that will enable lower costs and larger markets will occur. If key technical breakthroughs don't occur, progress along the curve may still happen, and may even exhibit an impressive learning rate, but will take a very long time to achieve.

Likewise, even a complete understanding of the factors that affect the slope of the learning curve leaves significant variability in how fast costs will fall. Many learning curves show different slopes at different times, with no obvious cause behind the change in improvement rates.[689] And just as different factories in the same industry (or even within the same company) may have substantially different production costs, different factories making the same thing can have different learning curve slopes. Lockheed's B-17 factory had a different learning curve slope than Boeing's, for example.[690]

Similarly, outside forces may disrupt the learning rate. Returning to nuclear power plants, frequent changes to safety regulations resulted in steadily increasing construction costs. Likewise, more stringent rules for coal plant emissions, among other regulatory changes, interrupted the multidecade decline in the price of coal power.[691] On a smaller scale, minor process disruptions can cause production costs to spike. In some cases, these spikes may quickly revert, as with the Boeing B-17, but in other cases the curve may reset to the new, higher level.

Moreover, the fact that learning curves typically use log-log plots can conceal a great deal of meaningful variation. In solar PV production, for instance, between 1976 and 2016, cost per watt declined around 23 percent for every doubling of production volume *on average*, but in any given year, costs differed from the predicted value by up to 150 percent.

More generally, although learning curves appear highly regular, with costs falling at a constant rate with every doubling, this phenomenon is less illuminating than it appears. Many—perhaps most—cost functions that both decrease with increased production volume and have diminishing returns will form a straight line on a log-log plot. In other words, if costs generally fall with increased production volume (which we should expect, for the reasons discussed in this book) and if returns are diminishing (which we should also expect), it will nearly always be possible to fit a learning curve to the data. This limits how much a learning curve can tell us about what's happening inside an improving process, since a learning curve can accommodate a variety of different relationships between production volume and cost. There's no deep reason why costs appear to fall in such a consistent fashion on a learning curve—this is simply an emergent property when you have both diminishing returns and improvement correlated with increasing production volume.

None of this is to say that learning curves aren't useful. They show that production volume is a relevant unit of experience and that returns to improvement diminish over time. They help us pin down the S curve, itself a useful conceptual tool for understanding technological progress, with some quantitative rigor. They allow us to quantify improvement rates, compare the progress of different technologies, and understand what helps and what hinders technological development and efficiency improvements. Learning curves are thus best thought of as a high-level map of process improvement. They offer a general picture and describe the major features of the landscape, but they can't provide detailed, turn-by-turn directions on how to navigate it or whether a given level of performance is reachable.

## Summary

Learning curves reveal the close coupling of production process efficiency and scale. When something is expensive and takes significant resources to produce, less of it will be produced. The things that are abundant are the things that are cheap. Making a process more efficient—changing the production method, altering the design of the product, reducing the costs of inputs, increasing the scale of production, making the process work more reliably and predictably, and cutting out steps or buffers—is how we get from expensive and scarce to cheap and ubiquitous. But just knowing the different types of process improvements doesn't tell us how, exactly, that change happens.

Learning curves can fill in some of this picture. As we make more of something, our skill improves, production problems are found and fixed, and potential process improvements are explored, which together cause costs to fall. As production volume increases, the market grows, which incentivizes inventors, entrepreneurs, and companies to develop and adopt better production technology and solve problems. And as production costs continue to fall, whether through experience or via some other mechanism, more and more potential uses for the product become feasible and production increases even more.

But this process isn't guaranteed. In some cases, we may not find process improvements, consigning us to an expensive product that's limited to niche uses, as with titanium. In other cases, the improvement process might be disrupted by product or process changes, resetting production at a higher cost level. If there are too many disruptions to the process or too much variation within it, this improvement process may not happen at all. For improvement to occur, some degree of stability is required, which is likely one reason why process improvements tend to take the form of a series of small enhancements rather than sudden leaps.

# 8

# Bundles, Chains, and Feedback Loops

A central aim of our investigation of efficiency has been to disaggregate the different mechanisms by which a production process can get cheaper and more efficient. In practice, however, these mechanisms—reducing variability, lowering the cost of inputs, decreasing the size of buffers, cutting out unnecessary steps, and taking advantage of economies of scale—will often occur together, with one process change triggering multiple types of improvements.

Consider Ford's Model T assembly line. By moving the work to the worker rather than having the worker move around the factory, the assembly line removed a large number of unnecessary steps. Because a conveyor belt kept parts off the floor, the assembly line eliminated many put-things-down and pick-things-up movements, as well as workers walking back and forth to retrieve parts.[692] Later on, raising the conveyor to the height of the worker eliminated many bend-down-and-straighten-back-up movements as well.[693] This elimination of steps allowed for increased economies of scale, because by cutting out unnecessary movements, workers could complete tasks in less time. For example, when the first Model T assembly line was introduced, the assembly time for magneto coils went from 20 minutes to just over 13 minutes.[694] This meant more units could be assembled in a single day, spreading labor and fixed production costs across a larger number of units. And because the assembly line moved work at a constant rate, it kept worker operations in sync. As historian David Hounshell notes in his study of American manufacturing, by "speed[ing] up the slow men and slow[ing] down the fast men," the assembly line "would bring regularity to the Ford factory, a regularity almost as dependable as the rising sun."[695] Keeping worker operations in sync eliminated demand mismatch between steps in the process, preventing inventory from accumulating between workstations and reducing the amount of work in process and its associated costs.

The development of machine-made interchangeable parts offers another example of how a single process change can trigger multiple improvements. By making every part similar enough to be interchangeable, many subsequent steps in the process could be removed. Assembly no longer required extensive fitting operations to get parts to fit together and there was less variation in assembly time, making the process more predictable. Ford achieved interchangeability by making extensive use of custom machine tools and specially designed jigs and fixtures: Workers could quickly place a part into a machine designed to accommodate it, rather than having to carefully measure and then place a part in the machine in just the right spot. This greatly simplified machining operations, cutting out unnecessary steps and reducing the amount of skilled (and thus expensive) labor required.[696]

In the case of both interchangeable parts and the assembly line, one change to the process triggered multiple simultaneous improvements. The same is true for many other types of process improvements. Redesigning a product to require fewer parts both reduces the number of assembly operations and improves quality, since fewer operations means fewer potential errors. Introducing a division of labor both cuts out unnecessary steps and enables the use of cheaper, less-skilled labor. Using standardized, off-the-shelf components allows for economies of scale (since standardized parts will be produced in much higher volumes than custom ones), eliminates the costs of designing and validating new parts, and reduces errors (since standardized parts will be better understood than never-before-used custom ones).

In this chapter, we'll explore when and how such improvements occur together. This generally happens in one of two ways. First, a single change to a process might result in improvements along several different axes. We'll call these *bundles of improvements*. Second, one type of improvement might unlock or make possible another type of improvement. We'll call these *chains of improvements*.

### Bundles of improvements

The most common case in which a single process change results in improvements along multiple axes is when a change is made to the production method. This is because the production method will influence, if not directly determine, every single aspect of the process, as we discussed in Chapter 1. How fast a process runs, how accurate it is, what inputs are required—these are all a function of what production method we choose, so a change in method can potentially change all these parameters at once. Replacing hand-forged nails with machine-cut nails reduced the input requirements (nail plate for nail rod), lessened the amount of labor required, decreased the variability of the process, and increased the rate of production, all at once. Similarly, replacing a manual process with a mechanized or automatic one can yield increases in speed and quality, reductions in labor cost, and economies of scale, all at the same time. The development of continuous automated mills to produce steel sheets, for example, both reduced costs and increased quality.[697]

Many changes that result in bundles of improvements are in fact special cases of this more general phenomenon of production method change. Cutting a step out of the process, for instance, can be thought of as a production method change that improves the process along several axes. Each process step requires some amount of inputs and will fail some percentage of the time, so removing a step from the process not only cuts out the inputs that step requires but also removes a potential source of failure, increasing process yield.[698] This is why pharmaceutical companies will often look for synthetic routes that can produce a drug in as few steps as possible. Since reaction yields are less than 100 percent at each step, which has a cumulative diminishing effect on the final output, decreasing the number of process steps can significantly increase overall yield.[699] Similar gains from reducing the number of process steps have been observed in semiconductor manufacturing.[700]

Reducing work in process can result in a similar bundle of improvements. As with value-adding process steps, work in process requires a scaffolding of supporting operations because it needs to be stored, transported, tracked, inspected, and insured. Therefore, reducing work in process not only reduces the waste of producing something that hasn't yet been sold but also decreases all of the supplementary costs and operations. Each item of inventory reduced can be thought of as a process step that has been eliminated, along with its supporting operations.

Standardization can also result in efficiency improvements along several axes, and it, too, can be thought of as a special case of production method change. Standardization changes some part of the process (say, how a part is made or how an operation is performed) to one that has greater scale, higher yields, and requires fewer ancillary operations and inputs (designing the new part, coordinating with people who aren't familiar with the new design, training people on how to interact with it, and so on).

Of course, a production method change won't necessarily enhance every aspect of a process. As we saw in Chapter 2, we typically have a palette of methods available for accomplishing a given task, and different methods may have different strengths and weaknesses, resulting in trade-offs rather than across-the-board improvements. But it's nevertheless possible, and common, for a change to the production method to yield multiple improvements at once.

## Chains of improvements
In addition to triggering improvements along many different axes, process improvements often unlock or enable further process improvements. For example, removing a bottleneck or constraint often enables additional advances that previously weren't feasible.

Take cement kilns, which produce cement by heating limestone and clay in a long, rotating tube. Historically, cement kilns had been manually controlled: An operator would monitor the kiln and adjust parameters such as temperature, rotation speed, and material feed rate to maintain the proper conditions inside the kiln. When manual kiln control was replaced by automatic, computer-based control starting in the 1950s, it became possible to build and operate much larger kilns because automatic processes—unlike human operators—could engage in the ceaseless monitoring and instant, incremental adjustments required to keep enormous kilns within tolerance. Automating kiln operations thus unlocked economies of scale by removing the bottleneck on control.[701]

More generally, smoothing the variation in a process and reducing its likelihood of failure is often a prerequisite for higher production volumes and, therefore, for economies of scale. A failure rate that is tolerable at low volumes may be untenable at higher volumes. Reflecting on the importance of process control methods at HP when the company decided to increase its production volume of printers, one engineering manager observed, "Before, when we produced at low volume, stopping the line was no big deal. But now inventory would pile up fast."[702] Similarly, the introduction of automatic control devices in chemical plants in the mid-20th century, along with the resulting reduction in errors and increased reliability, enabled much larger-scale operations.[703]

It's also common for one step in the production process to be the result of variation and lack of control in a previous step. Prior to interchangeable parts, imprecision in part fabrication required fitting steps in which gunsmiths would manually file and adjust parts so they would fit together. Likewise, metal parts are often made by casting or forging, but imprecision in these methods means that cast or forged metal parts often require additional machining steps to cut away extra metal and turn them into finished parts.[704] Beyond requiring extra fitting or adjusting steps, such variation also leads to an accumulation of inventory buffers between steps, necessitating additional steps to maintain, move, and monitor those buffers. In each case, by reducing variation at one step in the process, subsequent steps can also be removed.

A process improvement can also enable additional improvements by making follow-on improvements easier to uncover. In the Toyota production system, reducing the size of inventory buffers made any disruption to the process immediately apparent and, therefore, easier to find and fix. A manufacturer of birdfeeders implementing the Toyota production system discovered reliability problems with their glue guns after lowering work-in-process buffers.[705] More generally, it is often only possible to notice subtle problems in a process once the more severe problems have been fixed.[706]

Because producers often have a wide array of production technologies to choose from, each with distinct benefits, drawbacks, and constraints, an improvement to a production process may make a different, lower-cost process feasible when it wasn't before. Automation, for example, is often predicated on first reducing upstream variation in the process so that human judgment is no longer required. In his book on mechanized assembly, Geoffrey Boothroyd notes that while "defective parts do not generally create great difficulties when assembly is by hand," a defective part fed into a machine may severely damage it. So mechanized assembly is only cost-effective above some minimum quality threshold.[707] Similarly, the invention of the flyball governor, which reduced variability in steam engine speed and made its motions more uniform, made it possible to use textile machinery to process cotton for high-quality thread.[708] In agriculture, the development of corn plants that matured at the same time and grew ears at a uniform height enabled the use of mechanical corn pickers.[709]

When one improvement triggers another, it also creates the possibility of long chains of improvements, where each improvement enables the next. Interchangeable parts made the Model T assembly line feasible, which made it possible to complete assembly operations quickly and predictably. The assembly line, in turn, allowed for greater production rates and economies of scale. And as scale increased, further improvements, such as the development of special-purpose machine tools, became possible.

The Toyota production system explicitly harnesses this potential. In fact, one way of thinking of the system is as a deliberately structured chain of improvements. In his book on the Toyota production system, Yasuhiro Monden diagrams the method by which it functions. (See Figure 27.) According to Monden, three basic chains are at work in the Toyota production system, all of which contribute to its fundamental goal of greater profits. The first is increased quality, enabled by autonomation (autonomous defect control), which prevents defective parts from disrupting a downstream process. This results in higher sales and lower service costs. The second is decreased workforce costs, made possible by being able to flexibly reallocate workers as demand changes. This reallocation is accomplished by cross-training workers and rearranging machines so one worker can service multiple machines.

The largest chain in Monden's schema is devoted to inventory reduction. By producing products just in time, work in process is kept to a minimum and inventory costs are reduced. Just-in-time production is achieved by steadily producing products at the rate of actual demand. If demand is for 3,000 units of product A per month, product A should be produced at a rate of 100 per day, or one every 4.8 minutes. To ensure equipment doesn't lie dormant, production of product A can be interleaved with the production of other models. If there is demand for 1,500 units of product B per month, or 50 per day, the line would produce A, A, B, A, A, B. This arrangement minimizes inventory accumulation and maximizes equipment use, but it requires reducing setup times (using SMED) so equipment can quickly change over to the correct model. It's also facilitated by placing machines close together, minimizing travel distances, and training workers to oversee multiple machines simultaneously and perform different sorts of tasks. Just-in-time production is also made possible by kanban, the control system we discussed in Chapter 6, which keeps inventory to a minimum between processes running at different rates.

So, by reducing setup times and travel distances, increasing part quality, cross-training workers, and capping the amount of inventory, Toyota's production process can produce at exactly the rate of demand, minimizing inventory and quickly adjusting as demand changes. This chain structure explains why guides to the Toyota production system emphasize that the system must be implemented in its entirety to achieve results. Simply picking and choosing elements—a little kanban here, a few work cells there—will not yield the desired outcomes.

Ideally, these sorts of improvement chains can be arranged to form a virtuous circle, in which one improvement triggers or enables another. In fact, economies of scale, as discussed in Chapter 4, can be seen as just such virtuous circles, because they're self-reinforcing. Increased production volumes results in lower costs, which, all else being equal, leads to more sales, which leads to higher production volumes and even greater economies of scale. By a similar mechanism, any process improvement that results in decreased costs can potentially enable greater economies of scale and trigger this sort of cycle. Conversely, because so many process improvements are predicated on some level of scale, and because higher production volume yields more experience, which enables more advances, increased production scale can trigger additional process improvements.

Because of these bundles, chains, and feedback loops, process improvement is typically a complex combination of many different mechanisms. A single change might result in improvements along several different axes; one of those improvements might then trigger additional improvements; and so on. To understand how these improvements tend to occur together, let's take a closer look at Ford's production of the Model T, a canonical example of a chain of efficiency improvements that occurred over many years.

*Case study: Ford and the Model T*
Ford's status as a large-volume car producer began with the predecessor to the Model T: the Model N, a four-cylinder, two-seater car initially priced at $500.[710] At the time, the average car in the US cost more than $2,000, and it seemed nearly unimaginable that a car with the capabilities of the Model N could cost so little. In 1906, the year the Model N was introduced, Ford sold 8,500 of them, making the automaker bigger than the next two biggest car producers, Cadillac and Rambler, combined.[711]

To produce such a huge volume of cars, Ford began to use many of the production methods it would develop more fully with the Model T. Many of the Model N's parts were made of vanadium steel, a strong, lightweight, durable steel alloy. Vanadium steel allowed for a lighter car (the Model N weighed only 1,050 pounds), and was "machined readily."[712] This was important because Ford also made increasing use of advanced machine tools that allowed it to produce highly accurate interchangeable parts.[713] In 1906, Ford advertised that it was

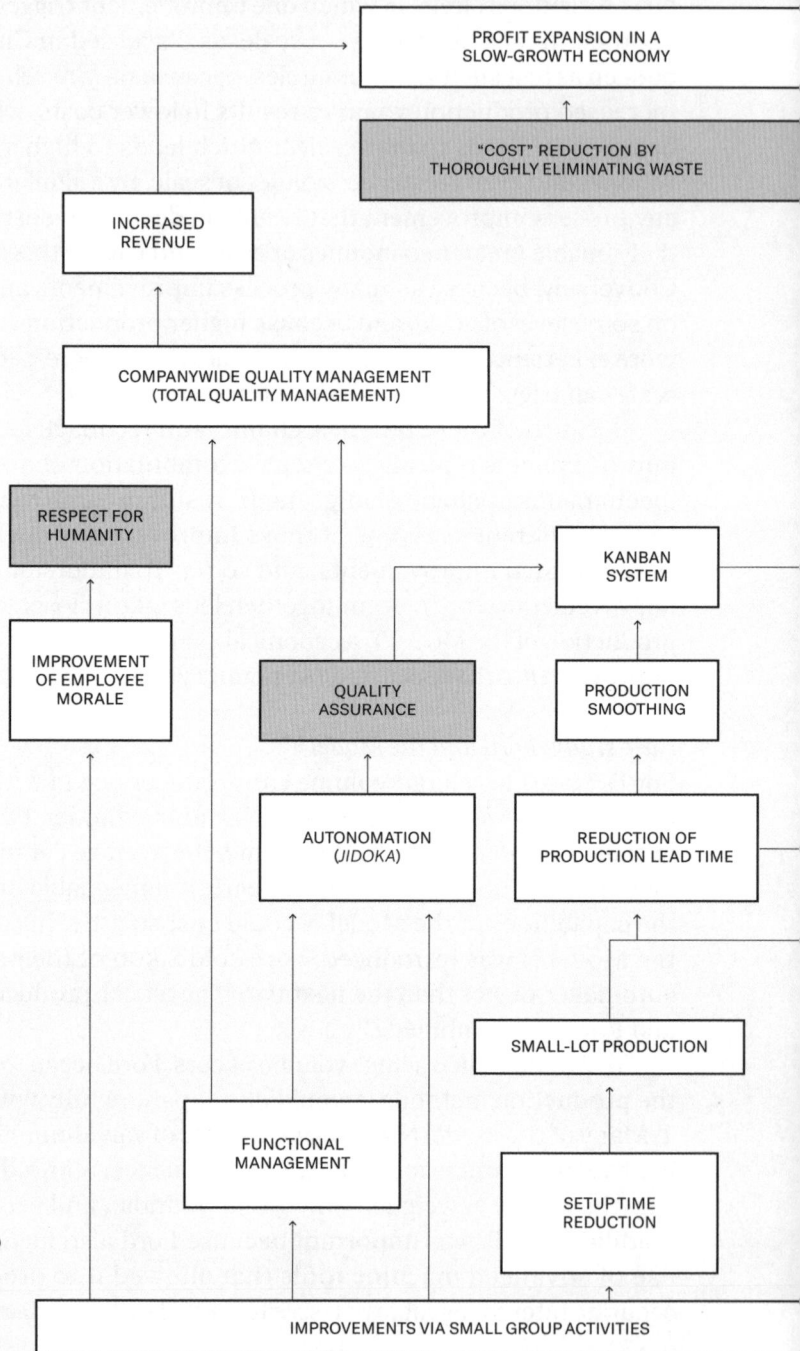

**Figure 27.** Chains of improvement in the Toyota production system. Reproduced by permission from Taylor & Francis Group through PLSclear, Monden, *Toyota Production System*, fig. 1.2, © 2012.

PROFIT EXPANSION IN A SLOW-GROWTH ECONOMY

"COST" REDUCTION BY THOROUGHLY ELIMINATING WASTE

INCREASED REVENUE

COMPANYWIDE QUALITY MANAGEMENT (TOTAL QUALITY MANAGEMENT)

RESPECT FOR HUMANITY

KANBAN SYSTEM

IMPROVEMENT OF EMPLOYEE MORALE

QUALITY ASSURANCE

PRODUCTION SMOOTHING

AUTONOMATION (*JIDOKA*)

REDUCTION OF PRODUCTION LEAD TIME

SMALL-LOT PRODUCTION

FUNCTIONAL MANAGEMENT

SETUP TIME REDUCTION

IMPROVEMENTS VIA SMALL GROUP ACTIVITIES

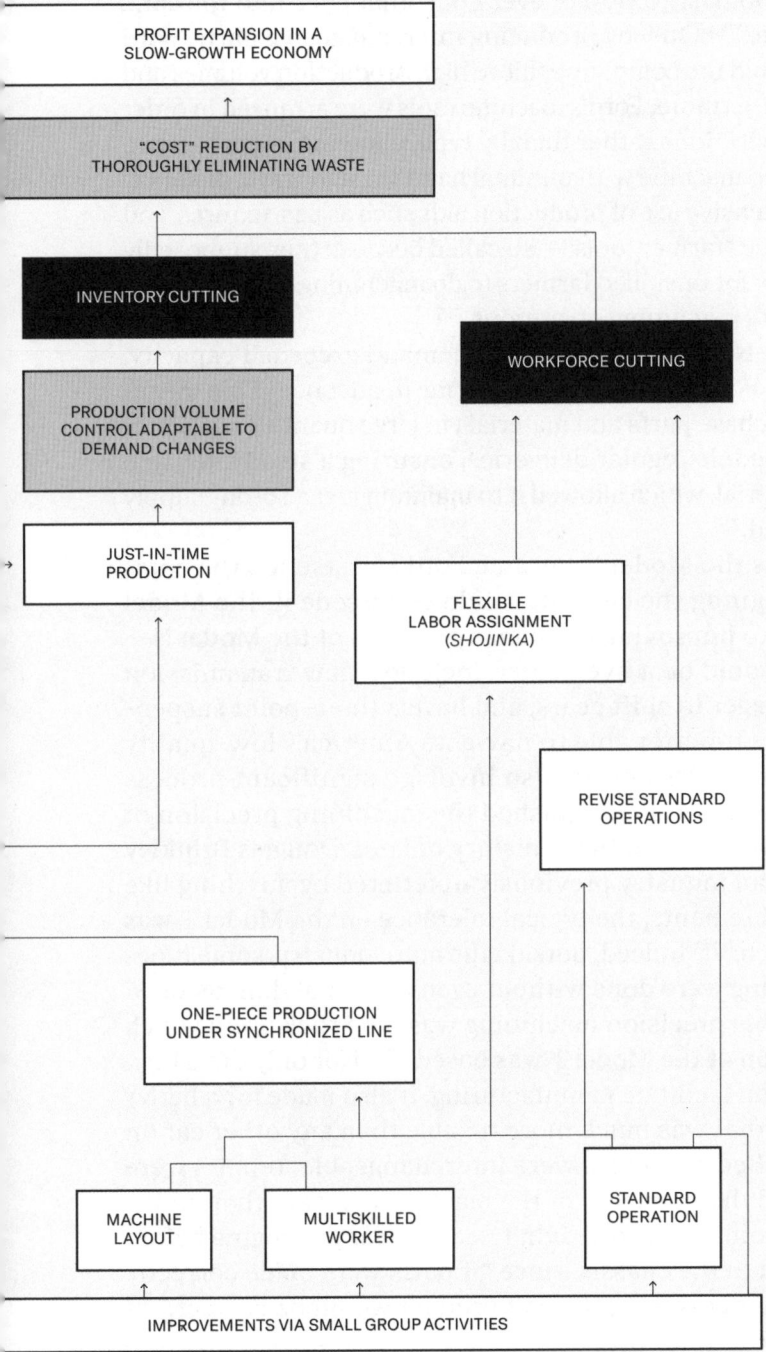

PROFIT EXPANSION IN A
SLOW-GROWTH ECONOMY

↑

"COST" REDUCTION BY
THOROUGHLY ELIMINATING WASTE

↑

INVENTORY CUTTING

WORKFORCE CUTTING

↑

PRODUCTION VOLUME
CONTROL ADAPTABLE TO
DEMAND CHANGES

↑

JUST-IN-TIME
PRODUCTION

FLEXIBLE
LABOR ASSIGNMENT
(SHOJINKA)

REVISE STANDARD
OPERATIONS

ONE-PIECE PRODUCTION
UNDER SYNCHRONIZED LINE

↑

MACHINE
LAYOUT

MULTISKILLED
WORKER

STANDARD
OPERATION

↑

IMPROVEMENTS VIA SMALL GROUP ACTIVITIES

"making 40,000 cylinders, 10,000 engines, 40,000 wheels, 20,000 axles, 10,000 bodies, 10,000 of every part that goes into the car... all exactly alike."[714] Only by producing interchangeable parts, Ford determined, could the company achieve high production volumes and low prices. Furthermore, Ford's machine tools were arranged in order of assembly operations rather than by type, allowing parts to move from machine to machine with minimal handling and travel distance. It also made extensive use of production aids such as jigs, fixtures, and templates. These "farmer tools"—so called because they supposedly made it possible for unskilled farmers to do machining work—greatly simplified Ford's machining operations.[715]

The Model N was so popular that demand exceeded capacity, which allowed Ford to plan production far in advance. This meant Ford could purchase parts and materials in large quantities at better prices and schedule regular deliveries, ensuring a steady, reliable delivery of material, which allowed it to maintain just a 10-day supply of parts on hand.[716]

But even as the Model N became Ford's bestseller, the company was designing the car that would supersede it: the Model T. In addition to improving upon many aspects of the Model N— the Model T would be a five-seater, include a new transmission that made it easier to shift gears, and have a three-point suspension that made it better able to navigate America's low-quality roads—the Model T's design also involved significant process improvements.[717] For one, it pushed the machining precision of the Model N even further. In his history of Ford, Douglas Brinkley notes that "in an industry previously unfettered by anything like exacting measurements, the typical tolerance on the Model T was 1/64th of an inch."[718] Indeed, outside the auto industry, some types of manufacturing were done without even the aid of dimensioned drawings.[719] This precision machining was "the rock upon which mass production of the Model T was based."[720] Not only did a high level of precision facilitate manufacturing, it also made for a better product—one that was much more reliable than any other car on the market.[721] Because parts were interchangeable, repairs were simpler.[722] And the precision of the machining meant that unlike most other automakers, Ford didn't need to test the engine before it was attached to the chassis, since "if parts were made correctly and put together correctly, the end product would be correct."[723]

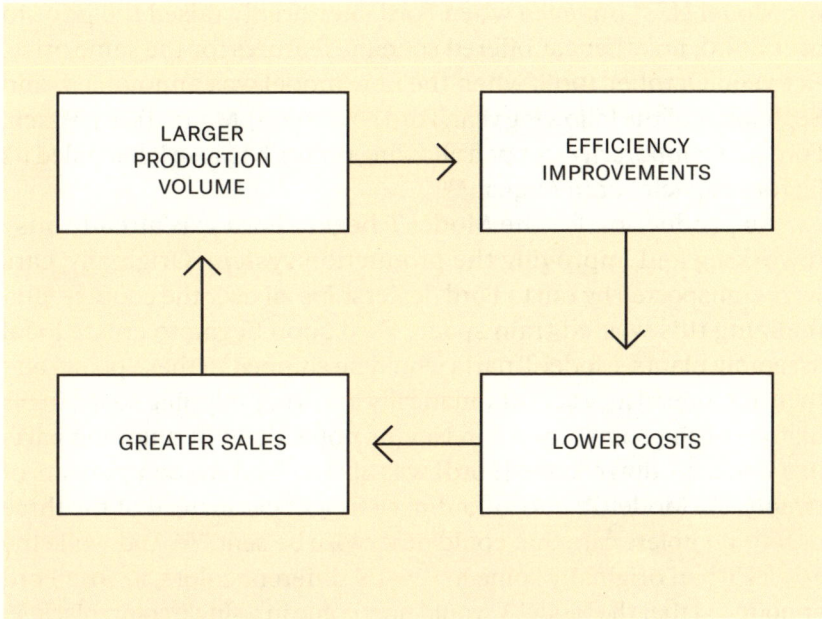

**Figure 28.** The improvement cycle triggered by economies of scale.

The Model T would ultimately cost around $100 a year to maintain, at a time when the maintenance of other cars cost $1,500 per year.[724]

At the time, most four-cylinder engines were cast either as four separate cylinders or two groups of two cylinders and then attached together, which required extra time and material. The Model T's engine block, on the other hand, was cast as a single piece.[725] Several components—the rear axle housing, transmission housing, and crankcase—were not made of the more customary cast steel but rather stamped steel, a then-novel technology for automobiles that was cheaper than casting.[726] Like any new technology, these production methods required time and effort to implement—it took 11 months of development to figure out how to produce the drawn-steel crankcase—but the manufacturing cost savings were worth the effort.[727] Over time, more and more Model T parts would be made from pressed steel, though the transmission housing itself was later changed to cast aluminum.[728]

The Model T was not the cheapest car on the road when it was introduced—at $850, it cost several hundred dollars more than the Model N.[729] But even when Ford later briefly raised the price to over $900, no other car offered so many features for the same price. Between October 1908, when the new model was announced, and September of the following year, Ford sold 10,607 Model Ts. By March, Ford had temporarily stopped allowing orders because it had filled its factory capacity until August.[730]

As production for the Model T began, Ford was already busy reworking and improving the production system. Originally, cars were transported by rail to Ford dealerships all over the country. But, realizing this wasted train space, Ford soon began to create local assembly plants. Model T parts would be shipped to these plants and then assembled into cars, dramatically lowering shipping costs. In his history of the company, Allan Nevins notes that "by shipping parts in a knocked-down state, [Ford] was able to load the components of twenty-six Model Ts into an ordinary freight car instead of the three or four complete cars that could otherwise be sent."[731] And while the Model T had originally come in several different colors, in 1912 Ford announced that the Model T would now come in a single color: black.[732]

The first Model Ts were assembled in Ford's Piquette Avenue plant in Detroit, which was built in 1904. But in 1910 it moved production to the new, larger Highland Park factory, also in Michigan, which was considered to be the best designed factory in the world.[733] At a time when electricity was still somewhat uncommon in manufacturing, electric motors drove mechanical belting and overhead cranes were used to move material.[734] At the Piquette Avenue plant, material came in on the bottom floor and final assembly was done on the top floor. But at Highland Park, material came in on the top floor and gradually moved down to assembly on the ground floor. To facilitate the movement of material, thousands of holes were cut in the floor, which allowed parts to move down through the factory through chutes, conveyors, and tubes.[735]

At Piquette Avenue, machine tool use had been extensive, but the machinery was largely general-purpose. With the decision to focus on a single model and the subsequent enormous increase in production volume, Ford began to buy or create dozens of special-purpose machine tools designed specifically for the Model T, such as a machine for automatically painting wheels and another for drilling holes in a cylinder block.[736] As with the farmer tools first introduced on the Model N, these special-purpose tools not only produced parts more cheaply but could also be operated by less-skilled machinists, reducing labor costs.

It was only the enormous production volumes of the Model T that enabled Ford to make such extensive use of special-purpose machinery.[737] Similarly, it was only by virtue of its large volumes that Ford could afford to purchase dedicated steel-stamping presses to churn out pressed-steel crankcases, which were cheaper and used less material than the cast iron employed by other manufacturers.[738]

Ford experimented with machinery continuously, and the factory was in a constant state of rearrangement as new machinery was brought online and old machinery was scrapped. In some cases, machines that were just a month old were replaced with newer, better ones. By 1914, Highland Park had 15,000 machine tools. As at other automakers, machinery was packed close together in the order that operations were performed. But Ford took this concept much further, sandwiching drilling machines and even carbonizing furnaces between heavy millers and press punches. Not only did this machine placement for material flow keep handling to a minimum, but the tight packing of machinery also prevented inventory from building up in the factory.[739]

Ford's process improvements weren't limited to new and better machine tools. The company constantly examined its operations to figure out how they could be done in fewer steps. In one case, a part being machined on a lathe required four thumbscrews to attach it to the lathe, each of which had to be twisted into place to position the part and untwisted to remove it. By designing a special spindle with an automatic clamp for the part, Ford reduced the time to perform the operation by 90 percent.[740]

The Model T itself was continuously redesigned to reduce costs. When the stamped-steel axle housing proved complex to manufacture, it was redesigned to be simpler. Brass carburetors and lamps were replaced by cheaper ones made of iron and steel. A water pump that was found to be extraneous was removed.[741] As a result of these constant improvements, Ford was able to continuously drop the price. By 1911, the cost of a Touring model had fallen to $780; by 1913, it had dipped to $600. And as costs fell, sales rose. In 1911, Ford sold 78,000 Model Ts. In 1912, it sold 168,000. And in 1913, it sold 248,000.[742]

Then, in 1913, Ford began to install the system that would become synonymous with mass production: the assembly line. Though gravity slides and conveyors had existed prior to 1913, Ford hadn't yet developed a systematic method for continuously moving the work to the worker during assembly. The first assembly line was installed in the flywheel magneto department. Previously, workers had stood at individual workbenches, each assembling an entire flywheel magneto. But on April 1, 1913, Ford replaced the workbenches with a single steel frame with sliding surfaces on top. Workers were instructed to stand in a designated spot and, rather than assemble an entire magneto, perform one small action, then slide the work down to the next worker, repeating the process over and over.[743]

The results spoke for themselves. Prior to the assembly line, it took a single worker an average of 20 minutes to assemble a flywheel magneto. With the assembly line, it took just over 13 minutes.[744]

Ford quickly found even more ways to improve the process. To prevent workers from having to bend over, the height of the line was raised several inches. Moving the work in a continuous chain allowed it to be synchronized, which sped up the slow workers and slowed down the fast ones to an optimal pace. Within a year, the assembly time for flywheel magnetos had fallen to five minutes.[745]

This experiment in magneto assembly was quickly replicated in other departments. In June 1913, Ford installed a transmission assembly line, bringing assembly time down from 18 minutes to nine. In November, the company installed a line for the entire engine, slashing assembly time from 594 minutes to 226 minutes. As with the flywheel magneto, further adjustments and refinements to the lines yielded even greater productivity gains. In August, Ford began to create an assembly line for the entire chassis. Its first attempt, using a rope and hand crank to pull along the car frames, dropped assembly time from 12.5 hours to just under six. By October, the line had been lengthened and assembly time had fallen to three hours. By April 1914, after months of experimentation, car assembly time had been cut down to 93 minutes.[746]

In addition to reducing assembly times, the assembly line decreased inventories. By moving the work continuously along the line, there was no opportunity for parts to accumulate in piles near workstations. The Highland Park facility kept enough parts on hand to produce 3,000 to 5,000 cars—just six to 10 days' worth of production. This was only possible through careful control of material deliveries and precise timing of the different assembly lines.[747]

As the assembly lines were installed, Ford continued to make other process improvements. Operations were constantly redesigned to require fewer production steps. One new machine reduced the number of operations required to install the steering arm to the stub axle from three to one.[748] An analysis of a piston rod assembly found that workers spent almost 50 percent of their time walking back and forth. When the operation was redesigned to reduce time spent moving about, productivity increased 50 percent.[749] A redesigned foundry that used molds mounted to a continuously moving conveyor belt not only increased assembly speed but also allowed the use of less-skilled labor in its operation.[750]

Meanwhile, Ford continued to tweak the design of the Model T. The body was redesigned to be simpler and less expensive to produce. Costly forged parts were eliminated by combining them with other components. Fastener counts were reduced.[751] By 1913, comparable cars to the Model T cost nearly twice as much as the Model T did.[752] And the price continued to fall. By 1916, the cost had dropped to just $360—a two-thirds reduction in just six years.

*The Model T and improvement chains*
With the Model T, Ford didn't just create a cheap, practical car. It built an efficiency engine. With high-precision machining, Ford was able to manufacture highly accurate parts that resulted in a better, more reliable car, required less work to assemble, and used less-skilled labor. This made the car inexpensive, which, along with its excellent design, resulted in sky-high demand. High demand and high production volume enabled Ford to make additional process improvements. It designed and deployed special-purpose machine tools—large fixed costs that were only practical at huge production volumes—which increased production rates and decreased labor costs. It set up dedicated assembly plants (also large fixed costs), which enabled substantial reductions in transportation costs. It built a new factory (another large fixed cost) specially designed to optimize the flow of production. It placed massive material orders, resulting in lower prices, lower inventory costs, and smoother material delivery, reducing variability and making maximum use of production facilities. And, as all these improvements drove down the cost of the car, demand for the Model T continued to rise, enabling Ford to improve its processes even more.

More generally, the high production volumes of the Model T made any process improvement incredibly lucrative. Even a small change had a big impact when multiplied over hundreds of thousands of Model Ts.

Consider, for instance, the effect of one minor improvement among many, the removal of a forged bracket:

> Presume that it took just one minute to install the forged brackets on each chassis. Ford produced about 200,000 cars in 1914. It would have taken 200,000 minutes, or better than 3,300 hours for the installation of these forgings. Each of these brackets was held in place with three screws, three nuts, and three cotter pins; that's six screws and nuts per car—1,200,000 of each! This saving does not take into account the cotter keys nor the brackets themselves. Each bracket had four holes which had to be drilled—1,600,000 holes—which took some time as well. If the screws alone were as cheap as ten for a penny, the savings on screws alone would have been $1,200![753]

Since any production step would be repeated millions of times, it was worth carefully studying even the smallest step for possible improvements. This resulted in an environment of continuous improvement, where processes were constantly experimented on, tweaked, ripped out, and replaced with better ones. Ford could, and often did, experiment with and create designs for its own machine tools, only later having a tool builder supply them.[754] And if new machines didn't work properly, Ford could afford to abandon the experiment. In 1916, a custom-designed piston-making machine that cost $180,000 to produce—$5 million in 2023 dollars—was "thrown into the yard" after repeated failures and replaced with simple lathes.[755]

The ultimate example of this environment of constant tinkering is the assembly line, which took years of experimentation to fully work out and required restructuring almost all of Ford's operations. By breaking down operations into a series of carefully sequenced steps and mechanically moving material through them, Ford was able to eliminate extraneous operations, reduce inventories, and increase production rates, enabling even lower costs and greater scale. This entire chain of improvements was *itself* made possible by the development of precision-machined parts.

The Model T would change the world, both by making the car a ubiquitous feature of American life and, more subtly but no less significantly, showing what could be achieved with large-volume production and a cascading chain of improvements.

## Summary

As we've investigated efficiency over the course of this book, two recurring themes have emerged. One is the importance of accumulating improvements over time: Even if no individual improvement is transformative, the collective effect of many of them together can be enormous. The gradual accumulation of improvements explains both technological development (as seen in technological S curves) and cost reduction (as seen in learning curves). And in this chapter, we've seen how the accumulation of improvements can be a self-perpetuating cycle, as one improvement makes further improvements possible, and how production systems, from Ford's Model T to the Toyota production system, have leveraged this phenomenon to drive down costs.

The other recurring theme is the importance of scale. Not only is scale a massive driver of efficiency in its own right, through phenomena like fixed cost spreading and geometric scaling, but it's also often a prerequisite for many other types of efficiency improvements, such as value engineering, DFMA, or automation, which require a sufficiently large output to justify the up-front costs. Although it's best to eliminate economies of scale where possible, in practice they will always be present in some form, and it will always be beneficial to take advantage of them.

These two themes are intimately related. As we saw in Chapter 7, scale makes it possible to accumulate large numbers of improvements, and more improvements make greater scale possible. Each additional item produced is another chance to find a better method of performing a task or to cut out a step that doesn't need to be there. Each increase in market size generates greater incentives to develop new, better technology. And each decrease in cost creates an even bigger market for a product, enabling even greater production volumes. With enough scale and enough accumulated improvements, production costs can be driven down by orders of magnitude.

We can see the importance of scale and accumulating improvements in two technological revolutions—one that has already changed the world and one that seems poised to. Over the past several decades, advancing computer and information-processing technology has transformed every aspect of everyday life. This has been driven not only by a small number of breakthrough inventions, such as the transistor and the integrated circuit, but by decade after decade of improvements to the semiconductor fabrication process that has steadily chipped away at the cost and size of computation. As transistors have become unimaginably small and cheap, they've been produced in increasingly incomprehensible volumes: Trillions of transistors are now made every second.[756] And, as we saw in Chapter 4, it's this enormous scale that has made it possible to continue to push the boundaries of the technology.

Huge volumes of cheap semiconductors have enabled another revolution, that of artificial intelligence. Modern AI models consist of neural networks trained on enormous amounts of data: the entire text corpus of the internet, countless hours of gameplay, millions of driving hours. By feeding a model millions of examples, the model gradually learns how to make more accurate predictions: what text is most likely to follow, what game moves will result in a win, what maneuvers to take to quickly and safely arrive at a destination. The availability of huge data sets, along with the requisite computation to train the networks, has made modern AI technology possible.

In this chapter, we've looked more deeply at the relationship between accumulated improvements and scale. As we saw with Ford's Model T, scale makes it possible to take chains of improvement and loop them around to create virtuous circles. Though Ford made the occasional major change in Model T production (like building a new factory), most production changes were relatively small—a different headlamp design, a new piston-making machine, a change to the height of the assembly line. But as the cycle of improvement ran for years, these small changes were transformative, both for the Model T and for the world. Such is the power of slow and steady efficiency improvements.

# 9

# Continuous Processes

The idea of a production process covers an enormous range of activities, from construction to manufacturing to agriculture and even, arguably, transportation. Different production processes, then, will vary widely depending on what they produce and how they produce it: Although a blacksmith, a corn farm, and an oil refinery all use production processes, the processes will obviously look quite different from one another.

But as production processes improve along the various axes described in previous chapters, they tend to converge on a common form: *continuous* processes. In a continuous process, input materials are continuously transformed into outputs. Everything flows smoothly from one step to the next without delays, rework, or the accumulation of large inventories between process steps. The process, once a series of discrete steps, begins to resemble a single, continuous transformation.

Consider papermaking as an example of this sort of process evolution. Paper is made by extracting cellulose fibers from plant matter and pressing the fibers into a thin sheet. In 18th-century Europe and America, papermaking consisted of a series of separate, time-consuming steps. First, papermakers would take cloth rags, cut them into squares of roughly equal size, and sprinkle them with lime to help them decompose. After being allowed to sit for days or weeks, the rags would be placed into a beating engine, a large tub filled with water with a rotating wooden drum inside it. Blades were mounted on the surface of the drum so that as the drum turned, the rags would be pulled against them and be cut up into a mass of cellulose fibers.[757] From there, the fiber mass would be brought to the vat room and placed in a large, water-filled vat. A vatman would dip wire molds into the mass of wet cellulose, collect a thin layer on the mold, and place it onto a sheet of felt. When several hundred sheets had been collected, they would be taken to a press, which would squeeze them all at once to remove the water, after which they would be brought to a loft to dry. After drying, the paper was coated with size (animal glue) to prevent it from absorbing ink and then either sold directly to customers or, for finer paper, pressed to give it a smooth finish.

This process was slow and labor-intensive. Outside of the beating engine, each step was done by hand, and a large amount of work in process accumulated between each step. After the rags were cut apart in the beating engine, the processed fibers would wait to be turned into sheets at the paper molds, after which they would wait again until there were enough sheets to run through the press.

That all changed in 1799 with the invention of the fourdrinier (named after the brothers who financed it), which mechanized the molding and pressing portions of the process.[758] In the machine, the rag fiber-water mix from the beating engine would flow onto a continuous wire belt, which would shake back and forth, imitating the motion of the vatman and causing the cellulose fibers to intertwine. As the cellulose moved along the belt, the water would drain from it. Then the pulp would continue through a series of rolls that squeezed out even more of the water and flattened the pulp into paper. The damp paper would then be collected into a large roll, which would be cut into individual sheets and placed in a loft for drying.

Following the invention of the fourdrinier, additional parts of the process were gradually mechanized. Steam dryers, which took the feed of damp paper from the machine and pulled it around a series of heated iron cylinders to dry it, first appeared in 1820. Sizing machines, which took the feed of dried paper and automatically coated it with size, began to appear in 1830.[759]

Not every step in the process was fully mechanized. In the rag room, beater room, and finishing room, mechanization encountered "intractable problems." It proved very difficult to build machines to duplicate the motions of the workers, and these steps remained labor-intensive. Nevertheless, the fourdrinier and its extensions massively reduced the cost of making paper, such that by 1850, machine-made paper was one-eighth the cost of handmade paper.[760]

The fourdrinier did more than mechanize large portions of papermaking; it also transformed papermaking into a continuous process. Whereas previously molding, pressing, drying, and sizing had been discrete steps in the process, they now happened continuously, with the output of one step feeding seamlessly into the next. This eliminated the large piles of work in process that had accumulated between steps, since the next step began as soon as the previous one had finished. Increasingly, the papermaking process began to resemble a single large machine, with every step working in sync.

This kind of continuous process is the natural outcome for a process that develops sufficiently along the various axes of improvement. A process where one step feeds seamlessly into the next without delay or buffering is one that necessarily operates at a high level of reliability, as any disruption will shut down the entire operation. The direct connection between process steps eliminates many non-value-adding steps, such as handling and storage operations, as well as any rework from process failures. Continuous processes often resemble (or are) one enormous machine, and manufacturers can achieve high production rates and large economies of scale by increasing the speed of this machine over time. Between 1867 and 1919, fourdriniers increased in speed from 100 feet per minute to over 700 feet per minute, and were also made wider, further increasing the volume of paper produced.[761]

More specifically, continuous processes *require* many of these improvements in order to be feasible. Reducing buffer inventory between process steps requires a highly reliable process; without it, process steps are constantly starved of inputs. Similarly, a continuous process in which the output of one step is immediately fed into the next requires a high level of reliability. And a continuous process not only enables large-scale production, but large-scale production is a precondition for the existence of a continuous process, because the expense of the equipment and the development effort needed to carefully sequence each step can typically only be amortized over massive production volumes. Because they often require highly developed production processes and enable vast production volumes, continuous processes are often the sign of a large, mature industry.

### Continuous versus batch processes

In the industrial improvement literature, most notably in discussions of lean manufacturing, continuous processes are often contrasted with batch processes. In a batch process, a large number of items are processed simultaneously in a discrete step. A heat-treating furnace might heat-treat dozens or hundreds of parts at the same time, or parts might be transported between workstations by forklifts that move hundreds of parts at once. In batch processes, work will often arrive at a step in the process and then wait around until enough work has accumulated so that a large batch can be processed at once. A part might sit for hours or longer in front of the heat-treating furnace until enough parts have accumulated for the process to run.

To illuminate the contrast between these processes, consider two different methods for delivering natural gas. Over land, natural gas is often transported via pipeline, which moves gas continuously between two points. The 1,679-mile-long Rockies Express, for example, brings natural gas produced in Colorado and New Mexico to the Midwest and the East Coast.[762] But over water, natural gas is moved in its liquid state via large tankers. In this case, natural gas is brought to a liquified natural gas (LNG) terminal and cooled until it becomes a liquid, which reduces its volume by a factor of roughly 600.[763] It is then loaded onto an LNG tanker, which moves the entire batch of gas at once. At its destination, the LNG is turned back into a gas. The pipeline is a continuous process: Gas is delivered constantly throughout the day, week, and month. The tanker, on the other hand, is a batch process: Gas is moved in one large batch, and no additional gas is delivered until another shipment arrives.

Batch processes are often considered inferior to continuous processes. One reason is that a batch process will always have more work in process than a continuous process with the same level of output. As an example, let's consider a modification of the pin factory example we discussed in Chapter 6. In this new factory, pin making is a two-step process. In the first step, lengths of wire are cut from a long coil. In the second step, the lengths of wire are turned into pins: Straightening, sharpening, adding the head, and painting all happen within a single machine. If each step in the process takes one second, then the factory's work in process will consist of two pins: the pins being worked on in each machine at any given time.

Now, let's replace the second machine with one that still takes one second to run but can process any number of pins at the same time (one pin, 10 pins, 100 pins...). If it's set to run 10 pins at once, the lengths of wire will no longer be processed immediately when they reach the machine. Instead, they'll wait at the machine until 10 pins have accumulated. As pins accumulate at the second machine, the level of work in process slowly increases until the machine runs, after which it starts accumulating again. Because pins now spend time waiting around, this modified process has an average work in process of 7.5 pins, compared to the two pins of the initial process. The two processes have the same level of output—an average rate of one pin per second—but the batch process results in a higher level of inventory waiting to be processed. And as batch size increases, so will the amount of work in process. If we double the batch size so the second machine is handling 20 pins at once, the average work in process rises to 12.5 pins, even though, as before, the overall rate of production hasn't changed.

This extra work in process has both direct and associated costs. Every piece of inventory in the process is an item that has been purchased but hasn't yet been sold, and it requires a scaffolding of support operations—storage, insurance, handling—whose costs can be significant. In his examination of Japanese manufacturing methods, Richard Schonberger compared Japanese and American factories making the same product, at the same volume, using the same amount of equipment. Because the Japanese factory didn't need space for work-in-process storage, it was one-third the size—and cost one-third as much to build—compared to its American counterpart.[764]

There are other downsides to batch processing beyond the costs of accumulated inventory. Batch processes greatly increase cycle time, or the time it takes for a piece of work to move through the process. In our original two-step pin factory, each pin went through the factory in two seconds. In the 10-pin batch factory, however, the average cycle time would be 12 seconds, a 500 percent increase. And with large batches, any process problem—a breakdown, an incorrect machine setting—will affect a huge amount of work at once. With a continuous process, by contrast, an error can be spotted as soon as the first bad piece is produced.

Smaller batches also require less expensive material-handling equipment. As Schonberger notes, while American-style factory strategy in the late 20th century was often to "make a bunch of parts on a fast machine, set them down somewhere, and periodically call for a forklift truck to move batches between workstations," Japanese factories, with their minimal amount of work in process, had much less use for such expensive equipment.[765]

These costs explain why industrial improvement systems like the Toyota production system and the theory of constraints advocate for reducing batch sizes as much as possible, ideally achieving one-piece flow, where every step processes a single item at a time at a steady rate matched to the level of overall demand. The smaller the batch size, the less work in process, and the lower the associated costs.

Nevertheless, batch processing has some benefits. For one, economies of scale are often associated with processing a large number of items at the same time. A larger-capacity heat-treating furnace might require less energy per item due to geometric scaling effects. Similarly, if a process has large setup costs or other fixed costs, a batch process may be able to spread those costs across a higher overall production volume.[766]

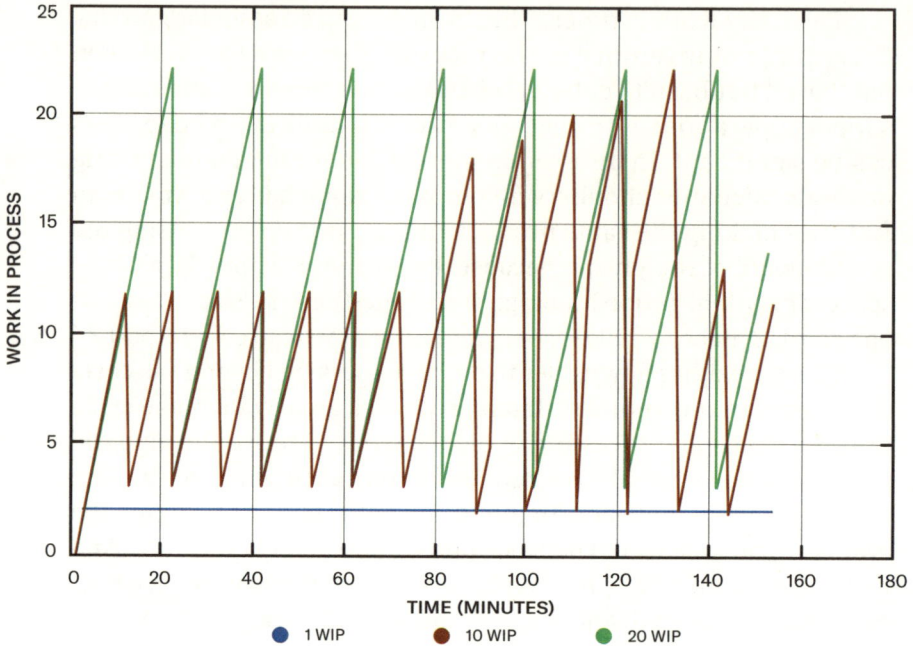

**Figure 29.** Work in process (WIP) over time with 1 WIP, 10 WIP, and 20 WIP machines.

A batch process is also often more flexible than a continuous one. A natural gas pipeline can only move gas between the two points connected by the pipeline. A liquefied natural gas carrier, on the other hand, doesn't have a fixed route and can move gas between any two LNG terminals. Similarly, batch processes that consist of several discrete pieces of equipment, such as individual machine tools, often have considerable flexibility in how that equipment is used. A lathe for producing part A this week can be used to produce part B next week and part C the week after that. This sort of process structure is sometimes called a job shop process.[767] A shop consists of multiple workstations that can be used to produce different products, and the material routing through the shop will change on a day-to-day or week-to-week basis depending on what needs to be produced. This process structure is common in industries where businesses sell small volumes of many different products. A continuous process, on the other hand, will often be designed to produce one specific product, and certain steps in the process will be mechanically coupled together. This sort of process, which is much less flexible, often can't easily adapt to producing different products.[768] While a manual glassblowing operation could make a variety of glass products, the Corning ribbon machine was only capable of making light bulb blanks.

The nature of batch processes yields an interesting dichotomy: Batch processes tend to be adopted both at low production volumes, where there's insufficient production volume to justify using continuous process equipment, and very high production volumes, where the economies of scale associated with batching are so great that it's worth the extra inventory costs. In chemical manufacturing, for example, batch production is often used for generating comparatively small volumes of chemicals, while large-volume chemical production typically uses continuous processes.[769] Similarly, large-volume coffee roasting is often done using continuous coffee roasters, where the beans are roasted as they move along a conveyor, whereas smaller, specialty producers that make a variety of different coffees will use smaller-batch roasters that roast a fixed quantity of beans all at once.[770] In both cases, the high cost of continuous process equipment means it can only be justified on high-volume products. In container shipping, however, the scale effects associated with higher-capacity ships are so great that ships now transport batches of tens of thousands of containers at once, as opposed to using a more continuous process (that is, a bigger number of smaller ships traveling more frequently). Similarly, in semiconductor fabrication, one of the most important developments was the invention of technology that could produce multiple transistors on a single silicon wafer at the same time, allowing for large-batch production and massive transistor production volumes. (Today, a single silicon wafer can hold trillions of transistors.) Indeed, as Christophe Lécuyer and David Brock observe in their history of Fairchild Semiconductor, it was the batch nature of its production processes that gave Fairchild its early successes, since batch processes "held the promise of markedly lowering the manufacturing cost of silicon transistors without sacrificing flexibility." Sequential methods, by contrast, required capital-intensive automation that would have locked them into a very narrow range of products.[771]

In certain cases, when output is great enough, the line between batch and continuous might blur: The process might be continuous or near-continuous, but each step deals with a large volume of work rather than a single item. Industrial juicing factories, for example, use juicers that process a large amount of fruit at once, but the whole process is automated and continuous.[772] Likewise, though the Bessemer process for steelmaking is a classic example of a batch process—the Bessemer converter is filled with several tons of liquid iron that is processed all at once when air is blown through it—in the late 19th century, large Bessemer steel mills began to resemble continuous processes. Iron from multiple blast furnaces would be pooled into a single large vat, which would be steadily replenished by fresh iron as it was fed into multiple Bessemer converters.[773]

Similarly, some steps in a production process may adopt batch processing while other steps may use continuous processes, depending on the economics and constraints of each portion of the operation. For instance, in the US, parking garages are often made from precast concrete, which is concrete that is cast into components—slabs, columns, beams—in a factory, trucked to the jobsite, and lifted into place by crane. The actual erection of the concrete components approximates a continuous, one-piece flow process: A crane lifts each piece into place one at a time, and builders aim to schedule deliveries so that a truck drives up with the next piece as soon as the crane has set the previous one without any waiting or wasted crane time. But the production of the precast pieces themselves is heavily batched. In the factory, pieces will be cast simultaneously on long casting beds that might hold a dozen pieces of precast or more, and large amounts of inventory will accumulate in the factory for weeks or months prior to construction rather than each piece coming off the casting bed and immediately being driven to the jobsite.

The use of different production styles at different points in the process can be traced to the economics and technological constraints of precast construction. On the jobsite, crane time is expensive and there's little room to store precast components (that is, work in process). Builders can minimize crane costs by carefully sequencing component delivery so each piece arrives precisely when it's needed. But because precast concrete takes, at minimum, several hours to produce (concrete is poured fluid and needs time to cure and reach strength), the rate of production in the factory can't match the rate of demand at the construction site, which might require a new piece every few minutes. Certain fixed costs of concrete production, such as the time and effort required for prestressing, are also better spread over large batches. For this reason, factory production takes place in batches of several pieces at a time and starts several months prior to building the garage, allowing construction to be continuous once it starts.

Other production methods, such as the Toyota production system, attempt to realize many of the benefits of a continuous process while also maintaining the flexibility of a batch system. By minimizing batch size (ideally achieving a batch size of one), grouping machines together in work cells, and reducing setup times to enable quick machine changeovers, the Toyota production system achieves close to a continuous flow. Work moves one piece at a time from station to station, with minimal downtime and waiting and no accumulated inventory. As with continuous process machinery, reliability is by necessity high and production is steady and constant, though not necessarily at large volumes (because the system is designed to produce at the rate of demand, not to maximize production volume).

But the system also manages to achieve these benefits while maintaining flexibility. Instead of using special-purpose machinery linked by fixed conveyors that move large volumes of material, the Toyota production system uses simpler, general-purpose machinery and moves smaller amounts of inventory by carts, wheeled racks, or other low-cost handling devices. Reduced setup time, accomplished using SMED methods, makes it possible for equipment to be quickly changed over to produce a new product, and cross-trained workers can be reallocated as demand shifts.

Another attempt to couple the benefits of continuous processing with those of batch production is group technology, a method for organizing small-lot production that might otherwise be done in a job shop-style operation. In group technology, workstations are organized into cells that perform similar types of work, with one machine feeding directly into the next to minimize travel time, much like the Toyota production system. Work is classified based on the types of operations it requires and work packages are sent to the appropriate cell, where they are worked on continuously. By grouping together parts that may have different shapes and sizes but which nevertheless require similar operations and quickly moving them through work cells dedicated to those types of operations, manufacturers can achieve many of the benefits of continuous production. Like the Toyota production system, group technology greatly decreases inventory levels, reduces the time material spends in process, and makes the process more predictable, all of which results in lower costs.[774] Also like the Toyota production system, group technology requires a highly reliable process to function. Following WWII, Japanese shipbuilders adopted group technology, along with the statistical control methods that made it feasible, enabling them to achieve much lower shipbuilding costs and capture a large share of the shipbuilding market.[775]

### Evolution toward continuous processes

Since the late 18th century, shortly after the beginning of the Industrial Revolution, a variety of industries have evolved toward continuous processes. Around 1785, not long before the creation of the continuous process papermaking machines we discussed earlier in this chapter, American inventor Oliver Evans built an automated continuous flour mill in Delaware.[776] Flour moved through the mill via a series of screws, conveyor belts, and bucket chains, ensuring that "no hand touched the grain except for loading and unloading the delivery carts."[777] The six workers employed at the mill spent most of their time closing barrels.[778] Evans described the design and operation of his mill in a book published in 1795, and similar mills soon became common in the US.[779]

Toward the middle of the 19th century, continuous process machinery began to appear for simple manufactured goods. Nail manufacturing was essentially continuous by the mid-18th century. Two-step machines powered by steam or waterwheel both cut the body of the nail and formed the head, while automatic feeding devices continuously fed nail plate into the machines.[780] In 1841, inventor John Howe developed a continuous automatic pin-making machine in which wire was fed into the machine and cut to the proper length, sharpened, and then headed.[781] In the 1860s and '70s, continuous mills and machines for manufacturing steel wire emerged, followed by automatic machinery for making barbed wire in 1875.[782]

Prior to the late 19th century, high-volume continuous process machinery would rarely have been economical for most goods, even if the technology had existed, because transportation technology limited the extent of the market. Beyond the high cost of transporting goods over long distances, the limitations of preindustrial transportation made it difficult to maintain a large, steady flow of material through the factory. As Chandler Jr. notes in *Scale and Scope*, "Transportation that depended on the power of animals, wind, and current was too slow, too irregular, and too uncertain to maintain a level of throughput necessary to achieve the potential economies of the new technologies."[783] But as the railroad developed in the second half of the 19th century, the costs of transportation over land declined dramatically and reliability greatly improved. This made it feasible to produce large volumes of goods in a single factory, and continuous automatic machinery became more common.

In the cigarette industry, for example, a single machine—the Bonsack machine, invented in 1881—performed every operation needed to produce a cigarette, including, eventually, placing them into packages.[784] In the 1870s, the Diamond Match Company developed continuous automatic machines that made and packed billions of matches.[785] Automatic screw-cutting machines also emerged in the 1870s. These were specially designed lathes with a series of different tools that could be rotated into place; bar stock would be fed into the machine, and a series of cams would rotate the appropriate tool into place, automatically cutting the features of the screw and producing a continuous output of screws.[786] And in the food-processing industry, continuous process mills not only for flour but also for oats, rye, barley, and other grains emerged by the end of the 19th century. In the case of oats, the continuous mills generated so much output that food processors were forced to invent a new product—breakfast cereal—to make use of it all.[787] Other industries that developed continuous processes in the late 19th century include soap, photographic film, sugar, beer, toilet paper, chewing gum, and candy.[788] In chemical manufacturing, Belgian chemist Ernest Solvay developed a continuous process for producing soda ash in 1872 that replaced the previous batch-based Leblanc process.[789]

Ford's development of the assembly line for the Model T is arguably the most famous example of a continuous manufacturing process. Prior to the Model T, most continuous processes were either for simple manufactured goods like nails, matches, screws, and cans, or undifferentiated bulk goods like flour, oats, and paper. With the Model T, Ford pushed continuous operations into previously unexplored realms, coordinating the efforts of tens of thousands of workers to produce a product with thousands of parts that fit precisely together. This style of production became known as mass production, a term that has become largely synonymous with the high-volume continuous production of complex manufactured goods.[790]

Outside of Ford, the trend toward continuous processes continued in the 20th century. In 1929, the trade journal *Chemical and*

*Metallurgical Engineering* noted that "continuous processing is every-where replacing batch handling as soon as it can be applied profit-ably."[791] In petroleum refining, cracking—splitting heavy crude oil into lighter hydrocarbons such as gasoline—was initially developed as the batch-based Burton process in 1913, but by the 1920s it had been superseded by continuous cracking methods like the Dubbs process.[792] And in ammonia synthesis, the continuous Haber–Bosch process was developed in 1909, which remains the primary source of nitrogen used in fertilizer today. These developments were often made possible by the invention of improved control systems and the spread of electric power, both of which enabled more precise control over equipment.[793]

Similarly, in metal production, a variety of developments toward continuous processes took place all along the value chain. In alumi-num manufacturing, the continuous Hall–Héroult process for pro-ducing aluminum from aluminum oxide, invented in 1886, replaced the Deville–Castner batch process.[794] And the continuous, automatic rolling of wide sheets of steel, attempted since the late 19th century, was perfected in the 1920s.[795]

Today, continuous processes continue to supersede batch pro-cesses. Even pharmaceutical manufacturing, which has traditionally relied on batch production methods, is beginning to adopt contin-uous process.[796]

In many cases, continuous process machinery produces goods in such large volumes that a minuscule number of facilities can manu-facture enough output for an entire country, or even the entire world. In 1885, just 30 Bonsack cigarette machines could produce enough cigarettes for the entire United States. By the 1980s, fewer than 15 ribbon machines made the entire world's supply of incandescent light bulb blanks. And when continuous process machinery for rolling steel sheets, which could produce 20 to 60 times more sheet per worker than previous technology, emerged, producers soon realized that "only a limited number of single continuous roll trains, each capable of mak-ing 150,000 to 300,000 tons per annum, could satisfy the country's needs."[797] More recently, in 2022, a severe shortage of infant formula occurred when a single plant that reportedly produced 20 percent of the United States' formula supply shut down.[798]

In many cases, the development of a continuous process marks the end of technological improvement. Incremental improvements may continue to accrue, but no further revolutionary changes take place. In S-curve terms, no additional overlapping S curves appear, and progress consists of going farther and farther on the top half of the S. After the ribbon machine, no further developments in incandescent bulb blank machinery occurred. Two hundred years after their invention, fourdriniers, now much improved, are still used for papermaking. The continuous Hall–Héroult and Haber–Bosch processes are still the primary method of producing aluminum and ammonia more than 100 years later, as is the 150-year-old Solvay process for soda production. Cars and other manufactured goods are still produced on assembly lines. In cases where a continuous process is superseded, it tends to be by another continuous process. Continuous cut-nail machines, for example, were eventually replaced by continuous wire-nail machines, which remain the primary method of producing nails today.

Not every process advances steadily toward a continuous transformation of inputs into outputs. In some cases, the advantages of batching push development away from continuity and toward larger and larger batches. In container shipping, for example, we have in some ways moved even further from a rapid, continuous transformation and toward the slower processing of big batches. Since 2007, ships have increasingly practiced slow steaming, or decreasing their speeds to reduce fuel costs.[799] In other cases, technological or product constraints can push a product toward a different sort of evolution. As in the case of precast concrete, limitations on the speed at which pieces can be produced and the incentive to minimize on-site time pushes factory production toward batches and huge inventory accumulations. On the other end of the spectrum, when production volumes are small or the goods require a considerable degree of customization, a continuous process may not be feasible. But absent these sorts of constraints, the push toward greater efficiency tends to result in processes getting more and more continuous.

**Summary**
Ford's invention of the assembly line and its adoption of mass production methods was a watershed moment in manufacturing. It allowed Ford to capture 50 percent of the American automobile market (that is, until its methods were mimicked by other auto manufacturers), and it revolutionized other industries as the assembly line spread to factories "producing foods, consumer appliances, household and

office equipment, tools, bicycles, toys, and games."[800] In his history of the assembly line, historian David Nye argues that it fundamentally changed the nature of consumption, and that "largely because of mass production, standards of living have probably risen more in the last 100 years than in any comparable period in human history."[801]

If anything, this understates the importance of mass production. Yes, mass production revolutionized the fabrication of large volumes of manufactured goods in the early to mid-20th century. But as we've seen in this chapter, the fundamental concept behind mass production—a smooth, continuous transformation of inputs into finished goods—has spread even further. Decades before Ford began to produce the Model T, continuous process machinery had transformed the production of other goods, from chemicals to foods to simple manufactured products, as part of an economic shift so vast it has been dubbed the Second Industrial Revolution.[802] And in the decades following the Model T, continuous process methods have slowly been transcending the limits of high-volume production, as manufacturers have discovered ways to adapt them to industries that require smaller volumes and more flexibility. In Japan, Toyota famously adapted Ford's methods into its own system of continuous, one-piece flow for what was, at the time, a lower-volume, higher-variety automobile market. Toyota's practices have since been adopted around the world in the form of lean methods. Even shipbuilding, the archetypical example of small-volume, highly customized, complex production, has been improved by continuous process methods. Group technology, designed to allow a smooth, continuous flow of material through a shipyard, is partly responsible for Japan conquering the shipbuilding market in the second half of the 20th century.

In industry after industry, adopting continuous processes has driven down costs and massively increased worker productivity. And while technological development tends to follow a series of overlapping S curves, as one technology is replaced by another, many continuous process technologies—fourdriniers, the ribbon machine, the assembly line—are so efficient that they continue to be used decades or even centuries after their introduction. In some ways, a continuous process marks the end of history for a given area of production—it's a form of production that has no obvious successor.

# 10

# Failures
# to Improve

The preceding chapters have laid out the different mechanisms by which production processes can become more efficient and what those improvements look like in practice. In other words, they explain how goods and services have become cheaper and cheaper over time. But goods and services don't always get cheaper. If we compare the price of various goods and services to overall inflation, we can see that many things have, in fact, become more expensive over time.

Figure 30 shows the relative price change of several consumer goods and services between 1978 and 2023, as well as the change in the US Consumer Price Index, a measure of overall inflation. During this period, televisions, toys, clothing, furniture, and other manufactured goods have all become cheaper, with their price increasing less than overall inflation. Some goods, such as televisions and toys, are cheaper even in nominal terms. Other goods and services, such as food, electricity, and housing, have mostly stayed the same, rising in cost at the same rate as inflation. And some things, like college tuition and medical services, have become much more expensive over time, rising in cost much faster than inflation.

Not every increase in cost can be attributed to a lack of efficiency improvement. In some cases, goods may be status or positional goods, where high cost and scarcity is the point. The price of a Rolex Submariner watch has risen twice as fast as overall inflation, but the purpose of a Rolex is to be an expensive luxury purchase, so lowering the price would be self-defeating.[803] Similarly, some people may deliberately purchase inefficiently produced goods for aesthetic reasons; preferring items that are artisanally crafted or handmade is an obvious example. In these cases, the inefficient production method is, in a sense, part of the product and can't easily be replaced by a cheaper production method.

In other cases, producers, governments, and other groups have successfully erected barriers to entry to the market, preventing competition and allowing sellers to maintain high prices. The high cost of medical care services in the US has many causes, ranging from the private insurance system to price opacity, but one reason is supply restrictions such as certificate-of-need laws, which require health care providers to prove there is a clear need for things like new hospitals or clinics before they are allowed to build them.[804]

But there are also cases where goods and services, despite being produced in highly competitive environments and despite deliberate efforts to improve production efficiency, fail to become cheaper over time. Housing costs, for instance, have only risen. Much of this is due to the cost of physically building a house, which has steadily increased over time. Indeed, between 1964 and 2021, the costs of new home construction increased almost 50 percent above inflation, as Figure 31 shows.

To be sure, there are many restrictions that limit the housing supply and would, therefore, be expected to drive up the price, from zoning and licensing restrictions to unions and opposition from local residents. Construction is also an incredibly competitive industry with extremely price-sensitive buyers. As we'll see in this chapter, there have been many attempts to find cheaper ways to build houses. Yet the costs of building a house in the US continue to rise.

What prevents a production process from getting cheaper over time? There are, broadly, two reasons this might happen. First, the process improvements discussed in the previous chapters might be blocked. It might not be possible to find better production methods, use cheaper materials, or make the process more reliable. If process improvements can't happen, costs won't fall. Second, a process might accumulate additional, uncontrollable costs over time, offsetting any efficiency improvements and resulting in steady or increasing costs even as the process becomes more efficient.

### Blocked paths to improvement

The limitations that prevent process improvement from happening fall into roughly three categories: technical limitations, political limitations, and market limitations.

Technical limitations occur when the fundamental nature of the process or product prevents or limits certain types of process improvement. For example, the physical properties of titanium make processing it fundamentally more expensive than other metals. Titanium dissipates heat slowly, which increases cutting tool wear and limits how fast it can be machined. And at high temperatures, titanium rapidly absorbs impurities from the air, which means that during processing, much of the material must be cut away to remove defects, resulting in significant yield loss.[805] In aerospace part fabrication, sometimes more than 95 percent of the metal is machined away and discarded.[806] This sort of trimming is waste in a lean manufacturing sense, since it's material that a customer doesn't value, so eliminating it and the process steps it requires would reduce manufacturing costs. But just because there's

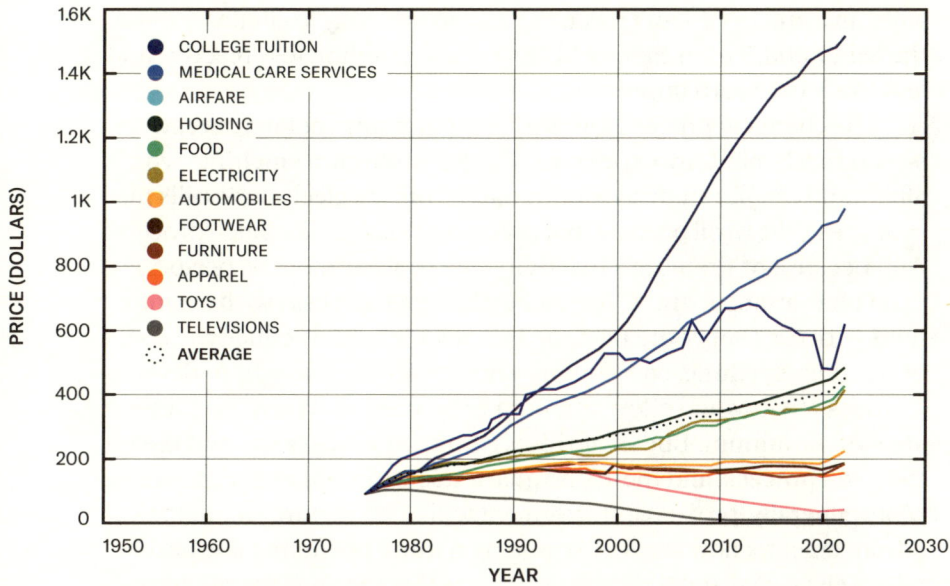

**Figure 30.** Price inflation of different goods and services over time. Data from US Bureau of Labor Statistics, "CPI Databases."

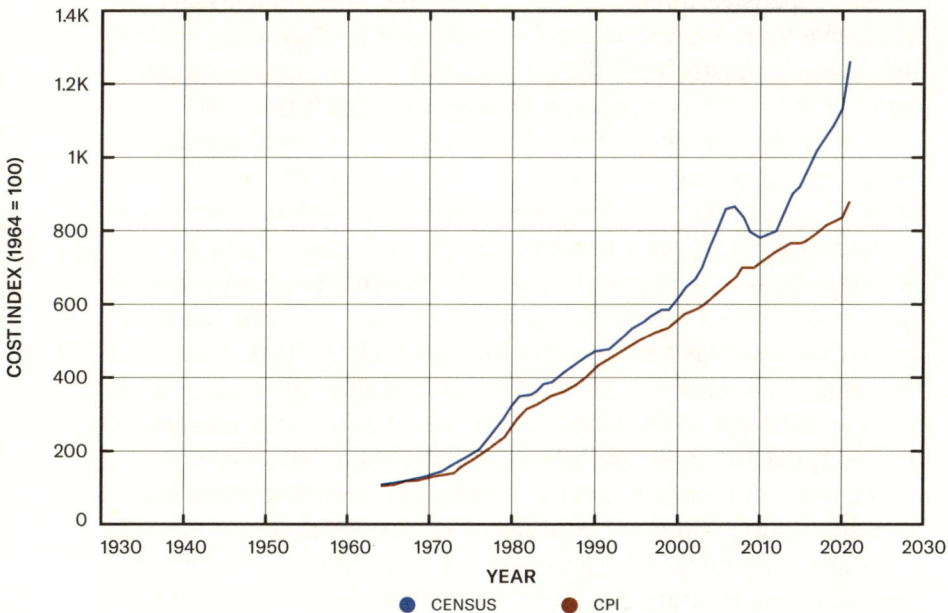

**Figure 31.** Comparing the Consumer Price Index to the price of a new home over time. Data from US Census Bureau and US Bureau of Labor Statistics.

waste in a process doesn't mean there's a *feasible* way to eliminate it—all the lean initiatives in the world can't make titanium less reactive and less likely to absorb impurities.

The chemical process for producing titanium metal from raw ore is also fundamentally expensive. Unlike aluminum smelting, where aluminum can be removed continuously from the electrolytic cells that produce it, the Kroll process that produces titanium is a batch process. And because of titanium's reactivity, the process must be performed in an inert atmosphere.[807] Replacing the batch process with a continuous process that could be done in open air would reduce the costs of titanium smelting, so for many years, producers sought to develop a process that would allow titanium to be used as widely as stainless steel or aluminum. But as of this writing, no process has surpassed the Kroll process, and there's no guarantee that such a process exists. More generally, process improvement will often require technological advancements: new ways of arranging natural phenomena to accomplish a given goal. But there's no guarantee that the universe will furnish a convenient, low-cost method of accomplishing a given task. Because of this, and because of titanium's physical properties, producing it remains expensive. Its cost has fallen over time, but it remains far more expensive than other structural metals such as aluminum or steel, and its use is mostly limited to applications in aerospace or medical implants, where its unique physical properties justify the high cost.[808]

The second category of efficiency blocker is political limitations. A better process may exist, but interested parties might have reasons to want to prevent it from being found or implemented. Government regulation, for example, can prevent the introduction of a better process by being overly prescriptive. Regulations may require or strongly incentivize a given production method, or may simply be written in a way that assumes a specific production technology will be used. For example, homes in the US are typically built to the requirements of the International Residential Code, a model code that most states have adopted for single-family home construction. This code contains a wide range of requirements the home must meet. The code defines minimum dimensions for concrete foundations and load-bearing metal studs; metal framing must use fasteners of a specific type, and ceramic tile must be attached with a specific backing material.[809] It is sometimes possible to get around these requirements (for example, by providing stamped engineering calculations or certified testing data that suggest an alternative will perform properly), but this adds cost and risk, since there's no guarantee a building official will approve it.

Prescriptive regulations can be contrasted with performance-based regulations, which stipulate that a product meet a certain standard of performance. In this model, as long as a product meets that performance standard, any production method is allowed.[810] In car manufacturing, crashworthiness is measured by subjecting cars to crash tests. Because crash tests don't mandate specific production methods, manufacturers have more freedom to change their product design and production methods. That's why, as we noted in Chapter 3, car manufacturers were able to shift from steel and aluminum stampings to large aluminum castings for the car body, a change that didn't require any updates to crash test regulations.[811] By comparison, the construction of tall timber buildings in the US only became possible after building codes were specifically changed to allow it.[812]

Political limitations can also take the form of opposition by stakeholders who stand to lose from the introduction of new technology. This can incentivize them to advocate for protectionist government regulation or threaten other measures, such as strikes, to try to prevent the introduction of new production methods. In the case of housing, building trade unions have often tried to limit the introduction of new building technology. Plasterers' unions lobbied for building code language that would forbid the use of drywall; plumbers' unions fought to stop codes from allowing PVC pipe; and building trades of all types have tried to prohibit the use of prefabricated construction.[813] Outside of housing, dockworkers' unions have tried to block the introduction of technology such as shipping containers and port automation.[814]

The third category of efficiency blocker is market limitations. If the market for a product is small and production volumes are low, taking advantage of economies of scale or process improvements that rely on them may not be possible and new, better production technology may not be developed. As we saw in Chapter 7, production volume determines how far an industry travels down the learning curve, so low-production industries will experience fewer cost-reducing improvements. High-end semiconductor manufacturing uses EUV lithography machines, which etch the semiconductor's minuscule features onto silicon wafers. The world's only supplier of these machines produces just 50 a year, putting them far out of range of any sort of mass production method. As a result, each machine takes months to

assemble and can cost up to $200 million.[815] Similarly, early proto-
types of a new car model, made before high-volume factories have
been tooled to manufacture it, are much more expensive to produce.
Each prototype for Ford's redesigned 1996 Ford Taurus cost on the
order of $400,000 to produce, 20 times as expensive as the produc-
tion version.[816]

For many products, as production costs fall, new markets open
up. Solar PV panels started out as an expensive aerospace technology
to provide power for satellites, and then for remote terrestrial power
stations such as offshore platforms or remote telephone repeaters.
But as the price of solar PVs fell, the market expanded to consumer
electronics, rooftops, and utility-scale power production.[817] With this
in mind, when high production costs are the result of market limitations, it may be possible to jump-start an industry by subsidizing
production until it has fallen far enough down the learning curve, as
happened with solar PVs and wind power.[818] But if there are technical limitations that prevent efficiency improvements or mean that a
product will always be more expensive compared to alternatives, this
jump-starting process will fail.

These different types of limitations on efficiency improvement—
technical, political, and market—will intersect in a variety of ways. For
instance, if the market for a product is small, there's less incentive to
search for better production technology. By contrast, a larger market
might also mean stronger market participants with the ability to shape
the regulatory environment in their favor.

## Blocked paths and stability

These different mechanisms often prevent efficiency improvements
by way of keeping the process from becoming more stable. As we saw
in Chapter 6, process improvement typically requires at least some
level of stability—the ability to make a similar product in the same way
each time. In a stable, repetitive production environment, producers
can take advantage of economies of scale, hunt down and eliminate
the causes of process failures, get rid of unnecessary steps, automate
production operations, and remove costly buffers from the process.
In an unstable environment where the product or process is constantly
changing, on the other hand, producers will find it difficult, if not
impossible, to accumulate improvements. The constant changes will
reset learning curves and prevent the alignment of different steps in

the process, resulting in costly buffers of time, material, or capacity. In an unstable production environment, not only will less automation be possible, but the process will also likely require more skilled, and thus more expensive, labor.

For these reasons, goods and services that continue to rise in cost are often those that are produced in highly unpredictable environments. In medical care, a patient can have an enormous number of different symptoms; a symptom may be caused by many possible diseases (or by nothing at all). And patients will vary in innumerable ways—age, gender, diet, size, physical fitness, stress, genetics—that can affect both the disease itself and the possible treatments. For this reason, a significant amount of medical care is often devoted to simply working out what to do. For each individual patient, a doctor may need to run several tests or try several treatments to see how a patient responds. Because of this variability, the process is difficult to automate or streamline and requires high-cost experts. In conventional manufacturing, by contrast, there's still a substantial figuring-out-what-to-do process at the outset, but it doesn't need to be repeated each time. Ford can figure out how to produce its Taurus once and then use that process to manufacture millions of nearly identical cars.

A variable environment also explains why repairing a car is proportionately so much more expensive than buying the same car new. Repairing even a small amount of damage affecting a tiny fraction of the parts can cost a large fraction of the cost of a new car. This is because a new car can be produced using a high-volume, streamlined, repetitive, highly automated process that has been designed to be as efficient as possible. Repairing a car, on the other hand, requires substantial effort merely to figure out which parts are damaged and need repair, and how. Removing and replacing the damaged parts is carried out almost entirely by expensive manual labor, since any given car might require a different repair process. (Even identical models may be damaged in different ways.) And while the production process is designed to make assembly of specific parts in a specific order as inexpensive as possible, removing and replacing any part once the car is complete may be much more difficult. For instance, accessing one part may require removing many others. As with medical services, the fundamentally variable nature of repairing damage to a car makes it difficult to substantially reduce its cost.

## Added production burdens

Just as a production process can accumulate improvements that enhance efficiency and reduce costs, it can also accumulate burdens that reduce efficiency and increase costs.

Perhaps the most common encumbrance a production process can accumulate is regulation. Over time, regulation can add costs to a production process, either by adding steps to the process itself or by requiring changes to the product. In car manufacturing, regulators have gradually added requirements for safety and fuel economy, requiring automakers to add catalytic converters to reduce emissions and airbags to prevent injuries during accidents, for example. One estimate suggests that between the 1960s and the 2010s, these sorts of requirements add $6,000 to $7,000 to the cost of a new car.[819]

Regulations can also add cost to the production process itself by requiring extra steps or equipment or restricting which methods can be used. For instance, the cost of coal-generated electric power began to increase in the 1970s when air quality regulations required coal plants to install scrubbing equipment to minimize emissions.[820] The National Association of Manufacturers likewise estimates that complying with US environmental regulations costs manufacturers on the order of $17,000 per employee per year.[821]

Another common burden is the increasing costs of inputs. As we saw in Chapter 4, input costs will sometimes rise over time due to diseconomies of scale. If local, easily accessible inputs are exhausted, production processes may need to use inputs that are farther away or of lower quality, adding production cost. Alternatively, input costs might increase because demand for a limited supply surges, thereby raising the price. Perhaps the most important input in this category is labor, which tends to get more expensive over time. This is due in part to a phenomenon known as Baumol's cost disease or the Baumol effect: As labor productivity in certain economic sectors increases, wages in that sector will tend to rise. This puts upward pressure on wages across the entire economy, including in areas that haven't experienced the same productivity growth. Increased productivity in a specific sector thus drives up the costs of labor across the entire economy, and tasks where the amount of labor can't easily be reduced will exhibit steadily rising labor costs. The canonical example is a string quartet. It takes just as much labor for a string quartet to play today as it did 100 years ago, but because other sectors of the economy have become much more productive and wages have risen significantly, it is proportionately much more expensive to hire a string quartet today than it was a century ago.[822]

It's important to note that nearly every industry must cope with these expanding burdens. Regulations tend to increase across the entire economy, and the Baumol effect will impact every part of the economy. Whether these burdens add to the final price of a product will depend on whether there are sufficient process improvements to overcome them. The magnitude of these burdens will also intersect with the technical, political, and market limitations on efficiency improvements we discussed earlier in this chapter. The impact of safety regulations on cost, for example, will depend partly on how easily a process can be automated. The less labor-intensive a process is, the smaller the impact safety regulations will have on cost, all else being equal. Likewise, the effect of rising labor costs is also determined in part by the extent to which a process can be automated.

*Case study: building construction*
To understand what these different burdens and limitations on efficiency improvement look like in practice, let's examine an industry that has repeatedly tried and failed to reduce production costs: building construction.

First, consider construction's costs. According to a variety of different cost indexes, which track the costs of construction over time, since the early 20th century the costs of construction in the US have almost always risen faster than inflation.

Table 3 shows the rate of increase of different measures of construction cost minus the rate of increase of the Consumer Price Index over different decades. In other words, it shows when, and by how much, construction costs have increased faster than inflation. We can see that in almost every decade, construction costs have risen faster than inflation, regardless of which measure of cost we use.

Construction cost indexes are somewhat abstract measures, but the measured costs of actual buildings display the same trend. In a 1946 article for *Fortune* magazine, a California builder listed his construction costs for a new housing development, which came in at around $5.69 per square foot (or $92 per square foot in 2022 dollars), including the builder's profits.[823] In 2022, the average cost per square foot of a new US home, excluding land, was $178 dollars—nearly

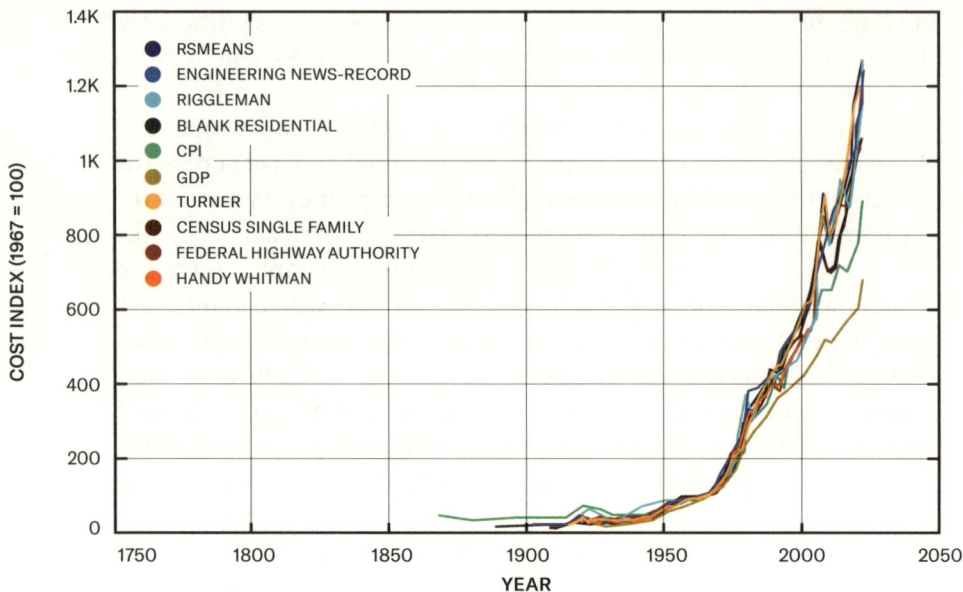

**Figure 32.** Increase of different construction cost indexes over time compared to different measures of inflation. Values have been adjusted so that 1967 = 100.

twice as much as in 1945—and costs in California were even higher.[824] Although there have been quality improvements in new homes over time—better windows, better insulation, better HVAC systems, and the like—they are not enough to account for this increase.

Productivity measures tell the same story. Between 1950 and 2019, labor productivity—the amount of output for a given amount of labor—in the US economy overall increased by nearly 200 percent. Labor productivity in construction, by comparison, was flat or even declining by some measures. Between 1962 and 2022, the number of labor hours it took to build one square foot of a single-family home increased from 85 hours in 1962 to 86.5 hours in 2022.[825]

This problem doesn't appear to be restricted to the US. Germany, Belgium, Sweden, Norway, and France all have similar levels of construction cost inflation. Between 2005 and 2023, construction costs in Europe as a whole increased at a similar pace to the US: just over 80 percent in the US and just under 70 percent in Europe, both higher than the corresponding inflation rate.[826]

A brief foray into the mechanics of the construction industry makes clear why costs haven't fallen: Many of the paths of process

| INDEX | TURNER | CENSUS SINGLE FAMILY | FEDERAL HIGHWAY AUTHORITY | HANDY WHITMAN | RSMEANS | ENGINEERING NEWS-RECORD | RIGGLEMAN | BLANK RESIDENTIAL |
|---|---|---|---|---|---|---|---|---|
| 1870–1879 | | | | | | | -0.1% | |
| 1880–1889 | | | | | | | 1.1% | |
| 1890–1899 | | | | | | | 1.7% | 1.1% |
| 1900–1909 | | | | | | | 0.6% | 1.5% |
| 1910–1919 | 6.2% | | 1.5% | 0.2% | | 2.4% | 2.3% | 0.6% |
| 1920–1929 | -2.2% | | -3.0% | -1.2% | | -0.3% | -0.3% | -0.2% |
| 1930–1939 | 3.3% | | -0.1% | 2.1% | | 3.5% | | |
| 1940–1949 | 2.5% | | 0.7% | 2.4% | | 2.2% | | |
| 1950–1959 | 1.3% | | -0.2% | 2.4% | | 2.8% | | |
| 1960–1969 | 1.5% | | 2.5% | 0.6% | 1.1% | 2.6% | | |
| 1970–1979 | 0.0% | 1.8% | 2.4% | 0.2% | 0.3% | 1.1% | | |
| 1980–1989 | 0.2% | -0.9% | -3.6% | -1.2% | -0.6% | -0.9% | | |
| 1990–1999 | 0.2% | -0.1% | 0.2% | -0.1% | -0.3% | 0.0% | | |
| 2000–2009 | 0.6% | 0.0% | 1.6% | | 1.9% | 1.1% | | |
| 2010–2019 | 2.2% | 2.1% | 1.4% | | 0.8% | 1.0% | | |

Table 3. Average annual increase of different construction cost indexes minus the rate of increase of the Consumer Price Index, by decade.

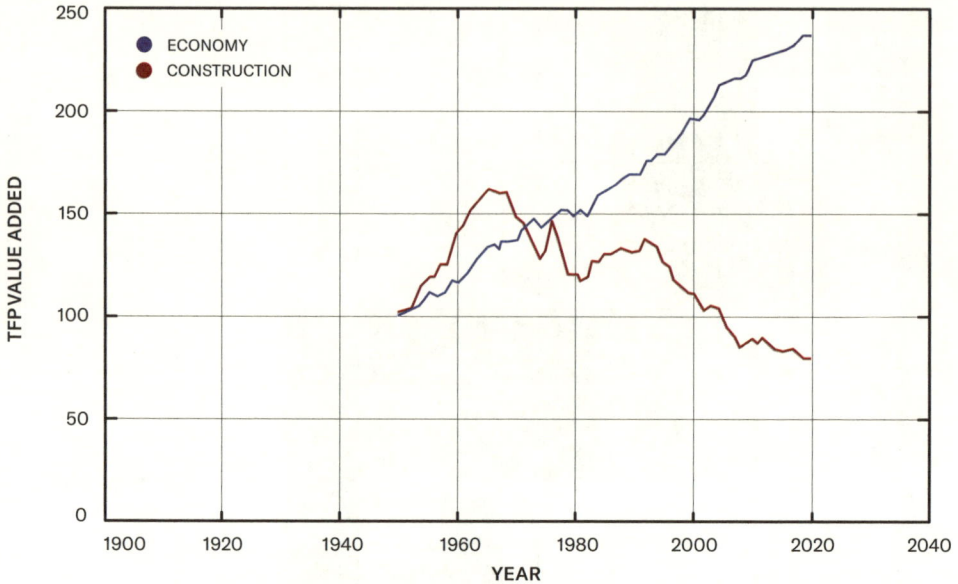

**Figure 33.** Productivity changes in construction compared to the economy overall. Data from Goolsbee and Syverson, "Path of Productivity."

improvement have been blocked, and over time construction has steadily accumulated burdens that add cost to the process.

First, let's look at the process barriers. For most modern goods, factory production has gradually replaced craft production and high-volume automated machinery has increasingly replaced physical labor. As we saw in Chapter 2, nails went from being made by hand with blacksmiths' tools to being manufactured using dedicated nail-making machines. Construction, however, remains largely craft-based. Buildings in the US are still mostly assembled on-site, by hand, by skilled tradespeople, just as they have been for hundreds of years. Compared to other industries, little of the work has been automated or mechanized. In home building, direct labor—the physical assembly tasks—can make up nearly 50 percent of the cost of building a home, and around 30 percent of the overall sales price.[827] In car manufacturing, by contrast, direct labor costs are on the order of 10 percent or less of a car's sales price.[828]

This lack of improvement isn't for lack of trying. There have been many attempts to push construction along the path other industries have followed, for example by having most or all of the work done in

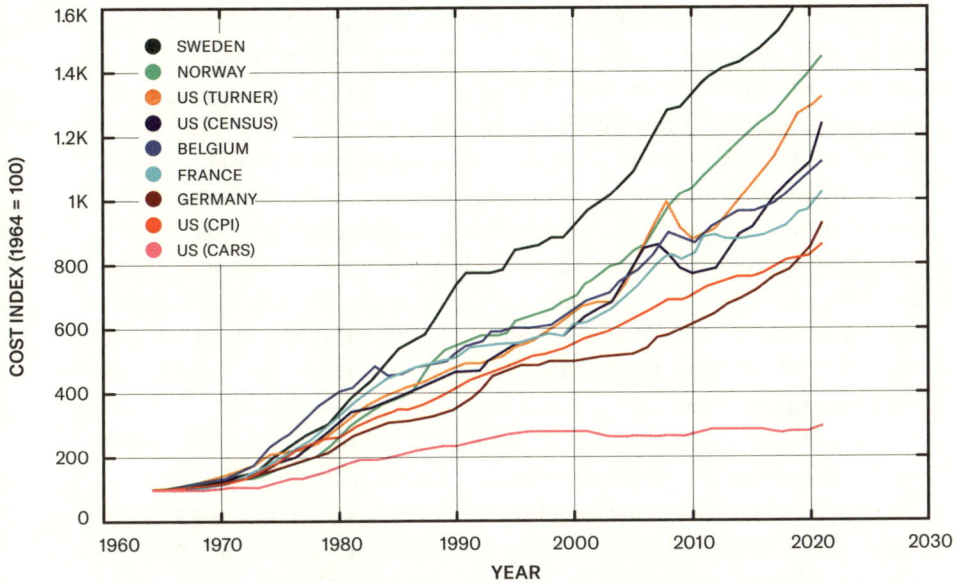

**Figure 34.** Changes in international construction costs over time.

factories rather than on-site. This is broadly known as prefabricated or modular construction. But while it's possible to make a building using prefabricated construction, it hasn't been able to substantially lower the costs of construction for single-family homes.[829] In the US, prefabrication currently makes up around 2 percent of new home construction, with another 10 percent or so going toward so-called manufactured or mobile homes, despite many attempts to adopt the practice more widely.[830] Companies such as National Homes, Gunnison Homes, and General Houses built hundreds of thousands of prefabricated homes in the 1950s, '60s, and '70s, and companies such as Lustron in 1947, Stirling Homex in 1967, and Katerra in 2015 spent hundreds of millions or even billions of dollars attempting to do for housing what Ford did for car manufacturing: massively reduce cost through mass production methods.[831] In 1969, the US's newly formed Department of Housing and Urban Development (HUD) launched Operation Breakthrough, an effort to kick-start a factory-built housing industry, and spent more than $70 million (nearly $500 million in 2023 dollars) developing and subsidizing innovative factory-built housing systems.[832]

None of these efforts were successful. The Operation Breakthrough systems all fell out of use after government subsidies were withdrawn. Lustron, Stirling Homex, and Katerra all declared bankruptcy, and National Homes, Gunnison Homes, and General Houses also went out of business. When Ford developed its mass production methods, the resulting product was so much less expensive that it swept away the previous craft process for car building (outside of high-cost luxury assemblers). Even if Ford had ultimately gone out of business, its production methods would have lived on in companies like GM or Chrysler. But none of the factory homebuilders—not even those that carved out successful businesses for decades, like National Homes—managed to put a serious dent in craft-process homebuilding in the US.

Prefabrication has always struggled to achieve cost parity with on-site construction, much less produce homes at a massively reduced cost. In the 1950s, the price for a house from National Homes was similar to, if not more expensive than, a conventional site-built home.[833] The same holds true today. In Sweden, prefabrication is much more common than in the US; over 80 percent of new detached single-family homes in Sweden use prefabricated construction.[834] But the average construction costs of a Swedish home are higher than the average costs of a home in the US, where houses are built almost entirely using conventional construction.[835]

Perhaps the biggest indicator of the challenge of reducing home construction costs using factory methods comes from Toyota. As we've seen throughout the previous chapters, Toyota's efficient car manufacturing methods and low production costs—the eponymous Toyota production system—are credited with the company's enormous success, and they've been adopted around the world in the form of lean methods. Toyota literally wrote the book on efficient manufacturing.[836] But cars aren't the only thing Toyota makes. It has several other lines of business, including a homebuilding division. Toyota Housing Corporation was founded in 1975 as part of an effort to find new applications for its manufacturing technology and expertise. Since then, Toyota has built thousands of prefabricated homes in Japan.[837] But despite 40 years of homebuilding experience and its wealth of manufacturing expertise, Toyota has not found a way to make prefabricated construction cheap and ubiquitous. In 2008, Toyota homes cost on the order of $200 or more per square foot, almost twice as much as an average new home in the US at the time.[838]

And while Toyota tops the list of the world's largest car manufacturers, it's only a small player in the Japanese housing market, which continues to make extensive use of conventional, on-site construction (though prefabrication is more popular in Japan than in the US). In other words, Toyota's expertise in low-cost, efficient manufacturing simply hasn't translated to home construction.

Even in the best-case scenario, the cost savings of factory production in construction have been lower than in other industries. Mobile homes (now referred to in the industry as manufactured homes) are factory-produced and can be less expensive than site-built construction in the US. The average manufactured home costs around 50 percent of what the average site-built home costs, due in part to factory efficiencies and in part to things like lower-end finishes, simpler foundations, and less strict (and more consistent) code requirements.[839]

But the further you move from the tiny-trailer form factor toward more conventionally sized and shaped homes, the narrower the gap in cost becomes. Small, single-section manufactured homes cost around 35 percent of an equivalent site-built home—a very large cost reduction. But these gains are only possible for small homes that don't need to be transported far, require only a limited amount of site work, and make limited use of additional building modules. Multisection homes built to a quality level similar to site-built homes cost around 70 to 80 percent of what a site-built home costs, including transportation.[840] And while 20 to 30 percent in savings is still substantial, it's much less than what other industries that moved from craft to factory production have achieved. Recall, for example, that when cotton thread production moved from craft to factory production in the late 18th and early 19th centuries, its cost fell by more than 90 percent.

Why haven't factory methods dramatically lowered the costs of construction the way they have for other types of manufacturing? Part of the explanation involves technical factors. Even a small building is large, which makes transporting it from a factory to the jobsite difficult and expensive. Builders either split the building into parts that are then stitched together on-site or ship huge, wide loads that require special transportation arrangements, such as specific routes, limited speeds and travel times, or escort vehicles. Both options add difficulty and expense compared to conventional construction. To survive transport, modules must be self-supporting, which can require additional structural supports, and tying them together means adding a great number of connectors and interfaces that aren't required by conventional construction. And either way there will be significant additional transportation costs compared to conventional construction. A building is expensive, but it's so large that its cost per pound or cubic foot is actually quite small. A dollar's worth of building is much heavier and requires much more physical space than a dollar's worth of cars, electronics, or appliances. In other words, the dollar density of a building is quite low. A single 20-foot shipping container, which is around 1,172 cubic feet in size, can hold on the order of $10 million worth of iPhones, but only around $30,000 worth of a single-family home. This means transporting a building is proportionately much more expensive than transporting something with higher dollar density. While this is especially true of transporting an entire building, which is mostly empty space, it's also true for building components. Even if a building is broken into individual panels that can be packed more tightly, the dollar density is still much lower than electronics or appliances. And as a building is broken down into a larger number of components, the amount of on-site labor, costly component interfaces, and probability of an assembly error all rise.

Moreover, high transportation costs not only add significant overhead but also limit the market a given factory can serve. The lower the dollar density of a product, the shorter the distance it can be transported cost-effectively. High dollar-density items, like semiconductors or hard drives, can be made in factories in Taiwan or Thailand and shipped around the world. Low dollar-density items, like gravel or toilet paper, have much shorter cost-effective shipping distances.[841] With large, heavy prefabricated building components, shipping distances are typically limited to a few hundred miles, the distance that can be driven in a single day.[842] Even bulk building materials, such as lumber and gypsum, have low enough dollar density that they tend to be used comparatively close to where they're produced. In the southeastern US, it's common to use southern yellow pine as a building material, whereas in the Pacific Northwest, it's more common to use Douglas fir. The limited cost-effective transportation distance means that it isn't possible to outsource construction to countries with lower labor costs and that the catchment area for any given prefab building factory is fairly low, reaching only a limited number of customers.

This small market size further reduces the possibilities for economies of scale in what is already a comparatively low-volume industry.

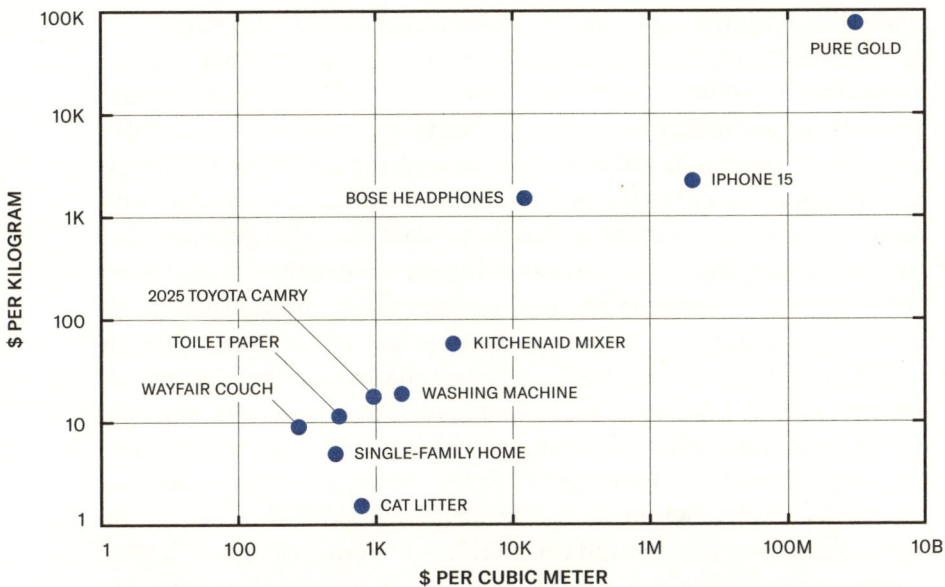

**Figure 35.** Dollar density (dollars per cubic meter and dollars per kilogram) of various goods.

There are only around 1 million single-family housing starts (houses on which construction has been started) in the US each year, compared to 15 million new cars and over 100 million new smartphones.[843] Even large building factories rarely make more than a few thousand modules a year, and many of them make much less than that. By comparison, a car assembly plant will typically make several hundred thousand cars a year—hundreds or even thousands of times more than a modular building factory.[844]

The small addressable market for any building factory is compounded by the high amount of variability in the construction process. Each building is constructed on a particular site, and each site will be somewhat different in size and shape, soil and climate conditions, buried infrastructure, access requirements, and more. Even a factory-produced building will require a significant amount of work on-site. In addition to the final module assembly, on-site workers will need to clear the site, run utilities, and lay the foundation. This will all be done outdoors, in weather conditions that can't be controlled and will vary from day-to-day.

Moreover, unlike most other manufactured goods, constructing a building requires a permit. There are an estimated 20,000 permitting jurisdictions in the US, each of which may have its own requirements.[845] Construction approval is ultimately at the discretion of the permitting reviewer, who is often forced to interpret unclear requirements language in the building code, as well as a zoning board, which may require changes to the design of a building based on community feedback or for other reasons. So even a single jurisdiction may wind up having varying requirements. This variability means that even the comparatively small volume of material going through a building factory will be far from uniform, which makes it much more difficult to use automation or repetitive, high-volume production equipment. Indeed, building factories will often simply duplicate conventional on-site methods rather than use any sort of factory-specific technology. A prefabrication building factory is sometimes little more than a large warehouse where skilled tradespeople build modules in the exact same way they would on-site. Even in comparatively highly automated factories, much of the work, such as drywall finishing, is still done manually.

Within a factory, the low dollar density of buildings rears its head again. Construction material is huge and bulky, which means it takes larger, more expensive equipment—bigger robot arms, wider conveyors—to handle a comparatively small value of product. It also means that it takes a lot of expensive automation, and expensive experts to maintain that automation, to duplicate what a relatively small amount of manual labor can do in the field. And these costs are spread over a relatively low volume of output.

Dollar density also restricts possible innovations in products or processes. Buildings use such a large volume of material that the material needs to be as cheap as possible, which makes it hard to substitute materials or processes that might be more attractive from a manufacturing perspective. Drywall, for instance, is reliably rated by prefab builders as one of the most troublesome materials because it's time-consuming and labor-intensive to hang and risks being damaged during transportation. But while several different interior finish materials could be used as substitutes, many of which are more attractive from a mechanized production standpoint, they are all far more expensive.[846] Dimensional lumber also has unattractive manufacturing qualities because it's a highly variable material that frequently warps or splits, but the fact that other materials are much more expensive makes it difficult to replace. A HUD study found that a house made from thin cold-formed steel studs, which are much more uniform than dimensional lumber studs, was around 14 percent more expensive than an equivalent one made from dimensional lumber.[847]

The cyclical nature of the construction industry—changes in economic conditions can have an enormous effect on the demand for new buildings—also reduces the viability of modular or factory production of housing, since such operations have trouble weathering downturns. This is exacerbated by the comparatively small addressable markets of building factories, which amplify variations in demand, since falling demand in one location can't be offset by rising demand somewhere else. Many modular construction companies and mobile-home builders went out of business during the recession of the 1970s. During the 2008 financial crisis, shipments of modular homes dropped by over 50 percent. Over a decade later, they have yet to recover to their pre-2008 level.[848] And while every industry is affected by downturns, because factory-built construction shows at best relatively modest savings (in the case of manufactured homes) or no savings at all (for other types of construction), these factories are still competing with conventional on-site builders, which have an easier time scaling back operations during slow periods given their much lower overheads.

The same factors that make it difficult for factory production to outcompete site-built construction in the US—high overheads, low production volumes, and high variability in the product and process—also make it difficult for construction to take advantage of economies of scale. In the absence of factories that can make use of large-volume, high-speed equipment, there are many fewer opportunities to spread costs or take advantage of geometric scaling effects from high-capacity machinery. Scale effects that do exist are often influence effects, such as having access to better financial terms or being able to negotiate better deals with suppliers, yet these are sufficiently slight that it's often possible for a small builder to be cost-competitive with a large builder. In fact, according to a 2005 study from Harvard's Joint Center for Housing Studies, the largest homebuilders (those building more than 10,000 homes a year) had slightly higher costs per square foot than small builders.[849]

Even if portable, high fixed-cost production equipment existed, the same factors would affect its on-site use outside of the factory setting. The variable nature of construction would make it difficult to ensure that the machine had a high degree of utilization (because it would be hard to know exactly when it would be needed and ensure that each new job lined up), and there would be high transportation costs associated with moving the equipment over long distances, as well as setup and teardown costs, which would make it hard to use for small projects.

For similar reasons, we see little vertical integration in the construction industry. Large homebuilders, such as Lennar and D.R. Horton, have few construction crews on staff and own little, if any, equipment. Instead, they subcontract out almost all their work to local crews. It's difficult for a construction company to have such a high volume of projects that they can have steady, reliable work for an in-house team of electricians, for instance. And unlike Toyota's flexible staffing, the skills and (often) the licenses required for different trades make it difficult to flexibly reallocate workers from one task to another. A drywaller can't simply switch to electrical work.

The high variability of home production also makes it difficult to accumulate incremental process improvements of the kind that help an industry fall down the learning curve and avoid the types of disruptions that reset the learning curve.[850] On large, repetitive projects, such as building skyscrapers with identical floors, learning effects do occur. Higher floors will be built faster and more efficiently than lower floors as workers learn the rhythm of the process. But these efficiency gains will evaporate once the project is complete or if it is interrupted.

This variability also makes introducing process improvements fraught. As discussed in Chapter 7, new technology inevitably has a shakeout period during which bugs are worked out. But without a high production volume over which to spread this cost or a controlled and consistent environment in which to tease out the causes of certain bugs, the introduction of a new process is quite risky. In general, construction projects have a fat-tailed, right-skewed distribution of outcomes: A project that goes well might be slightly under budget, but a project that goes poorly can be double, triple, or other multiples more than the projected cost. So the potential upside of a novel production method is relatively low, whereas the potential downside is extremely high.[851] As a result, rational risk aversion means many builders will resist adopting novel construction technology that dramatically alters the production process, in contrast to a company like Ford, which was constantly rearranging the Model T production process, adopting brand-new, never-before-tried machines, and disposing of them if they didn't work. This has the effect of dampening incentives to develop novel construction technology. The result is that construction technology tends to improve within the existing paradigm. In S-curve terms, construction methods tend to stay on the same S rather than being superseded by a new method with its own S curve.

Beyond these effects, other aspects of the industry make reducing the costs of construction difficult. Outside of labor costs, building materials make up the largest percentage of construction costs. But here, too, it's hard to lower costs. Bulk building materials, which are already factory-produced, are made as cheaply as anything civilization makes in mass or volumetric terms. It's not obvious that there's significant potential for more reduction here. Moreover, moving a huge amount of solid material around will always incur significant transportation costs, and it's difficult to use fewer or lighter building materials because doing so often requires compromising on functionality or contravenes the aesthetic preferences of buyers, who often favor heavier, more luxurious materials such as ceramic tile and hardwood. It can also be hard to decrease material usage without adding labor cost. For example, it's often possible to reduce the amount of concrete in a building by using complex formwork shapes or inserts, but these typically require more labor to install, a trade-off that's often not worth it.[852]

The variability of site conditions and permitting requirements, the lack of precisely specified drawings, and relatively low volume also make it less viable for the construction industry to implement DFMA and other design optimization efforts. The cost to design most products is many times the cost of the product itself. Car manufacturers will spend billions of dollars developing a new model, thousands of times more than the cost of the car it ultimately sells. But design costs are often a minuscule fraction of the overall cost of a building, on the order of 5 percent or less. Although a building has many more parts than a car, many fewer people will work on its design and much less will be spent on it. It's still possible to value-engineer building designs, but without a large production volume over which to spread the design costs, building design efforts are typically much more limited. This often means that building elements are oversized and, therefore, more expensive compared to what might be absolutely necessary.

Then there's the fact that many construction tasks have historically been resistant to automation. Bricklaying, for example, seems

| ITEM | COST PER CUBIC FOOT |
|---|---|
| LOW-PRICED COMMODITIES | |
| Coal | $ 3.15 |
| Corn | $ 3.35 |
| Urea | $ 7.15 |
| Iron ore | $ 7.83 |
| Potatoes | $ 8.77 |
| Gasoline | $ 19.00 |
| Polyvinyl | $ 32.40 |
| BUILDING MATERIALS | |
| Lumber | $ 5.54 |
| Poured concrete | $ 5.93 |
| Drywall | $ 10.89 |
| 10" hollowcore plank | $ 12.36 |
| Brick and mortar | $ 15.59 |

Table 4. Cost per cubic foot of various low-priced commodities and building materials.

like it would be a perfect task for automation. It's physically taxing and highly repetitive; even a small house will use thousands of identical bricks. But there have been more than a century's worth of attempts to develop bricklaying machines, all of which have had limited success at best as of this writing. The machines have struggled to handle mortar, required making maps of the location of each individual brick, and had high setup and teardown costs. One bricklaying robot was found to be economical only for large, uniform sections of brick wall.[853] More generally, construction materials and methods have typically evolved for ease of use by human workers, and what's easy for a human is not necessarily easy for a machine. More often than not, mechanizing a task involves changing it into something that's easy for a machine to do, like how a car uses wheels rather than mechanical legs. But without being able to dramatically rework production methods—which, for the reasons discussed in this chapter, is difficult in the construction industry—it's often infeasible to have a machine perform a task the way a human would do it.

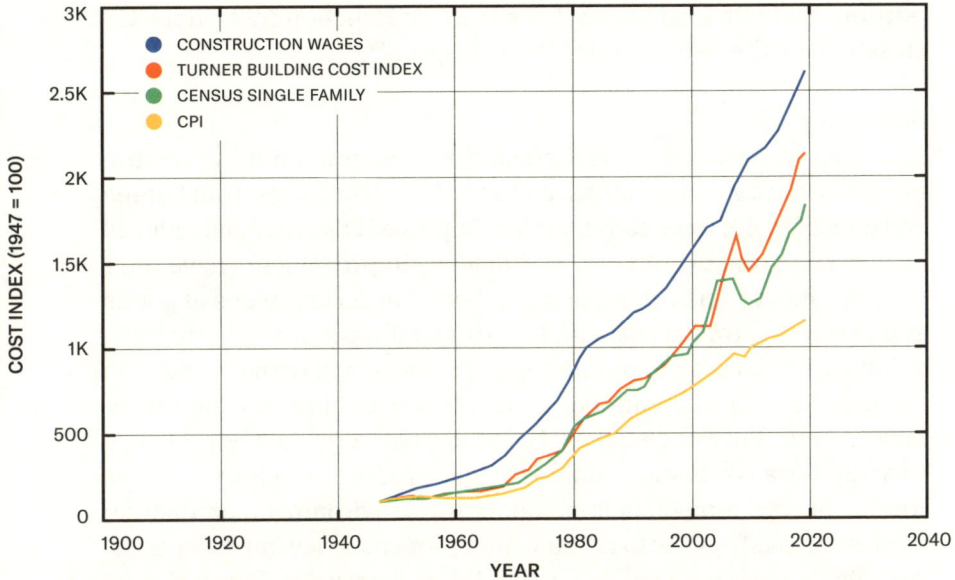

**Figure 36.** Construction wages versus construction cost indexes versus the Consumer Price Index. Construction cost indexes increase at about half the rate of construction wages. This is roughly what we would expect to see if construction costs were roughly half labor and half material, labor was not getting more productive over time, and material costs were rising at the rate of overall inflation.

In addition to all these blocked paths of process improvement, construction has accumulated additional burdens over time. Regulations have become stricter in a number of domains. Building codes are longer and more complex than ever before and include more stringent requirements for things like energy efficiency, safety requirements are more cumbersome, and environmental regulations increasingly require impact studies and mitigation measures to prevent negative environmental effects. The effect of such regulation has been particularly pronounced in nuclear power plant construction, which has steadily grown more expensive over the years.

Finally, construction is encumbered by increasing labor costs. As we've seen, the sector has been limited in its ability to mechanize or automate tasks, so there has been little to no reduction in the amount of labor required to construct homes. This lack of improvement, combined with the fact that construction is difficult to outsource to low-labor-cost countries, means it's subject to Baumol's cost disease. In fact, if you adjust for the fact that construction wages are only a portion of construction costs, the rise in construction costs in the US closely track the rise in construction wages.[854]

## Summary

Efficiency improvements have created enormous amounts of wealth, brought abundance to billions, and saved countless lives from famine and infection. But the steady march of improved efficiency and reduced cost is a narrow sort of victory. Efficiency improvements have their greatest, most transformative impact only on certain types of goods and services—things that can be produced repetitively in massive volumes, like cars, semiconductors, or steel, or where the output can be spread over a large number of people, like commercial aircraft or power plants. But a huge swath of goods and services can't be produced this way. Because they are limited to small production volumes or have irreducible variation in their manufacturing environment, production in these areas struggles to take advantage of efficiency improvements. And efficiency improvements can only be taken so far if the universe doesn't supply the right sort of phenomena to be leveraged. Titanium is strong, light, corrosion-resistant, and one of the most abundant elements in Earth's crust. But if we can't find some way around its reactive chemical nature when smelting and machining it—and so far, we haven't—it can only get so cheap.[855]

Cost-reducing improvements aren't the only forces at work in the efficiency landscape. Production processes are sociotechnical systems, embedded in human organizations and subject to the vagaries of human decisions, which are often guided by considerations other than cost. These decisions can and do influence the trajectory of production process development, by adding costs in service of achieving other goals like increasing safety or reducing environmental harms, or by limiting the sorts of improvements that can be made.

Some of the most important parts of modern civilization—building and infrastructure construction, medical care, education—are ones where many efficiency improvements have been screened off and costs have steadily risen. While efficiency improvements have reduced the cost of manufacturing penicillin by several orders of magnitude, many other aspects of the medical system have not been so fortunate. The cost of developing a new drug has increased by a factor of 14 between the 1970s and the 2010s in inflation-adjusted terms, in part because of the rising costs of clinical trials.[856] Similarly, incomes for physicians and nurses have more than tripled in inflation-adjusted terms between 1960 and 2016.[857] So while efficiency-improving techniques such as route design have lowered the cost of one step in the medical care value chain, other steps have gotten much more expensive.

As we saw in the previous chapter, efficiency-improving techniques are slowly being adapted for use in production environments outside of repetitive, high-volume manufacturing. But there's much work to be done. Creating a world where more people benefit from more abundance will require pushing efficiency improvements further into areas that have historically resisted them.

# Conclusion

# The Future
# of Production

The different mechanisms for improving production we've discussed in this book can transform a process that is slow, unreliable, labor-intensive, and expensive into one that is fast, predictable, and cheap. And these transformations, applied across the globe in every type of industry, have been world changing. Cheap, abundant food has vastly reduced the incidence of famine. Cheap, abundant light has given people more hours in the day. Cheap, abundant drugs like penicillin have greatly reduced the amount of suffering and death in the world. Cheap, abundant electricity has made it possible for people to have an endless number of modern conveniences, from kitchen appliances to smartphones. Process efficiency—the continuous effort to produce goods and services more cheaply—has given us a world of incredible abundance.

Each of the mechanisms of efficiency improvement we've looked at in this book—technological improvement, reducing the cost of inputs, economies of scale, removing unnecessary steps, and reducing variability and eliminating buffers—have been critical for this transformation. The Industrial Revolution in Britain was triggered by technological improvements in textile manufacturing and steam engine design. Reducing input costs, in the form of strategies like DFMA, has become table stakes for achieving cost-effective manufacturing. The Second Industrial Revolution in America was triggered by exploiting economies of scale. Cutting out unnecessary steps is the cornerstone of lean methods, an industrial improvement system that has broadened beyond its origins in the Japanese auto industry to nearly every type of manufacturing. Variability reduction and process control helped make Japanese manufacturers the envy of the world in the late 20th century, enabling them to conquer industries like shipbuilding with astonishing rapidity.[858] Because these mechanisms all interact, producing bundles, chains, and virtuous circles of improvement, none of them can be ignored. All have been critical to the project of improving efficiency.

And yet, some of these mechanisms loom larger than others. When we take a 30,000-foot view at where and how efficiency has improved, one recurring theme is the importance of scale. Economies of scale are not only a major driver of efficiency improvements in and of themselves, but they are also a gating mechanism that unlocks other efficiency improvements by making it possible to amortize the large fixed costs that such improvements often require over a sufficiently large output. To return to our bulb blank example from Chapter 1, the

Corning ribbon machine was able to continuously produce blanks at an incredibly low cost per bulb, but the machine itself took years to develop and remains extremely expensive to build and operate. It was only by virtue of its enormous output—the millions of bulbs it could make in a single day—that the costs of developing, building, and operating it could be recouped. Likewise, strategies like value engineering, DFMA, and automation all have significant fixed costs that need to be spread over a sufficiently large production volume.

More generally, improving the efficiency of a process has historically been dependent on repeating the same process over and over again. When we can't repeat a process, or when there's significant variation in the production environment or in the product itself, many paths toward increased efficiency are blocked. Without repetition, we can't take advantage of economies of scale. A constantly changing process or a highly variable environment is harder to predict and control, making it more difficult to increase yield and decrease waste. It's also harder to automate or mechanize in such circumstances, which limits our ability to reduce labor costs. This is one reason why services like medical care, car repair, and construction, which rely on skilled labor that can flexibly adapt to highly variable conditions, remain expensive. Finding ways to make these things cheap will require escaping the constraints of scale to get the benefits of a continuous process—that is, a constant transformation of inputs into a finished product without waiting, buffering, failures, or wasted actions—without needing to repeat that process over and over again.

Some types of process improvement are inextricably coupled with repetition. Taking advantage of phenomena like geometric scaling will always push producers toward larger production volumes. But in other cases, the coupling between efficiency and scale is more contingent. The development of continuous processes has historically been most easily achieved with a carefully orchestrated arrangement of workers and special-purpose equipment, which is costly to build and operate and, therefore, requires the cost to be spread over a very large output. It takes time, effort, and expense to make the steps in a process work together in sync, a result that can often only be achieved with dedicated equipment and justified with large production volumes.

But what if we could reduce that time and effort? What if it was quick and easy to figure out the order of steps in a process and arrange them so that the transformation is swift and steady? And what if this process didn't require special equipment but could use flexible, adaptable machinery able to produce a variety of different items? Then it would be possible to achieve a smooth continuous process that worked in a highly variable environment, bringing the benefits of efficiency improvement to domains that have traditionally resisted it.

**Flexible equipment and information processing**
With the right technology—in particular, information-processing technology capable of flexibly responding to a variable environment and sequencing the appropriate actions in response—this sort of improvement would be possible.

Consider the printing press, which creates text by coating movable metal type with ink and then pressing that type against a piece of paper. The first presses could print a single sheet at a time, using a large screw to press the type against the paper. By the mid-19th century, these had evolved into rotary presses, which fed a roll of paper through rotating drums that had the inked metal type mounted to them.[859] Rotary presses could print continuously, producing thousands of pages per hour. But the fixed cost of setting up the rolls with the movable metal type meant that, like other continuous process machinery, rotary presses were best suited for producing texts in very high volumes.

Now consider a modern inkjet printer. Unlike movable metal type, the inkjet printer works by moving a printhead back and forth, shooting tiny jets of ink onto a piece of paper. By varying the location and color of the ink, an inkjet printer can produce any type of document or image. And unlike the printing press, which requires physically rearranging the metal type to produce a new design, an inkjet printer can print a different document every time simply by feeding it a different set of electronic instructions that tell the printhead where to move. The inkjet printer is, therefore, a much more flexible production technology than the rotary press.[860]

The inkjet printer and other similar desktop printing technologies make it possible to cheaply produce printed material in very small volumes—as small as a single document. They don't eliminate economies of scale; it's still cheaper to produce large print runs with high-volume printing presses rather than inkjet printers.[861] But they make it possible to use a low-cost continuous production process in a highly variable environment, printing many different documents in small quantities. And while a modern printing press is incredibly expensive, an inkjet printer is cheap enough that an individual consumer can afford one.

What would it take to adapt this model of production more broadly? How could it be applied to car repair or medical care or construction? For production technology to be able to flexibly adapt to a highly variable environment, there are two broad requirements. First, it requires equipment that is physically capable of responding flexibly—that can take a broad range of possible actions, or a small number of actions that can be combined to produce a wide variety of outputs. The inkjet printer, for instance, has a small number of possible actions: It can move the printhead relative to the page on the X and Y axis, and it can shoot a tiny dot of several different colors (typically cyan, magenta, yellow, and black). But by combining individual colors in different amounts and placing jets of color in distinct locations, different texts or images can be created.

Beyond inkjet printers, highly flexible production equipment with many degrees of freedom already exists in a range of industries. Robots, which can be programmed to take many possible actions and use many possible tools, have been in commercial use since the 1960s. General Motors began using Unimate, the first industrial robot, in 1961 to remove parts from a die-casting machine. Shortly thereafter, other car manufacturers began using Unimates for tasks such as workpiece handling and automated welding.[862] Today, robots are widely used across a variety of industries, from robotic welders in car factories to robotic vehicles in distribution centers and robotic cranes in seaports. And in machining, flexible production equipment has been in use for even longer. Milling machines, which can cut metal into a variety of shapes by moving it against a rotating metal cutter, have been in use since the early 1800s.[863] Modern milling machines are capable of manipulating the workpiece and the cutting tool to cut from different directions without human intervention, and they often have built-in tool changers so they can quickly and automatically change out cutting tools.[864] By pairing these with robot arms that load and unload the machine, it's possible to create an automated machining cell where a workpiece is loaded, machined, and unloaded without any manual labor at all.

The second requirement for flexible, automated production is that the equipment has some method of determining the sequence of the proper actions to take—an information-processing component that can, based on data from the environment, break a desired output into a series of achievable production steps. Unfortunately, in contrast to the first requirement, the information-processing capabilities of existing equipment lag much further behind. Historically, mechanized or automated processes have only been able to respond to the environment in limited ways or not at all. A mechanical corn picker, for example, doesn't try to figure out where the corn is and pick it—the plant is simply fed into a mechanism that the corn can't pass through, which filters it out from the rest of the material. As we discussed in Chapter 6, automatic control systems can adjust production equipment in response to variation in the environment, but these systems can typically only make relatively narrow adjustments, such as keeping a certain measurement, like temperature, at a specified level. More flexible, higher-level adaptation and control tends to require a human in the loop, as we saw in our discussion of statistical process control and the Toyota production system.

These control systems, both manual and human, exist in the context of a process that has already been figured out and arranged. They're of no help in determining what the process should be: how to map a desired product to a series of operations and how those operations should be structured. Welding, for example, is often automated; robots are more than capable of performing many welding tasks, and automating a welding process can greatly reduce production costs. But most welding robots aren't yet capable of figuring out what should be welded.[865] They must be explicitly instructed by people. In some cases, this will be done by a person physically moving the robot through the path it should take, which the robot can then repeat.[866]

Even when it is possible to automate the process of deciding what sequence of actions to take, there's still the problem of figuring out what should be produced in the first place. An inkjet printer can automatically translate a file into a series of printhead movements that will reproduce the document on a sheet of paper, but the printer can't author the document for you. Producing something with flexible production equipment, therefore, often has the same sort of time-consuming and labor-intensive figuring-out process we saw with other types of services, such as medical care.

In 2023, electrical engineer Heath Raftery documented the process of trying to 3D print a replacement plastic part for an inexpensive tent. The part was used to hold the rods of the tent together, and Raftery went through many rounds of design iteration as he attempted to print a part that could hold the rods together successfully. After many hours and many prototype prints, Raftery eventually found that while the design he had converged on worked when the tent was put together, it failed when trying to stow the tent in its storage case, which put different forces on the plastic part. As he wistfully reflected, "lurking within some forgettable $50 Kmart capitalist wonder is the carefully applied expertise of designers and engineers."[867] Raftery undoubtedly spent hundreds, perhaps thousands, of times the cost of the plastic part itself in trying to recreate its design. It's a powerful illustration of the problem of scale and repetition: Only over a large production volume can we economically spread the time and effort required to design something to successfully perform a task, even if 3D printing or a similar technology is *physically* capable of producing it quickly and cheaply. Without better information-processing technology, flexible and adaptable production technology won't reduce costs any more than cheap penicillin reduced the costs of medical care. The cost of the process will still be weighed down by the expensive human expertise in the loop.

We're likely to see significant advances in this sort of technology. Indeed, we've already made considerable progress. As computers have gotten more powerful, it's become possible to simulate certain aspects of the real world in software, running experiments and observing the results entirely in silicon. This has made it feasible to rapidly explore a much wider design space. Once upon a time, a car's safety profile could only be determined by crashing an actual model of the car and observing how it behaved, an expensive and restrictive limitation. Since the work of designing the car would need to be mostly done before the crash test, any post-crash redesign would be extremely expensive. But now crashes can be simulated in software with reasonable accuracy, making it possible to quickly and easily create successive car designs (although this hasn't completely eliminated the need for real-life crash testing).[868] Similarly, advances in computational fluid dynamics have greatly expanded how much aerospace design can be done in the computer. It's increasingly possible to predict the behavior of aircraft components using software, reducing the need to build and test models in wind tunnels.[869] Similar models are being used in manufacturing to simulate a production line and determine things like flow rates, bottlenecks, and capacity issues without having to build an actual factory. More generally, as computers become more powerful and software tools become increasingly intelligent, it will become possible to use methods from operations research and other disciplines to investigate larger and larger design spaces—machine designs, factory layouts, new technological possibilities—by swiftly simulating and sorting through thousands or millions of different configurations and selecting the best ones. Better computers and smarter software will accelerate the figuring-out process, requiring less feedback from the real world to produce robust, functional designs. Combined with flexible production technology, this will enable highly efficient production at smaller and smaller scales.

The obvious question is, what are the limits of these advances? Will it ever be possible for product and technological development to take place entirely in silicon? Will there be software or AI that can spit out a design for a highly efficient machine or process—a new type of titanium smelting, a much more efficient solar PV cell, a fully DFMA'd electric car and the factory to produce it—without needing real-world feedback and the painstaking, step-by-step improvement that currently characterizes technological development and process improvement? Could we simply start at the top of the S curve and avoid the messy, expensive process of climbing it? It's hard to say. There are, after all, countervailing forces that might thwart a completely simulated or algorithmic development process—limits to what we can accomplish without building something and seeing how it works in the real world.

Computational fluid dynamics (CFD) provides an example. Increasingly accurate computer simulations of fluid flow have made it possible to do much aircraft design work, which previously required physical mockups and wind-tunnel testing, in software. This not only greatly reduces the cost and accelerates the pace of trial-and-error testing, but also makes it viable to explore a much wider design space by experimenting with shapes and fluid flows that would be impractical or impossible to test in a wind tunnel. CFD is based on a set of equations known as the Navier–Stokes equations, which essentially translate Newton's second law of motion (force equals mass times acceleration) to the flow of fluids, although the burdens of computation often mean that other simplified models of fluid flow are used. The ultimate goal of aerospace CFD is to "fly the Navier–Stokes equations"—to digitally model an entire aircraft and completely simulate its behavior in flight, using the Navier–Stokes equations to accurately model how moving air will behave.[870]

However, completely simulating the flow of air over an aircraft without modeling simplifications is estimated to require an increase in computing power some 20 orders of magnitude. One researcher estimated in the year 2000 that such computing power might be available in the year 2080, assuming Moore's law continued unabated, but 15 years later they weren't confident it would happen in the 21st century, if ever.[871] The problem is that of turbulent flow, or the tendency of fluids to engage in chaotic, irregular changes in certain conditions, which requires an enormous degree of precision to simulate accurately.

Even direct simulation with Navier–Stokes might not be enough to simulate fluid flow with complete accuracy. There's still an open question known as the smoothness problem: Simply put, it's not clear whether the Navier–Stokes equations will in some cases "blow up" and produce nonsensical answers, such as fluids with infinite velocities. This would mean that in some cases the Navier–Stokes equations can't accurately model the behavior of fluids in the real world.[872] The smoothness problem has been deemed important enough that it was designated as one of the seven Millennium Prize Problems by the Clay Mathematics Institute, which has offered a $1 million reward for its solution.[873]

CFD is an example of how, despite great strides in software improvement, there are still major barriers to creating completely accurate computer models that would obviate the need for testing and feedback in the real world. Similar barriers likely exist for other sorts of simulation. As we observed in Chapter 6, there are innumerable factors at work in any production process or technological implementation: tiny gravitational changes, minute fluctuations in air currents and temperature, and so on. As technology shrinks, these small effects become more important, as in the case of semiconductor fabrication. Sometimes, slight perturbations or changes in parameters can have large impacts: the so-called butterfly effect, studied by a branch of science and mathematics known as chaos theory. These sorts of effects may ultimately limit how effective a simulation can be and how much it can accurately predict without relying on the slow, messy process of building something and testing it in the real world. On the other hand, algorithmic breakthroughs, massive increases in computing power, and other technological developments may make these concerns appear quaint in the future. As AI capabilities accelerate and increasingly large computing clusters are built to train them, it's a question to which we may soon have the answer.

This, then, is one of the great challenges for the future of production: developing manufacturing and information-processing technology that can flexibly respond to an uncontrolled, variable environment. We need to automate not only the production process but also the upstream design and decision process—for example, mechanization that works not only in the controlled confines of a factory but also in the variable conditions of a construction site. This sort of automation technology, coupled with continuing progress in flexible, high-degree-of-freedom production equipment that can quickly produce something different when fed different instructions, could free us from the yoke of Baumol's cost disease and push efficiency improvements and cost reductions into the labor-intensive services that have so far resisted getting cheaper.

What might a process like this look like? Let's examine a hypothetical future construction process in which steadily advancing technology has made production both highly flexible and highly efficient.

*Case study: home construction of the future*
The process starts with a homebuyer, John, who has acquired a parcel of land on which he'll build his new home. He logs in to the home design software provided by the homebuilder he's hired and gets to work. The software already has access to a government-supplied high-fidelity map of the land parcel, which includes topographic information, soil properties, rainfall patterns, existing infrastructure, local code requirements, and other relevant information. To this, John adds basic details about his family's lifestyle: kids, pets, working locations and commute distances for him and his wife, income. He also has the option of uploading a preference map derived from his and his family's various social media accounts, which will help the software figure out what everyone likes and what people similar to him have chosen.

From there, the software begins to generate and sort through millions of possible home configuration options. It takes into account variables like current and future predicted climate, likely lifestyle trajectories (will they need a home office? bedrooms for kids? a mother-in-law suite for an aging parent?), urban development patterns (will this window still have a good view if an apartment building gets built on the corner?), and so on. It analyzes typical family traffic patterns and room uses, with modifications for the specific needs of John's family (for instance, his son plays the drums and needs a place to practice that won't disturb everyone else).

The software starts by producing simple, high-level simulations, using algorithms to minimize factors like travel distance and maximize natural light exposure. Out of millions of possible options, it selects some promising candidates. It then starts to produce more detailed simulations, running a high-fidelity virtual family through a variety of situations to detect issues like traffic flow (is there a backup at a bathroom? do people bump into each other in the kitchen?), home features (are there enough electrical outlets where people need them? would it be helpful to have a bigger shower with a second showerhead? would a pot filler over the stove get used?), and materials (is more durable carpet needed in the upstairs hallway to handle the traffic?). The software ends up with several promising options for home layouts, along with several color and material palettes that would pair nicely with them.

John could simply pick one of these software-selected choices, but he opts to use them as a starting point for further iteration. He tweaks a few options ("The workshop is nice, but it needs to be bigger"; "Let's do a herringbone pattern instead of hexagons for the bathroom tile"), and the software modifies the options accordingly and generates several new ones based on John's feedback. The program includes cost sliders that can be toggled to upgrade or downgrade various finishes. If John bumps up the cost slider by $10,000, it can show potential upgrades and make recommendations ("For another $10,000, the best bang for your buck is 10-foot ceilings; the second best is nicer kitchen floors"). John continues to tweak the room layouts but doesn't stray too far from the software's recommendations, since deviating too much from the suggested parameters requires manual review by an architect at the homebuilder's headquarters. As changes are made, the total cost is instantly updated. The software defaults to home designs that keep construction costs low, such as arranging bathrooms to share plumbing walls, but this can also be changed so that cost is given less weight in the design algorithms.

John finishes the design in a few hours—most of which is spent mulling over various options, as the software updates nearly instantaneously—and then sends his selection to the homebuilder. They arrive on-site first thing the following morning to start construction.

First to show up is an automated foundation installation truck. To minimize the amount of excavation and site work required, the home will sit perched on a foundation of large steel screws, which are drilled into place by a boom mounted to the truck. Once that's done, the truck leaves, having already been routed to another job across town, and the actual home construction begins. Now a second truck arrives on-site, this one packed with a dozen humanoid robots and various pieces of construction equipment. (While a humanoid shape isn't always the most efficient design for a robot, it tends to be a useful shape for construction, since so much of the work is done in spaces sized for humans.) The robotic construction crew wastes no time. The home's structure will consist almost entirely of thin sheets of steel, bent into the proper shape by truck-mounted sheet-bending equipment: steel box sections for the heavy beams that span foundation posts, lighter C-shaped joists for the girders, steel roof trusses, and so on. Each one is fabricated on the truck, moved into position by a pair of robots, and quickly welded in place with welding guns mounted to the robots' arms.

The walls, floors, and roof consist of steel triple-sandwich panels. These are made by taking a sheet of steel and placing a layer of expanding foam insulation on top of it, followed by another sheet of steel. Because the steel is so thin and the insulation expands so much, the truck can store a surprisingly large amount of both, but it will receive automatic deliveries of new coils of sheet steel and liquid insulation when it starts to run low. On top of the steel panels is a cavity to run wiring, plumbing, and micro-HVAC ducts, followed by one final sheet of steel. The outer layers, which will be exposed to the elements, use a special corrosion-resistant steel, and the panels have specially bent edges so they can lock tightly together.

Once the foundation beams and girders are in place, the truck begins to produce the panels. The home design software has already broken John's design into an optimal panel layout and generated the fabrication information and installation sequence for each panel. The bending and sheet-steel cutting equipment can make panels of any possible two-dimensional shape; curved, three-dimensional panels are also possible, but they're a little more expensive and harder to make using portable equipment, so they tend to be reserved for higher-end homes. For each panel, the truck-mounted fabrication equipment cuts a sheet of steel, sprays the insulation, positions the middle layer of steel, and installs wiring, plumbing, and HVAC ducts, and anything else that needs to go into it (light switches, thermostats, and the like), followed by the third sheet of steel. Once the panels have been made, they're carried into place and welded into position by the robots. The panel fabrication is rapid and continuous, but the truck can only work so fast, so the truck started making the panels the night before, right after John submitted his order.

Panel installation proceeds quickly, and in a matter of hours the structure is in place, along with insulation, electrical wiring, plumbing, and most of the HVAC ducts. This is followed by window and door installation. A truck full of prefabricated windows and doors arrives on-site, and as soon as the robots change their arm-mounted welding guns for window installation tools, they get to work. John has wisely chosen standard-sized windows and doors, which can be drawn from stock already waiting at the homebuilder's headquarters. If he'd wanted custom-made doors and windows, next-day construction wouldn't have been possible. Within an hour, the windows and doors are in place and the house is watertight.

Right now, every surface in the house is steel. But the panels have been fabricated with attachment points to receive special finish panels, which can be anything from engineered wood and carpet to tile and stone (all synthetic but designed to mimic the real thing). Since John chose standard materials, installation begins immediately following delivery from the factory. The panels slot into place and lock together, creating seamless interior, exterior, and roof finishes. John mostly went with the software's recommendations, but if he wants to upgrade in the future, it will be as easy as taking off the existing finish panels—which the homebuilder will buy back, albeit at a significant discount—and slipping on new ones. In an hour, all the finishes have been installed. As with every other task, the robot's software has automatically translated the installation sequence ("First this panel goes in, then that one, then this one") into a series of specific actions ("Pick up this panel, orient it like this, walk forward 40 feet") that are executed smoothly and seamlessly and choreographed with the work of every other robot. If something unexpected should happen—if a finish panel breaks or doesn't fit properly—the task sequence is automatically updated, with barely a blip in the smooth, continuous installation of material.

As the finish panels are being placed, the rest of the equipment is arriving on-site: HVAC units, water heaters, appliances, toilets,

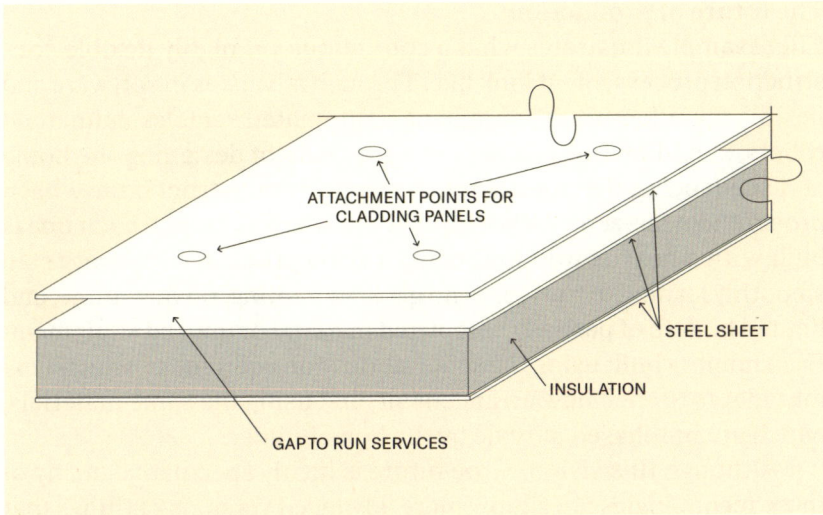

**Figure 37.** A triple-sandwich building panel.

preassembled kitchen islands, cabinetry, lighting fixtures, and more. These items are too bulky to be stored at the homebuilder's headquarters, so the company has placed orders for standard units at suppliers around the city, and an automated truck picked them up earlier this morning. These too are quickly installed by the humanoid robots, which have once again switched out their end effectors for a new set of tools. Then comes the landscaping and outdoor work, which John has opted to keep minimal: just some paving stones for the front walk and the driveway.

When the landscaping work is done, the truck returns to the homebuilder's headquarters and begins to fabricate panels for another house that will be built the next day. No home inspection is needed; pictures and other data were automatically recorded during the fabrication process, and as installation took place, a variety of sensors (visual, thermal, and so on) monitored progress to ensure that everything was connected and working properly. Before leaving, the robots check each switch, cabinet door, faucet handle, toilet flush, and every other home feature, recording the result. The data is fed into the local jurisdiction's automatic checking software, which instantly verifies that everything is working and signs off on the home. John can move into his new house whenever he wants. The entire process, from starting design to being handed the keys, has taken less than 24 hours.

## The future of production

This example illustrates what a continuous yet highly flexible construction process might look like. Through advances in software and flexible manufacturing equipment—automated vehicles, humanoid robots, steel-bending equipment—the costs of designing the home and sequencing the operations to smoothly construct it have been brought to near-zero. Although every house is different, each one is built with a near-continuous construction process, with every step smoothly leading to the next—no gaps, no waiting, no downtime, and limited buffers of partially completed material or unused equipment. Each home is built using the same fabrication equipment, which flexibly adapts to each new arrangement, and using the same materials, which are purchased at scale by the homebuilder.

Although this vision of the future is highly speculative, many of these technologies are simply more advanced versions of things that already exist. Construction already makes use of steel foundation screws (called helical piles), as well as metal-insulated sandwich panels, prefabricated doors and windows, and portable steel-bending equipment. Design software that can quickly generate an architectural model and translate that model into fabrication information is in development, as are humanoid robots. This future may not be as speculative, or as far off, as it seems.

To be sure, many other factors are responsible for high construction costs. Zoning restrictions and other burdensome land use rules can drive up the costs of housing even if the construction process itself has been radically improved; indeed, in cities like San Francisco, the cost of land can approach 80 percent of the price of a home. And we've skipped over many of the difficulties of getting such a system in place: revamping the permitting and inspection system, creating computer-interpretable building codes, overcoming trade union opposition. This speculative case study isn't intended to lay out a specific path for the future of construction but to show how even a highly variable and uncontrolled production environment can, with sufficiently advanced technology, make use of efficient continuous process production methods.

We can imagine similar technologies that make domains like health care and education more efficient. In health care, advancements in AI and biological simulation may one day greatly reduce the time and effort to develop new drugs by being able to discover new compounds and accurately predict their effects without running time-consuming and costly human trials.[874] In education, curriculum software that tests a student's capabilities and creates a tailored sequence of explanations and example problems that will predictably improve their skills could greatly reduce the time a learner spends struggling to understand new material.

More generally, we are perhaps in the very early stages of developing information-processing technology that makes it possible to push efficiency improvements into the high-variability environments that have long resisted them. As of this writing, AI's capabilities are advancing rapidly. AI can vastly outperform humans at chess, Go, and nearly every other game, flexibly determining which actions to take to achieve a purpose in narrow environments more tractable than the real world. Self-driving cars, which are gradually being deployed to more cities, are capable of navigating an uncontrolled, real-world environment. Large language models can answer questions, create software, write fiction, perform well on standardized tests, and do numerous other things that previously only a human could do. If this technology can become general-purpose enough and be coupled with general-purpose physical automation (like a low-cost humanoid robot), it has the potential to radically change areas of the economy that have been resistant to efficiency improvements.

This sort of technology wouldn't eliminate economies of scale or the types of high-volume production that can take advantage of them. But it would enable dramatically new production possibilities, bringing the kinds of world-changing improvements we've seen in finished goods and agriculture to new domains, such as housing and medical care, where greater efficiency—and greater abundance—is needed.

While we can expect flexible production and information-processing technology to open new avenues for efficiency improvement, it's critical that we continue to study and understand what makes efficiency improvements possible and what prevents them from occurring. This book has attempted to lay out a high-level map, but many details of the territory remain to be filled in. Often, issues of efficiency and productivity are studied only at a high level of abstraction, as a series of lines on graphs going up or down over time. This can make improvements in efficiency and productivity feel inevitable or like the product of broad, historical forces or government policies that most of us have little ability to influence. But nothing about efficiency is inevitable. Every line trending upward, every drop in cost, every additional ounce of efficiency we can squeeze from a bundle of inputs is the product of deliberate effort—of thousands of workers, engineers, factory managers, and line supervisors redesigning products, rearranging factories, testing and exploring new ways to do things. The better we understand how efficiency improvement happens, the more fruitful our efforts to improve standards of living will be—letting us build a better world for ourselves, our children, and future generations.

# A
# Notes

## INTRODUCTION

1 Jeffrey S. Sartin, "Infectious Diseases During the Civil War," *Clinical Infectious Diseases* 16, no. 4 (April 1993): 580–84, https://doi.org/10.1093/clind/16.4.580; Peter Neushul, "Science, Government and the Mass Production of Penicillin," *Journal of the History of Medicine and Allied Sciences* 48, no. 4 (1993): 371–95, https://doi.org/10.1093/jhmas/48.4.371.

2 Neushul, "Mass Production of Penicillin"; Robert Gaynes, "The Discovery of Penicillin—New Insights After More Than 75 Years of Clinical Use," *Emerging Infectious Diseases* 23, no. 5 (May 2017): 849–53, https://doi.org/10.3201/eid2305.161556.

3 Gaynes, "Discovery of Penicillin."

4 Christoph Wittmann and James C. Liao, eds., *Industrial Biotechnology: Products and Processes* (John Wiley & Sons, 2017), 16.

5 David Wilson, *In Search of Penicillin* (Knopf, 1976), 158–59.

6 Neushul, "Mass Production of Penicillin."

7 Gaynes, "Discovery of Penicillin"; Michael L. Shuler and Fikret Kargi, *Bioprocess Engineering: Basic Concepts* (Prentice Hall, 2002), 4.

8 Wittmann and Liao, *Industrial Biotechnology*, 17–18; Neushul, "Mass Production of Penicillin"; C.P. van der Beek and J.A. Roels, "Penicillin Production: Biotechnology at Its Best," *Antonie van Leeuwenhoek* 50, no. 5–6 (November 1984): 625–39, https://doi.org/10.1007/BF02386230.

9 Neushul, "Mass Production of Penicillin."

10 Neushul, "Mass Production of Penicillin"; Richard I. Mateles, *Penicillin: A Paradigm for Biotechnology* (Candida Corporation, 1998), 21–22.

11 Neushul, "Mass Production of Penicillin."

12 Richard Felix, "Making Penicillin in Milk Bottles: A Tale of Wartime Antibiotic Manufacture," *The Pharmaceutical Journal* (May 25, 2018), https://pharmaceutical-journal.com/article/opinion/making-penicillin-in-milk-bottles-a-tale-of-wartime-antibiotic-manufacture; Mateles, *Penicillin*, 11.

13 Shuler and Kargi, *Bioprocess Engineering*, 6.

14 Neushul, "Mass Production of Penicillin."

15 Gaynes, "Discovery of Penicillin."

16 Neushul, "Mass Production of Penicillin"; Mateles, *Penicillin*, 22.

17 Neushul, "Mass Production of Penicillin"; Wittmann and Liao, *Industrial Biotechnology*, 17.

18 Gaynes, "Discovery of Penicillin."

19 Neushul, "Mass Production of Penicillin";

van der Beek and Roels, "Penicillin Production."

20 Arthur E. Humphrey and S. Edward Lee, "Industrial Fermentation: Principles, Processes, and Products," in *Riegel's Handbook of Industrial Chemistry*, ed. J.A. Kent (Springer, 1992), 916–86.

21 Neushul, "Mass Production of Penicillin."

22 van der Beek and Roels, "Penicillin Production."

23 Joseph Gottfried, "History Repeating? Avoiding a Return to the Pre-Antibiotic Age" (third year paper, Harvard Law School, 2005).

24 Gottfried, "History Repeating?"; W.A. Adedeji, "The Treasure Called Antibiotics," *Annals of Ibadan Postgraduate Medicine* 14, no. 2 (December 2016): 56–57.

25 Matthew I. Hutchings et al., "Antibiotics: Past, Present, and Future," *Current Opinion in Microbiology* 51 (October 2019): 72–80, https://doi.org/10.1016/j.mib.2019.10.008.

26 Robert C. Allen, "Economic Structure and Agricultural Productivity in Europe, 1300–1800," *European Review of Economic History* 4, no. 1 (April 2000): 1–25, https://doi.org/10.1017/S1361491600000125.

27 Guido Alfani and Cormac Ó Gráda, "The Timing and Causes of Famines in Europe," *Nature Sustainability* 1, no. 6 (June 2018): 283–88, https://doi.org/10.1038/s41893-018-0078-0.

28 Stanley Lebergott, "Labor Force and Employment, 1800–1960," in *Output, Employment, and Productivity in the United States after 1800,* ed. Dorothy S. Brady (NBER, January 1966), 117–204, https://www.nber.org/books-and-chapters/output-employment-and-productivity-united-states-after-1800/labor-force-and-employment-1800-1960.

29 "Ag and Food Sectors and the Economy," USDA Economic Research Service, last updated April 19, 2024, https://www.ers.usda.gov/data-products/ag-and-food-statistics-charting-the-essentials/ag-and-food-sectors-and-the-economy/.

30 USDA, *Crop Production Historical Track Records* (National Agricultural Statistics Service, April 2024), https://downloads.usda.library.cornell.edu/usda-esmis/files/c534fn92g/6t055511z/9593wh83b/croptr24.pdf.

31 USDA, *Crop Production.*

32 Prabhu L. Pingali, "Green Revolution: Impacts, Limits, and the Path Ahead," *PNAS* 109, no. 31 (July 2012): 12302–8, https://doi.org/10.1073/pnas.0912953109.

33 Alex de Waal, "Ending Mass Atrocity and Ending Famine," *The Lancet* 386, no. 10003 (October 2015): 1528–29, https://doi.org/10.1016/S0140-6736(15)00480-8.

34 Susanna Harris, "Flax Fibre: Innovation and Change in the Early Neolithic: A Technological and Material Perspective," in *Textile Society of America 2014 Biennial Symposium Proceedings*, Los Angeles, California, September 10–14, 2014, https://digital-commons.unl.edu/tsaconf/913/; Bernardo Chessa et al., "Revealing the History of Sheep Domestication Using Retrovirus Integrations," *Science* 324, no. 5926 (April 2009): 532–36.

35 Virginia Postrel, *The Fabric of Civilization* (Basic Books, 2020), 48–50.

36 Postrel, *Fabric of Civilization*, 49; "Wool in Ancient Rome," Dulcia's Roman Closet, Elizabeth Brooks, https://sites.google.com/view/dulciasromancloset/roman-textiles/wool-in-ancient-rome.

37 Postrel, *Fabric of Civilization*, 51.

38 Constance H. Berman, "Women's Work in Family, Village, and Town after 1000 CE: Contributions to Economic Growth?," *Journal of Women's History* 19, no. 3 (2007): 10–32, https://doi.org/10.1353/jowh.2007.0052; Postrel, *Fabric of Civilization*, 61.

39 Postrel, *Fabric of Civilization*, 48.

40 Craig Muldrew, "'Th'ancient Distaff' and 'Whirling Spindle': Measuring the Contribution of Spinning to Household Earnings and the National Economy in England, 1550–1770: Spinning and the English Economy," *The Economic History Review* 65, no. 2 (May 2012): 498–526, https://doi.org/10.1111/j.1468-0289.2010.00588.x.

41 Joel Mokyr, ed., *The British Industrial Revolution: An Economic Perspective* (Westview Press, 1999), 168; Gregory Clark, "The British Industrial Revolution, 1760–1860," UC Davis, Spring 2005, https://faculty.econ.ucdavis.edu/faculty/gclark/ecn110b/readings/ecn110b-chapter2-2005.pdf.

42 Clark, "British Industrial Revolution."

43 Andreas W. Engelhardt, "Textiles 2025," *International Fiber Journal*, April 2, 2020, https://www.fiberjournal.com/textiles-2025/.

44 Muldrew, "Spinning and the English Economy."

45 Postrel, *Fabric of Civilization*, 70.

46 Richard Kirwan and Sophie Mullins, eds., *Specialist Markets in the Early Modern Book World* (Brill, 2015), 331–50.

47 Eltjo Buringh and Jan Luiten van Zanden, "Charting the 'Rise of the West': Manuscripts and Printed Books in Europe, A Long-Term Perspective from the Sixth through Eighteenth Centuries," *The Journal of Economic History* 69, no. 02 (June 2009): 409, https://doi.org/10.1017/S0022050709000837; Thomas Kelly, *Early Public Libraries: A History of Public Libraries in Great Britain before 1850* (Library Association, 1966), 16; "Powell's City of Books," Powell's City of Books, Powell's, 2023, https://www.powells.com/locations/powells-city-of-books.

48 Kelly, *Early Public Libraries*, 14.

49 James Moran, *Printing Presses: History and Development from the Fifteenth Century to Modern Times* (Faber, 1973), 19–20.

50 Fernand Braudel, *Civilization and Capitalism, 15th–18th Century, Volume I: The Structures of Everyday Life: The Limits of the Possible* (University of California Press, 1992), 399; Stephan Füssel, *Gutenberg* (Haus Publishing, 2020), 31.

51 Füssel, *Gutenberg*, 38.

52 "200–250 Sheets per Hour Remains the Standard for Hand-Press Output," Jeremy Norman's History of Information, Jeremy M. Norman, accessed September 12, 2024, https://www.historyofinformation.com/detail.php?id=39.

53 Robert Hoe, *A Short History of the Printing Press and of the Improvements in Printing Machinery from the Time of Gutenberg Up to the Present Day* (Robert Hoe, 1902), 17; Füssel, *Gutenberg*, 37.

54 Buringh and van Zanden, "Rise of the West."

55 "Historic Trends in Book Production," AI Impacts, February 7, 2020, Gregory Clark, https://aiimpacts.org/historic-trends-in-book-production/.

56 Gregory Clark, "Lifestyles of the Rich and Famous: Living Costs of the Rich Versus the Poor in England, 1209–1869," UC Davis, August 2004.

57 Buringh and van Zanden, "Rise of the West."

58 For steel, see Bret Devereaux, "Collections: Iron, How Did They Make It? Part II, Trees for Blooms," A Collection of Unmitigated Pedantry, September 25, 2020, https://acoup.blog/2020/09/25/collections-iron-how-did-they-make-it-part-ii-trees-for-blooms/. For shoes, see Laura Quirk and Matthew Beaudoin, "If the Shoe Fits… : Analysis of Nineteenth-Century Shoes," *KEWA* 11, no. 4–6 (2012): 27–32, https://www.academia.edu/1771754/If_The_Shoe_Fits_Analysis_of_Nineteeth_Century_Shoes. For shipping, see Chapter 5 of this book. For watches, see Liza Picard, *Dr. Johnson's London: Coffee-Houses and Climbing Boys, Medicine, Toothpaste and Gin, Poverty and Press-Gangs, Freakshows and Female Education* (St. Martin's Press, 2014). In the late 1700s, a watch cost more than twice a sailor's monthly wages.

59 Forgeng, *Daily Life in Medieval Europe*

(Greenwood Press, 1999), 33.

60 Braudel, *Civilization and Capitalism*, 283–85.

61 Charlie Giattino, Esteban Ortiz-Ospina, and Max Roser, "Working Hours," Our World in Data, January 15, 2024, https://ourworldindata.org/working-hours.

62 "Agricultural Productivity in the United States," USDA Economic Research Service, last updated January 12, 2024, https://www.ers.usda.gov/data-products/agricultural-productivity-in-the-u-s/.

63 Jeremy Greenwood and Gokce Uysal, "New Goods and the Transition to a New Economy," *SSRN Electronic Journal*, October 7, 2003, https://doi.org/10.2139/ssrn.439381.

64 James M. Jasper, *Nuclear Politics: Energy and the State in the United States, Sweden, and France* (Princeton University Press, 1990), 45–49.

65 "How Is Uranium Made into Nuclear Fuel?," World Nuclear Association, 2003, https://world-nuclear.org/nuclear-essentials/how-is-uranium-made-into-nuclear-fuel.

66 "Too Cheap to Meter: A History of the Phrase," US Nuclear Regulatory Commission, last updated September 24, 2021, https://www.nrc.gov/reading-rm/basic-ref/students/history-101/too-cheap-to-meter.html.

67 Jessica R. Lovering, Arthur Yip, and Ted Nordhaus, "Historical Construction Costs of Global Nuclear Power Reactors," *Energy Policy* 91 (April 2016): 371–82, https://doi.org/10.1016/j.enpol.2016.01.011.

68 "Lazard's Levelized Cost of Energy Analysis—Version 15.0," Lazard, October 2021, https://www.lazard.com/media/sptlfats/lazards-levelized-cost-of-energy-version-150-vf.pdf. The World Nuclear Association states that "nuclear power is cost-competitive with other forms of electricity generation, except where there is direct access to low-cost fossil fuels." "Economics of Nuclear Power," World Nuclear Association, last updated September 29, 2023, https://world-nuclear.org/information-library/economic-aspects/economics-of-nuclear-power.aspx.

69 Julian Lincoln Simon, *The Ultimate Resource* (Princeton University Press, 1981), 89; Richard P. Hallion, ed., *The Hypersonic Revolution: Case Studies in the History of Hypersonic Technology, Vol. 2, From Scramjet to the National Aero-Space Plane (1964–1986)* (Air Force History and Museums Program, 1998), 1033.

70 "Cost of Space Launches to Low Earth Orbit," Our World in Data, 2023, https://ourworldindata.org/grapher/cost-space-launches-low-earth-orbit.

71 The electric arc furnace was invented in 1878, but in the early 20th century its costs were "very rarely lower than the open hearth." In 1955, electric-arc steel made up around 7 percent of American steel production compared to 90 percent open-hearth. But by the 1970s, electric arc furnaces were replacing open-hearth furnaces, and in the mid-1990s Clayton Christensen reported that electric-arc-powered mini mills were "the most efficient, lowest-cost steel makers in the world." William T. Hogan, *Economic History of the Iron and Steel Industry in the United States* (Heath, 1971), 414, 418–19, 1520, 1527, 1531; Clayton M. Christensen, *The Innovator's Dilemma: When New Technologies Cause Great Firms to Fail* (Harvard Business School Press, 1997), 80.

72 Lance A. Ealey, *Quality by Design: Taguchi Methods and U.S. Industry* (ASI Press, 1988), 50.

## CHAPTER 1

73 The basic concept behind the light bulb—running electric current through a filament of carbon inside a vacuum—was first suggested in 1838 by Belgian professor Ambroise-Marcellin Jobard, and there were at least 20 unsuccessful attempts to create an incandescent electric light prior to Edison. Robert Friedel, Paul Israel, and Bernard S. Finn, *Edison's Electric Light: The Art of Invention* (Johns Hopkins University Press, 2010), 128; Manning Eyre, "The Incandescent Lamp," *Electrical World* XXV, no. 1 (1895): 9–13.

74 Arthur Aaron Bright, *The Electric-Lamp Industry: Technological Change and Economic Development from 1800 to 1947* (Macmillan, 1949), 209.

75 This and other Edison companies would later merge to become Edison General Electric, which would eventually merge with the Thomson–Houston Company to form General Electric. Maury Klein, *The Power Makers* (Bloomsbury Press, 2008), 292–96; Friedel, Israel, and Finn, *Edison's Electric Light*, 136.

76 Gilbert Scott Ram, *The Incandescent Lamp and Its Manufacture* (Electrician Printing and Publishing Company, 1893), 103–8.

77 Frederick William Hodkin and Arnold Cousen, *A Textbook of Glass Technology* (D. Van Nostrand, 1925), 94. In the Leblanc process, either sodium or potassium salt would be combined with sulfuric acid to produce saltcake, which would then be heated with coal and

limestone in a furnace to produce sodium or potassium carbonate.

78    "Watchmaking," in *The Encyclopedia Americana: A Universal Reference Library Comprising the Arts and Sciences… Commerce, Etc., of the World*, eds. Frederick Converse Beach and George Edwin Rines (Scientific American Compiling Department, 1905); Leon Pratt Alford, *Production Handbook* (Ronald Press Company, 1944), 1339.

79    Thomas Thwaites, *The Toaster Project: Or A Heroic Attempt to Build a Simple Electric Appliance from Scratch* (Chronicle Books, 2012), 21.

80    Leonard E. Read, "I, Pencil," *The Freeman*, December 1958, 32–37.

81    US Tariff Commission, *Incandescent Electric Lamps: A Survey of the Production of Incandescent Electric Lamps and Glass Bulbs in the Principal Producing Countries, Processes of Manufacture, Marketing, and Industrial Agreements, with Special Reference to International Trade and Tariff Considerations* (US Government Printing Office, 1939), 10.

82    Ram, *Incandescent Lamp*, 106.

83    Ram, *Incandescent Lamp*, 107; US Tariff Commission, *Incandescent Electric Lamps*, 10.

84    Friedel, Israel, and Finn, *Edison's Electric Light*, 136.

85    Davis Dyer and Daniel Gross, *The Generations of Corning: The Life and Times of a Global Corporation* (Oxford University Press, 2001), 56.

86    Ram, *Incandescent Lamp*, 105–6.

87    Ram, *Incandescent Lamp*, 106.

88    Ram, *Incandescent Lamp*, 107.

89    Ram, *Incandescent Lamp*, 106.

90    Bright, *Electric-Lamp Industry*, 4; Friedel, Israel, and Finn, *Edison's Electric Light*, 182.

91    Bright, *Electric-Lamp Industry*, 4, 269.

92    American Society of Mechanical Engineers, "The Corning Ribbon Machine for Incandescent Light Bulb Blanks," American Society of Mechanical Engineers, 1983, https://www.asme.org/wwwasmeorg/media/resourcefiles/aboutasme/who%20we%20are/engineering%20history/landmarks/81-corning-ribbon-machine.pdf.

93    James H. Gardiner, ed., *The Chemical News and Journal of Industrial Science* (Chemical News Office, 1921), 166; US Tariff Commission, *Incandescent Electric Lamps*, 10.

94    ASME, "Corning Ribbon Machine."

95    Michelle Padilla, "Running the Ribbon Machine: Stories from the Team," Corning Museum of Glass, January 9, 2018, https://blog.cmog.

org/2018/01/09/running-the-ribbon-machine-stories-from-the-team/.

96    ASME, "Corning Ribbon Machine."

97    ASME, "Corning Ribbon Machine."

98    Bright, *Electric-Lamp Industry*, 130, 204–205.

99    Friedel, Israel, and Finn, *Edison's Electric Light*, 132.

100   Bright, *Electric-Lamp Industry*, 349.

101   Bright, *Electric-Lamp Industry*, 349–50.

102   Bright, *Electric-Lamp Industry*, 350.

103   Bright, *Electric-Lamp Industry*, 351.

104   Bright, *Electric-Lamp Industry*, 269, 363.

## CHAPTER 2

105   Nail rod, in turn, had its own production process. Nail rod was made at a slitting mill, where water-powered blades would cut iron bars or sheets into individual rods. R.F. Tylecote, *A History of Metallurgy* (Institute of Materials, 1992), 105.

106   Amos John Loveday, "The Cut Nail Industry 1776–1890: Technology, Cost Accounting and the Upper Ohio Valley" (PhD diss., Ohio State University, 1979), 17–22.

107   Loveday, "Cut Nail Industry," 22.

108   Loveday, "Cut Nail Industry," 11; Mapelli et al., "Analysis of the Nails from the Roman Legionary at Inchtuthil," *Steel Research International* 79, no. 7 (2008): 569–76, https://doi.org/10.1002/srin.200806168.

109   Daniel E. Sichel, "The Price of Nails Since 1695: A Window into Economic Change," *Journal of Economic Perspectives* 36, no. 1 (February 2022): 125–50.

110   Loveday, "Cut Nail Industry," 23–27, 47.

111   Loveday, "Cut Nail Industry," 24. In Amos Loveday's survey of the cut-nail industry, he describes a mill with 12 to 14 nail machines that produced 20,000 kegs per year, or roughly 420 pounds of nails per machine per day (a keg being 100 pounds). Per economist Daniel Sichel, there were roughly 70 nails to a pound, or about 29,500 nails produced per machine per day. Sichel, "Price of Nails"; Loveday, "Cut Nail Industry," 85.

112   Early cut-nail machines required the iron to be heated before cutting, but by the mid-19th century machines were cutting cold iron. Krysta Ryzewski and Robert Gordon, "Historical Nail-Making Techniques Revealed in Metal Structure," *Historical Metallurgy* 42, no. 1 (2008): 50–64.

113   Sichel, "Price of Nails."

114   J. Bucknall Smith, *A Treatise Upon Wire,*

*Its Manufacture and Uses, Embracing Comprehensive Descriptions of the Constructions and Applications of Wire Ropes* (Offices of Engineering, 1891), 19.

115 Sichel, "Price of Nails."

116 An S-shaped progression shouldn't be particularly surprising. Since technological performance is capped by zero at the bottom and some maximum level at the top (nothing can increase forever), any graph of performance on a wide enough time frame can be fit with an S curve.

117 Early in the development of radio, for example, it was expected to be used primarily for point-to-point communication, and its potential utility for broadcast communication was largely ignored. Nathan Rosenberg, *Studies on Science and the Innovation Process: Selected Works* (World Scientific, 2010), 337.

118 In practice, the initial stages of development, where the technology doesn't work, or doesn't work well enough to be commercialized, are often omitted, and graphs of performance simply show the top portion of the S curve, where the technology is improving and then slowing down.

119 Barrie Charles Blake-Coleman, *Copper Wire and Electrical Conductors: The Shaping of a Technology* (Harwood Academic Publishers, 1992), 127; Bence Jones, *The Royal Institution: Its Founder and Its First Professors* (Longmans, Green, 1871), 278; Henry Schroeder, *History of Electric Light* (Smithsonian Institution, 1923), 5.

120 Schroeder, *History of Electric Light*, 12–14, 17–18.

121 Friedel, Israel, and Finn, *Edison's Electric Light*, 91.

122 Schroeder, *History of Electric Light*, 14.

123 Friedel, Israel, and Finn, *Edison's Electric Light*, 92–93.

124 P.A. Redhead, "History of Vacuum Devices," National Research Council, May 1999, https://cds.cern.ch/record/455984/files/p281.pdf.

125 Friedel, Israel, and Finn, *Edison's Electric Light*, 92.

126 "The Edison Electric Light," *Journal of the Franklin Institute* 109, no. 3 (1880): 145–56.

127 Bright, *Electric-Lamp Industry*, 115–98.

128 Argon wasn't produced in industrial quantities until after WWI. Bright, *Electric-Lamp Industry*, 133.

129 Bright, *Electric-Lamp Industry*, 328.

130 Maxime F. Gendre, "Two Centuries of Electric Light Source Innovations," Eindhoven University of Technology, January 2003, https://www.einlightred.tue.nl/lightsources/history/light_history.pdf.

131 Don Klipstein, "The Great Internet Light Bulb Book, Part I," 1996, https://donklipstein.com/bulb1.html#eff.

132 Michael Scholand, "Max Tech and Beyond: Fluorescent Lamps," US Department of Energy, April 1, 2012, https://doi.org/10.2172/1047754.

133 Lisa Zyga, "White LEDs with Super-High Luminous Efficacy Could Satisfy All General Lighting Needs," Phys.org, August 31, 2010, https://phys.org/news/2010-08-white-super-high-luminous-efficacy.html.

134 Scholand, "Max Tech and Beyond."

135 Kevin Lane et al., "Lighting," IEA, last updated July 11, 2023, https://www.iea.org/reports/lighting.

136 E. Charles Vivian, *A History of Aeronautics* (Harcourt, Brace and Company, 1921); "Powering Up," Wright Brothers Aeroplane Company, accessed January 1, 2024, https://www.wright-brothers.org/History_Wing/History_of_the_Airplane/Century_Before/Powering_up/Powering_up.htm.

137 John Uri, "120 Years Ago: The First Powered Flight at Kitty Hawk," NASA, December 14, 2023, https://www.nasa.gov/history/120-years-ago-the-first-powered-flight-at-kitty-hawk/.

138 "Ernst Heinkel," San Diego Air & Space Museum, accessed September 6, 2024, https://sandiegoairandspace.org/hall-of-fame/honoree/ernst-heinkel.

139 "Grumman F8F-2 Bearcat 'Conquest I,'" Smithsonian National Air and Space Museum, accessed September 6, 2024, https://airandspace.si.edu/collection-objects/grumman-f8f-2-bearcat-conquest-i/nasm_A19770989000.

140 Scott Germain, "World's Fastest Piston-Power Airplane," *Smithsonian Magazine*, August 2018, https://www.smithsonianmag.com/air-space-magazine/worlds-fastest-piston-airplane-180969509.

141 "Setting Records with the SR-71 Blackbird," Smithsonian Air and Space Museum, July 28, 2016, https://airandspace.si.edu/stories/editorial/setting-records-sr-71-blackbird.

142 Rias J. van Wyk et al., "Permanent Magnets: A Technological Analysis," *R&D Management* 21, no. 4 (October 1991): 301–08; Vaclav Smil, "The Vacuum Tube's Power Law," *IEEE Spectrum*, February 2019, 23; Christensen, "Exploring the Limits of the Technology S-Curve, Part I," *Production and Operations Management* 1, no. 4 (1991): 334–57, https://doi.org/10.1111/j.1937-5956.1992.tb00001.x; Ron Adner and Rahul Kapoor, "Innovation Ecosystems and the Pace of Substitution: Re-Examining Technology S-Curves," *Strategic*

Management Journal 37, no. 4 (April 2016): 625–48, https://doi.org/10.1002/smj.2363; Jim Keller, "Moore's Law Is Not Dead," Berkeley EECS, September 18, 2019, video, 64 min., 52 sec., https://www.youtube.com/watch?v=ol-G9ztQw2Gc; Gerald James Holton, Thematic Origins of Scientific Thought: Kepler to Einstein (Harvard University Press, 1975); Roger C. Reed, The Superalloys: Fundamentals and Applications (Cambridge University Press, 2006), 5; Jim Taylor and Abijit Biswas, "Profile of Deep Space Communications Capability," NASA Jet Propulsion Laboratory, August 2015, https://descanso.jpl.nasa.gov/performmetrics/profileDSCC.html; Douglas A. Irwin and Peter J. Klenow, "Learning-by-Doing Spillovers in the Semiconductor Industry," Journal of Political Economy 102, no. 6 (1994): 1200–1227; Redhead, "History of Vacuum Devices."

143   Around 100 BC, the mathematician and engineer Hero of Alexandria described an aeolipile, a metal vessel containing water that would be heated, directing steam through a narrow spout. Such machines were used to open temple doors and perform other small mechanical tasks.

144   The Newcomen engine was predated by Thomas Savery's engine, a steam-powered pump patented in 1698. Richard L. Hills, Power from Steam (Cambridge University Press, 2003), 16, 20.

145   Smil, Energy and Civilization: A History (MIT Press, 2018), 237.

146   There were a variety of other improvements to the Newcomen engine over the 18th century. While the cylinder in the original engine was made from brass, by 1720 it began to be replaced with a cheaper cast-iron cylinder. Operating the engine required opening and closing a series of valves, which was initially done manually, but mechanisms were later added to open and close them automatically. Experimenting with the amount of pressure in the cylinder eventually revealed that a lower vacuum (7 or 8 psi below atmospheric pressure) worked better than the higher vacuum (9.4 psi below atmospheric pressure) Newcomen originally used. These improvements didn't impact thermal efficiency but were important for other performance metrics, such as cost per horsepower. John Paul Murphy, "Energy, Mining, and the Commercial Success of the Newcomen Steam Engine" (PhD diss., Northeastern University, 2012), 168–75; G.N. von Tunzelmann, Steam Power and British Industrialization to 1860, 17–18 (Clarendon Press, 1978); E.A. Forward, "The Early History of the Cylinder Boring Machine," Transactions of the Newcomen Society 5, no. 1 (January 1924): 24–38, https://doi.org/10.1179/tns.1924.004.

147   Matthew A. Carr, "Thermodynamic Analysis of a Newcomen Steam Engine," The International Journal for the History of Engineering & Technology 83, no. 2 (July 2013): 187–208, https://doi.org/10.1179/17581206 13Z.00000000024.

148   Murphy, "Newcomen Steam Engine," 161.

149   Hills, Power from Steam, 59.

150   Hills, Power from Steam, 64.

151   R.U. Ayres, "Technological Transformations and Long Waves," Technological Forecasting and Social Change 37, no. 1 (March 1990): 1–37.

152   Ayres, "Technological Transformations and Long Waves"; John L. Enos, Petroleum Progress and Profits: A History of Process Innovation (MIT Press, 1962), 246–51; Tibor Vaško, ed., The Long-Wave Debate (Springer–Verlag, 1987), 83; David J. Bricknell, Float: Pilkingtons' Glass Revolution (Crucible, 2009); Michael Cook, "The LeBlanc Soda Process: A Gothic Tale for Freshmen Engineers," CEE 32, no. 2 (Spring 1998), https://journals.flvc.org/cee/article/view/123287; Philip Anderson II, "On the Nature of Technological Progress and Industrial Dynamics" (PhD diss., Columbia University, 1988); "Aluminum," in Encyclopedia of Physical Science and Technology, ed. Robert A. Meyers (Academic Press, 2001); Harry N. Holmes, "The Story of Aluminum," Journal of Chemical Education 7, no. 2 (February 1930): 232, https://doi.org/10.1021/ed007p232.

153   Although a metric might also be a summary measure that combines several different factors, such as cost-per-unit output.

154   Gendre, "Electric Light Source Innovations."

155   Later process improvements would blunt this advantage. The eventual replacement of the trip hammer with the Burden rotary squeezer in the production of iron plate changed the direction of the grain in the plate, which allowed the resulting cut nails to be clinched. Loveday, "Cut Nail Industry," 89–90.

156   von Tunzelmann, Steam Power and British Industrialization, 75, 79.

157   Murphy, "Newcomen Steam Engine," 164–65.

158   Robert H. Todd, Dell K. Allen, and Leo Alting, Fundamental Principles of Manufacturing Processes (Industrial Press Inc., 1994), 7–16.

159   Li et al., "Review of Wire Arc Additive Manufacturing for 3D Metal Printing," International Journal of Automation Technology 13, no. 3 (May 2019): 346–53, https://doi.org/10.20965/ijat.2019.p0346. There are, of course, cases in which the old methods do get completely replaced. The

Bessemer process for making steel has been completely replaced by modern methods, for example.

160 Peter Priess, "Wire Nails in North America," *Bulletin of the Association for Preservation Technology* 5, no. 4 (1973): 87, https://doi.org/10.2307/1493446.

161 Christensen, *Innovator's Dilemma*, 63.

162 Christensen, *Innovator's Dilemma*, 63–69.

163 Otto Mayr and Robert C. Post, eds., *Yankee Enterprise: The Rise of the American System of Manufactures*, 63–103; Noble, *Forces of Production* (Alfred A. Knopf, 1984).

164 This directionality doesn't necessarily mean that the development of a new product will follow this pattern. While it's often the case that a company starts with an idea for a product, which gets turned into a prototype, at which point a production line is built to produce it, in some cases the product design comes first. For instance, synthetic fibers were often developed before a use for them was found. In the case of Gore-Tex, the production method of rapidly pulling strands of polytetrafluoroethylene was developed first. More generally, the functionality, product design, and production process will often coevolve, with changes to one resulting in changes to the others. David A. Hounshell and John Kenly Smith Jr., *Science and Corporate Strategy: DuPont R&D, 1902–1980* (Cambridge University Press, 1988); John Unland, *Warm, Dry, and Comfortable: The Story of Developing the Gore-Tex Brand* (John Unland, 2019), 5–20.

165 Douglas Brinkley, *Wheels for the World: Henry Ford, His Company, and a Century of Progress* (Penguin Publishing Group, 2004), 181.

166 *Ford Dealer and Service Field* (Trade Press Publishing Company, 1927).

167 Alfred Sloan, *My Years with General Motors* (Crown, 1990), 235–36; Hounshell and Smith Jr., *Science and Corporate Strategy*, 138–46.

168 Rosenberg, *Studies on Science*, 79.

169 R.E. Webb and W.M. Bruce, "Redesigning the Tomato for Mechanized Production," *Yearbook of Agriculture, 1968: Science for Better Living*, (US Department of Agriculture, 1968), 103–07, https://www.cabidigitallibrary.org/doi/full/10.5555/19691601886

170 Thomas Burnell Colbert, "Iowa Farmers and Mechanical Corn Pickers, 1900–1952," *Agricultural History* 74, no. 2 (2000): 530–44.

171 USDA, "Recent Trends in GE Adoption" USDA Economic Research Service, last updated July 26, 2024, https://www.ers.usda.gov/data-products/adoption-of-genetically-engineered-crops-in-the-u-s/

recent-trends-in-ge-adoption/.

172 Rosenberg, *Studies on Science*, 75.

173 Robert Grieve, *The Cotton Centennial, 1790–1890: Cotton and Its Uses, the Inception and Development of the Cotton Industries of America, and a Full Account of the Pawtucket Cotton Centenary Celebration* (J.A. & R.A. Reid, 1891), 23–27; Brooke Hindle and Stephen Lubar, *Engines of Change: The American Industrial Revolution, 1790–1860* (Smithsonian, 1986), 61.

174 Hounshell and Smith Jr., *Science and Corporate Strategy*, 76–98.

175 Craig S. Galbraith, "Transferring Core Manufacturing Technologies in High-Technology Firms," *California Management Review* 32, no. 4 (July 1990): 56–70.

176 Jeffrey K. Liker, *Becoming Lean: Inside Stories of U.S. Manufacturers* (CRC Press, 1997), 409–57.

177 Richard E. Caves and David R. Barton, *Efficiency in U.S. Manufacturing Industries* (MIT Press, 1990), 2.

178 Gabriel Szulanski, *Sticky Knowledge: Barriers to Knowing in the Firm* (Sage, 2003), 12.

179 Morris Chang, "Can Semiconductor Manufacturing Return to the US?" Brookings Institution, April 19, 2022, video, 42 min., 20 sec., https://www.youtube.com/watch?v=Nw-CWYcag5RE. On a personal note, I once worked for a company that had recently opened a new production facility that was never able to achieve the results of our existing plants. After a few years, the plant closed.

180 G.R. Hall and R.E. Johnson, "Transfers of United States Aerospace Technology to Japan," in *The Technology Factor in International Trade*, ed. Raymond Vernon (NBER, 1970) 325, 333, 337; Lars Bruno and Stig Tenold, "The Basis for South Korea's Ascent in the Shipbuilding Industry, 1970–1990," *The Mariner's Mirror* 97, no. 3 (January 2011): 201–17. Tenold et al., "International Transfer of Tacit Knowledge."

181 Robert J. Thomas, *What Machines Can't Do: Politics and Technology in the Industrial Enterprise* (University of California Press, 1994).

182 Liker, *Becoming Lean*, 290. Note that we don't always see such difficulty. For example, Eli Whitney's cotton gin, which dramatically reduced the amount of effort required to de-seed cotton, was so easily duplicated that Whitney made little to no money from it and he spent much of his life embroiled in patent infringement suits. Hindle and Lubar, *Engines of Change*, 81.

183 Irving Brinton Holley Jr., *Buying Aircraft:*

Matériel Procurement for the Army Air Forces, United States Army in World War II: Special Studies (Center of Military History, United States Army, 1964).

184   John K. Brown, "Design Plans, Working Drawings, National Styles: Engineering Practice in Great Britain and the United States, 1775–1945," *Technology and Culture* 41, no. 2 (2000): 195–238.

185   Hall and Johnson, "United States Aerospace Technology."

186   Pisano, "Learning-before-Doing."

187   John H. Meursinge, "Practical Experience in the Transfer of Technology," *Technology and Culture* 12, no. 3 (1971): 469–70. https://doi.org/10.2307/3103001.

188   Hindle and Lubar, *Engines of Change*, 68–69.

189   Paul A. David, "The Dynamo and the Computer: A Historical Perspective on the Modern Productivity Paradox," *The American Economic Review* 80, no. 2 (May 1990), 355–61.

190   J.Y. Kang et al., "Old Methods Versus New: A Comparison of Very Large Crude Carrier Construction at Scott Lithgow and Hyundai Heavy Industries, 1970–1977," *The Mariner's Mirror 101*, no. 4 (October 2, 2015): 426–57, https://doi.org/10.1080/00253359.2015.1085707.

191   Donald MacKenzie and Graham Spinardi, "Tacit Knowledge, Weapons Design, and the Uninvention of Nuclear Weapons," *American Journal of Sociology* 101, no. 1 (1995): 44–99.

192   David J. Jeremy, *Transatlantic Industrial Revolution: The Diffusion of Textile Technologies Between Britain and America, 1790–1830s* (MIT Press, 1981), 30.

193   Marcie J. Tyre and Eric von Hippel, "The Situated Nature of Adaptive Learning in Organizations," *Organization Science* 8, no. 1 (February 1997): 71–83, https://doi.org/10.1287/orsc.8.1.71.

194   Douglas Michael O'Reagan, "Science, Technology, and Know-How: Exploitation of German Science and the Challenges of Technology Transfer in the Postwar World" (PhD diss., UC Berkeley, 2014).

195   Thomas, *What Machines Can't Do*, 153.

196   Tenold et al., "International Transfer of Tacit Knowledge."

197   Christina Coleman, personal conversation with the author, 2022.

198   D.J. Teece, "Technology Transfer by Multinational Firms: The Resource Cost of Transferring Technological Know-How" *The Economic Journal* 87, no. 346 (June 1977), 242–61, https://doi.org/10.2307/2232084.

199   Thomas J. Misa, *A Nation of Steel* (Johns Hopkins University Press, 1995), 5–15.

200   Bricknell, *Float*, 60–61.

201   Michael S. Malone, *The Big Score: The Billion Dollar Story of Silicon Valley* (Stripe Press, 2021), 234.

202   Chris J. McDonald, "The Evolution of Intel's Copy EXACTLY! Technology Transfer Method," *Intel Technology Journal* Q4 98, 11–17.

203   Jeremy, *Transatlantic Industrial Revolution*, 51–52.

204   Marc Levinson, *The Box: How the Shipping Container Made the World Smaller and the World Economy Bigger* (Princeton University Press, 2016), 101–27; "US Labour Unions Warn against Use of Autonomous Cargo Vessels," *Maritime Gateway*, September 14, 2021, https://www.maritimegateway.com/us-labour-unions-warn-against-use-of-autonomous-cargo-vessels/.

205   Joel Mokyr, *The Lever of Riches: Technological Creativity and Economic Progress.* (Oxford University Press, 1992), 178–79.

206   Galbraith, "Transferring Core Manufacturing Technologies."

207   Szulanski, *Sticky Knowledge*, 29; Liker, Becoming Lean, 423.

208   Kim B. Clark and Robert H. Hayes, "Exploring the Sources of Productivity Differences at the Factory Level," in *The Uneasy Alliance: Managing the Productivity-Technology Dilemma*, eds. Kim B. Clark et al. (Harvard Business School Press, 1985), 168.

209   Christensen, "Limits of the Technology S-Curve."

210   James M. Utterback and Linsu Kim, "Invasion of a Stable Business by Radical Innovation," in *The Management of Productivity and Technology in Manufacturing*, ed. Paul R. Kleindorfer (Springer US, 1985), 113–51.

211   Paul Gipe, *Wind Energy Comes of Age* (Wiley, 1995), 96–114.Veers et al., "Grand Challenges in the Science of Wind Energy," *Science* 366, no. 6464 (October 2019): eaau2027, https://doi.org/10.1126/science.aau2027.

212   Rebecca Henderson, "Of Life Cycles Real and Imaginary: The Unexpectedly Long Old Age of Optical Lithography," *Research Policy* 24, no. 4 (July 1995): 631–43, https://doi.org/10.1016/S0048-7333(94)00790-X.

213   Margaret B.W. Graham and Bettye H. Pruitt, *R&D for Industry: A Century of Technical Innovation at Alcoa* (Cambridge University Press, 1990), 331–79.

214   Ben Thompson, "Intel Unleashed, Gelsinger on Intel, IDM 2.0," Stratechery, March 24, 2021, https://stratechery.com/2021/intel-unleashed-gelsinger-on-intel-idm-2-0/.

215 General Leslie R. Groves, *Now It Can Be Told: The Story of The Manhattan Project* (Hachette Books, 2009), 94.

216 "Operation Warp Speed: Accelerated COVID-19 Vaccine Development Status and Efforts to Address Manufacturing Challenges," US Government Accountability Office, February 2021, https://www.gao.gov/assets/gao-21-319.pdf.

217 Sichel, "Price of Nails," 5.

218 J.P. McGeer, "Hall-Héroult: 100 Years of Processes Evolution," *Journal of Metals* 38 (1986), 27–33, https://doi.org/10.1007/BF03257618.

219 Kevin H. R. Rouwenhorst et al., "Plasma-Driven Catalysis: Green Ammonia Synthesis with Intermittent Electricity," *Green Chemistry* 22, no. 19 (2020): 6258–87, https://doi.org/10.1039/D0GC02058C.

220 Alan Turing's model of a computer that could perform any computation, the Turing machine, was based on a human calculator following a specific set of calculation steps. Jonathon Keats, "How Alan Turing Found Machine Thinking in the Human Mind," *New Scientist*, June 29, 2016, https://www.newscientist.com/article/mg23130803-200-how-alan-turing-found-machine-thinking-in-the-human-mind/.

221 Hyunchul Kim et al., "Redundancy Resolution of a Human Arm for Controlling a Seven DOF Wearable Robotic System," *Annual International Conference of the IEEE Engineering in Medicine and Biology Society* (2011): 3471–74, https://doi.org/10.1109/IEMBS.2011.6090938; Thomas Feix et al., "The GRASP Taxonomy of Human Grasp Types," *IEEE Transactions on Human-Machine Systems* 46, no. 1 (February 2016): 66–77, https://doi.org/10.1109/THMS.2015.2470657.

222 There were, in fact, attempts to build nail machines that used mechanical hammering mechanisms, but these weren't successful.

223 James Bright, *Automation and Management* (Plimpton Press, 1958), 29.

224 "Light Bulb Mass Production Process. Last Incandescent Lamp Factory in Korea." *All Process of World*, April 25, 2022, video, 12 min., 4 sec., https://www.youtube.com/watch?v=v9Psw6_5lHg.

225 Aude Billard and Danica Kragic, "Trends and Challenges in Robot Manipulation," *Science* 364, no. 6446 (June 2019): eaat8414. https://doi.org/10.1126/science.aat8414. Of course, this is changing. As software and sensors evolve, machines are becoming better able to process and respond to an increasing amount of environmental variation and information.

The self-driving car, which is not in widespread use as of this writing but has nevertheless been tested and deployed successfully, is the obvious example of machines' increasing ability to respond to an uncontrolled environment.

226 Billard and Kragic, "Robot Manipulation."

227 Elena Karpova, Grace I. Kunz, and Myrna B. Garner, *Going Global: The Textile and Apparel Industry* (Bloomsbury Publishing USA, 2021), 208.

228 Brian Potter, "Why Did Agriculture Mechanize and Not Construction?" Construction Physics, June 16, 2021, https://www.construction-physics.com/p/why-did-agriculture-mechanize-and.

229 Earle Buckingham, *Production Engineering* (John Wiley & Sons, 1942), 163.

230 Jeremy, Transatlantic Industrial Revolution, 30; Robert B. Gordon, "Who Turned the Mechanical Ideal into Mechanical Reality?" *Technology and Culture* 29, no. 4 (October 1988): 744, https://doi.org/10.2307/3105044; Padilla, "Running the Ribbon Machine."

231 Of course, the technology might also develop in ways that don't directly improve production efficiency, like getting new features or being adapted for use in new environments.

## CHAPTER 3

232 McMaster-Carr, accessed October 12, 2023, https://mcmaster.com/.

233 A simplified example: If a car has 10,000 parts and each part has two possible designs, there are $2^{10000}$, or $10^{3010}$ unique varieties of car, compared to around $10^{82}$ atoms in the universe. And this greatly underestimates the possible varieties of car, as there are many more than two types of part, and they can be combined in many possible ways.

234 von Tunzelmann, *Steam Power and British Industrialization*, 17.

235 Rhys Jenkins, "Ironfounding in England, 1490–1603," *Transactions of the Newcomen Society* 19, no. 1 (January 1938): 35–49, https://doi.org/10.1179/tns.1938.002; A.E. Musson, "Joseph Whitworth and the Growth of Mass-Production Engineering," *Business History* 17, no. 2 (July 1975): 109–49, https://doi.org/10.1080/00076797500000018.

236 Bright, *Electric-Lamp Industry*, 125, 207.

237 Jack M. Walker, ed. *Handbook of Manufacturing Engineering* (Marcel Dekker, 1996): 3.

238 Buckingham, *Principles of Interchangeable*

Manufacturing (Industrial Press, 1921); J.R. Fawcett, *Designing for Mass Production: An Introduction* (Sir Isaac Pitman and Sons Ltd, 1939); Herbert Chase, *Handbook of Designing for Quantity Production* (McGraw-Hill, 1944).

239  American Management Association, *Production Planning and Control for the Mass Production Plant: Mass Production Series 1–7* (Literary Licensing, LLC, 2013).

240  Frank George Woollard, *Principles of Mass and Flow Production* (Mechanical Handling, 1954), 54.

241  Brown, "Design Plans, Working Drawings, National Styles."

242  John Henshaw, *Liberty's Provenance: The Evolution of the Liberty Ship from Its Sunderland Origins* (Pen & Sword Books, 2019), 78–80.

243  Hamid R. Parsaei and William G. Sullivan, eds., *Concurrent Engineering: Contemporary Issues and Modern Design Tools* (Springer US, 1993), 28–37.

244  James Brown, *Value Engineering: A Blueprint* (Industrial Press, 1992), 1–4.

245  Brown, *Value Engineering*, 2–3; *US Army Materiel Command, Inspections and Investigations: Value Analysis* (The Command, 1965).

246  Lawrence D. Miles, *Techniques of Value Analysis and Engineering* (McGraw-Hill, 1961), 2.

247  Jennifer Lloyd, "Instructional Bulletin No. 16-01 Value Engineering Policies and Procedures," State of Tennessee Department of Transportation, February 25, 2016, https://www.tn.gov/content/dam/tn/tdot/documents/InstructionalBulletinNo_16_01.pdf; Georgia Department of Transportation, "Value Engineering FY2022," 2022, https://www.dot.ga.gov/InvestSmart/SSTP/AttachmentD-ValueEngineering-SB200.pdf.

248  Later versions of Bolz's handbook were retitled *Production Processes: The Productivity Handbook*, as the term "producibility" fell out of fashion. Defense Technical Information Center, *Engineering Design Handbook: Design Guidance for Productivity*, US Army Materiel Command, August 1971, https://apps.dtic.mil/sti/tr/pdf/AD0890839.pdf; T.C. Kuo and Hong-Chao Zhang, "Design for Manufacturability and Design for 'X': Concepts, Applications, and Perspectives," in *Seventeenth IEEE/CPMT International Electronics Manufacturing Technology Symposium*, "Manufacturing Technologies—Present and Future" (1995): 446–59, https://doi.org/10.1109/IEMT.1995.526203.

249  Roger William Bolz, *Production Processes:*

*The Productivity Handbook* (Industrial Press, 1981), 1–5.

250  Bolz, *Production Processes*, 1–7.

251  A book of their findings was published in 1968. See Geoffrey Boothroyd and A. H. Redford, *Mechanized Assembly* (McGraw-Hill, 1968). Boothroyd, *Product Deisgn for Manufacturing and Assembly*, 2.

252  Geoffrey Boothroyd, Peter Dewhurst, and Winston A. Knight, *Product Design for Manufacturing and Assembly* (Taylor & Francis, 2011), 4.

253  P.G. Leaney and Gunter Wittenberg, "Design for Assembling: The Evaluation Methods of Hitachi, Boothroyd and Lucas," *Assembly Automation* 12, no. 2 (February 1992): 8–17, https://doi.org/10.1108/eb004359.

254  Anthony Bourdain, *Kitchen Confidential: Adventures in the Culinary Underbelly* (Bloomsbury, 2013), 209.

255  Saul I. Gass, *An Illustrated Guide to Linear Programming* (McGraw-Hill, 1970), 29, 50–56; Mikael Rönnqvist, David Martell, and Andres Weintraub, "Fifty Years of Operational Research in Forestry," *International Transactions in Operational Research* 30, no. 6 (2023): 3296–3328, https://doi.org/10.1111/itor.13316.

256  Jagjit Singh, *Great Ideas of Operations Research* (Dover Publications, 1972), 165–75.

257  SMED trades setup time for meta-setup time: You reduce the time spent on setups by spending time up front figuring out how to make the setups shorter, a process which might take weeks or months. Therefore, the time spent on SMED is only worth it if you plan on needing to do enough setups to amortize the cost. In the limit switch example above, it wouldn't be worth spending the time and effort to install five different switches if the changeover only needed to be performed on rare occasions.

258  Shigeo Shingo, *A Study of the Toyota Production System: From an Industrial Engineering Viewpoint*, trans. Andrew P. Dillon (Productivity Press, 1989), 52.

259  Shigeo Shingo, *A Revolution in Manufacturing: The SMED System*, trans. Andrew P. Dillon (Productivity Press, 1985), xxii.

260  This is also often true in manufacturing, where the initial prototype will be made using different, more expensive methods than the final production version. For example, preproduction car prototypes, which are made before the factory is built, can cost 20 times as much as the factory-produced version.

261  Ian Houson, ed., *Process Understanding: For Scale-Up and Manufacture of Active*

Ingredients (Wiley-VCH, 2011); Gary P. Pisano, "Knowledge, Integration, and the Locus of Learning: An Empirical Analysis of Process Development," *Strategic Management Journal* 15, no. S1 (1994): 85–100, https://doi.org/10.1002/smj.4250150907.

262 A. John Blacker and Mike T. Williams, eds., *Pharmaceutical Process Development: Current Chemical and Engineering Challenges* (Royal Society of Chemistry, 2011),102.

263 Philip Cornwall, Louis J. Diorazio, and Natalie Monks, "Route Design, the Foundation of Successful Chemical Development," *Bioorganic & Medicinal Chemistry* 26, no. 14 (August 2018): 4336–47. https://doi.org/10.1016/j.bmc.2018.06.006.

264 George W. Oliphant, *An Application of Some Industrial Engineering Principles to an Electrical Maintenance Organization* (Union Carbide Nuclear Company, 1957), 47–48.

265 Oliphant, *Electrical Maintenance Organization*, 34.

266 This is obviously related to obtaining cheaper labor or other inputs in different geographies: Do you do so by expanding your firm to the new location or outsourcing that aspect of the business to a local supplier?

267 Loveday, "Cut Nail Industry," 6–8, 117–131.

268 Susan Helper, "Strategy and Irreversibility in Supplier Relations: The Case of the U.S. Automobile Industry," *Business History Review* 65, no. 4 (1991): 781–824, https://doi.org/10.2307/3117265.

269 Kathryn Rudie Harrigan, "Formulating Vertical Integration Strategies," *The Academy of Management Review* 9, no. 4 (October 1984): 638–52, https://doi.org/10.2307/258487.

270 Ronald Coase, "The Nature of the Firm," *Economica* 4, no. 16 (November 1937), 386–405, https://doi.org/10.1111/j.1468-0335.1937.tb00002.x.

271 Dr. L.J. Hart-Smith, "Out-sourced profits—the Cornerstone of Successful Subcontracting," presented at the Boeing Third Annual Technical Excellence (TATE) Symposium St. Louis, Missouri, February 14–15, 2001, https://techrights.org/wp-content/uploads/2022/06/2014130646.pdf; Richard A. Bettis, Stephen P. Bradley, and Gary Hamel, "Outsourcing and Industrial Decline," *Academy of Management Perspectives* 6, no. 1 (February 1992): 7–22, https://doi.org/10.5465/ame.1992.4274298; Dan Wang, "How Technology Grows (a Restatement of Definite Optimism)," July 24, 2018, https://danwang.co/how-technology-grows/.

272 Nicholas Argyres, "Evidence on the Role of Firm Capabilities in Vertical Integration Decisions," *Strategic Management Journal* 17, no. 2 (February 1996): 129–50, https://doi.org/10.1002/(SICI)1097-0266(199602)17:2<129::AID-SMJ798>3.0.CO;2-H.

273 Of course, there are reasons other than cost minimization that may govern the make-or-buy decision, such as getting access to a wider pool of talent. Kenneth L. Deavers, "Outsourcing: A Corporate Competitiveness Strategy, Not a Search for Low Wages," *Journal of Labor Research* 18, no. 4 (December 1997): 503–19, https://doi.org/10.1007/s12122-997-1019-2.

274 Richard N. Langlois and Paul L. Robertson, "Explaining Vertical Integration: Lessons from the American Automobile Industry," *The Journal of Economic History* 49, no. 2 (June 1989): 361–75, https://doi.org/10.1017/S0022050700007993.

275 Helper, "Supplier Relations."

276 Alfred D. Chandler Jr., *Giant Enterprise: Ford, General Motors, and the Automobile Industry: Sources and Readings* (Harcourt, Brace & World, 1964), 10.

277 Langlois and Robertson, "Explaining Vertical Integration," 366.

278 Warren Brown, "America's Auto Industry Geared Up a Century Ago," *Washington Post*, July 9, 1996, https://www.washingtonpost.com/archive/1996/07/10/americas-auto-industry-geared-up-a-century-ago/4cbae9ab-136a-4c70-8890-992a50681ac4/; Chandler Jr., *Giant Enterprise*, 3–4.

279 Langlois and Robertson, "Explaining Vertical Integration."

280 Ford: "We make nothing for the sake of making… If those who sell to us will not manufacture at prices which, upon investigation, we believe to be right, then we make the articles ourselves." Allan Nevins, *Ford: The Times, the Man, the Company* (Scribner, 1954), 203, 230; Harold Katz, "The Decline of Competition in the Automobile Industry, 1920–1940" (PhD diss., Columbia University, 1970), 123–25. ProQuest (7023446).

281 Nevins, *Ford*, 220. This sort of "defensive" vertical integration to ensure steady material flow was common in the early 20th century and practiced by other notable companies, such as Standard Oil and Sears. See Chandler Jr., *Strategy and Structure: Chapters in the History of the Industrial Enterprise* (MIT Press, 1962).

282 Katz, "Competition in the Automobile Industry," 123–24.

283 George J. Brown, "The US Automobile Industry: Will It Survive Increasing International

Competition?" Defense Technical Information Center, April 22, 1991, https://apps.dtic.mil/sti/tr/pdf/ADA237808.pdf.

284  Ford R. Bryan, *Rouge: Pictured in Its Prime* (Wayne State University Press, 2003), 108.

285  Bryan, *Beyond the Model T: The Other Ventures of Henry Ford* (Wayne State University Press, 1997), 118–20.

286  "Ford Model T Original Prices." Mitch Taylor's Ford Model T.net, accessed September 13, 2024, https://www.fordmodelt.net/model-t-ford-prices.htm. As the car market evolved, however, consumers increasingly wanted greater variety, styling, and performance than the Model T could deliver. Over the course of the 1920s, Ford lost market share to other manufacturers, notably General Motors. In 1921, Ford owned 55 percent of the automobile market compared to GM's 12 percent, but by 1925 the gap had closed to 40 percent and 20 percent, respectively. As it lost market share and its production volumes fell, Ford's large investments in vertically integrated production began to look like a curse. The high fixed costs meant that when output fell beyond a certain point, Ford lost money on each car sold. Ford's struggles show that efficiency isn't everything—it's only valuable when coupled to a product that people want to buy. A cheap product that no one wants to buy may end up being incredibly expensive.

287  Katz, "Competition in the Automobile Industry," 130.

288  Katz, "Competition in the Automobile Industry," 130; "Responsible Material Sourcing," Ford, accessed September 13, 2024, https://corporate.ford.com/social-impact/sustainability/responsible-material-sourcing.html.

289  Alfred D. Chandler Jr., *Scale and Scope: The Dynamics of Industrial Capitalism* (Belknap Press, 1990), 207–08.

290  Chandler Jr., *Giant Enterprise*, 3.

291  Chandler Jr., *Scale and Scope*, 208.

292  William J. Abernathy, *The Productivity Dilemma: Roadblock to Innovation in the Automobile Industry* (Johns Hopkins University Press, 1978), 37.

293  Chandler Jr., *Scale and Scope*, 127–36.

294  Richard Preston, *American Steel: Hot Metal Men and the Resurrection of the Rust Belt* (Prentice Hall Press, 1991), 79.

295  Chandler Jr., *The Visible Hand: The Managerial Revolution in American Business* (Belknap Press, 1977).

296  Langlois, "The Vanishing Hand: The Changing Dynamics of Industrial Capitalism," *Industrial and Corporate Change* 12, no. 2, (April 2003): 351–385, https://doi.org/10.1093/icc/12.2.351.

297  Deavers, "Outsourcing."

298  Strictly speaking, it's usually not the case that the inputs in different places are identical and can be swapped out with no other changes. Different labor pools may speak different languages, for example, which means documentation or manuals may need to be translated.

299  Daniel Soyer, ed., *A Coat of Many Colors: Immigration, Globalism, and Reform in the New York City Garment Industry* (Fordham University Press, 2005), 28.

300  Innovation Resource Center, "Evolution of the Clothing Industry in the United States," March 1, 2023, https://web.archive.org/web/20230301133836/https://irc4hr.org/resource/evolution-of-the-clothing-industry-in-the-united-states/.

301  Philip H. Knight, *Shoe Dog: A Memoir by the Creator of Nike* (Scribner, 2016), 285–289, 312, 341–352, 374–375; "Nike Manufacturing Map," Nike, accessed January 6, 2023, https://manufacturingmap.nikeinc.com/.

302  Sarah Monks, *Toy Town: How a Hong Kong Industry Played a Global Game* (PPP Company, 2010), 3–4, 23–24; Stig Tenold, "The Declining Role of Western Europe in Shipping and Shipbuilding, 1900–2000," in *Shipping and Globalization in the Post-War Era: Contexts, Companies, Connections*, ed. Niels P. Petersson, Stig Tenold, and Nicholas J. White (Springer International Publishing, 2019), 9–36.

303  Douglas W. Rae, *City: Urbanism and Its End* (Yale University Press, 2003), 13–21.

304  Binyamin Appelbaum, "American Companies Still Make Aluminum. In Iceland," *The New York Times*, July 1, 2017, https://www.nytimes.com/2017/07/01/us/politics/american-companies-still-make-aluminum-in-iceland.html; Petur Blondal, "Icelandic Aluminium Production Is Nearly 900,000 Tonnes, the Second Biggest in Europe after Norway," interview by ALCircle, April 17, 2023, https://www.alcircle.com/interview/detail/92549/icelandic-aluminium-production-is-nearly-900-000-tonnes-the-second-biggest-in-europe-after-norway-petur-blondal-managing-director-icelandic-associatio; Congressional Research Service, "U.S. Aluminum Manufacturing: Industry Trends and Sustainability," October 26, 2022, https://sgp.fas.org/crs/misc/R47294.pdf. The US does, however, make much more aluminum once recycled aluminum is taken into account.

305  Willard Wesley Cochrane, *The Development of American Agriculture: A Historical Analysis* (University of Minnesota Press, 1993), 184.

306 Peter Temin, *Iron and Steel in Nineteenth-Century America: An Economic Inquiry* (MIT Press, 1964), 20; Smil, *Still the Iron Age: Iron and Steel in the Modern World* (Butterworth-Heinemann, 2016), 10.

307 James McIntyre Camp and Charles Blaine Francis, *The Making, Shaping and Treating of Steel* (Carnegie Steel Company, 120), 200.

308 Rae, *City*, 9, 52–53.

309 "The AWAC Business: Global Business," Alumina Limited, archived October 25, 2006, https://web.archive.org/web/20061025041242/http://www.aluminalimited.com/index.php?s=awac_biz&ss=global&p=global_op#Wagerup.

310 "Where Are Manufactured Homes Built?" Clayton Homes, accessed January 12, 2024, https://web.archive.org/web/20220707164045/https://www.claytonhomes.com/studio/where-clayton-manufactured-homes-are-built/. This occurs on a smaller scale in lean manufacturing optimizations, where equipment is arranged to minimize travel distance.

311 Smil, *Still the Iron Age*, 35, 37.

312 G.W. Josephson, *Iron Blast-Furnace Slag Production, Processing, Properties, and Uses* (US Government Printing Office, 1949), 3.

313 Josephson, *Iron Blast-Furnace Slag Production*, 3.

314 Kenneth C. Curry, "Iron and Steel Slag," *US Geological Survey*, January 2020, https://pubs.usgs.gov/periodicals/mcs2020/mcs2020-iron-steel-slag.pdf.

315 Pierre Desrochers, "Victorian Pioneers of Corporate Sustainability," *Business History Review* 83, no. 4 (2009): 703–29, https://doi.org/10.1017/S000768050000088X.

316 Vernon W. Ruttan, *Technology, Growth, and Development: An Induced Innovation Perspective* (Oxford University Press, 2001), 287–89.

317 Desrochers, "Victorian Pioneers of Corporate Sustainability."

318 Cook, "LeBlanc Soda Process."

319 A.H. Hooker, "Niagara Falls Power and American Industries: The Chemical Industries," Transactions of the American Electrochemical Society, Twenty-Ninth General Meeting, Washington, DC, April 27–29, 1916; Minnesota Legislature House of Representatives, *Journal of the House of Representatives, During the Twenty-Fourth Session of the Legislature of the State of Minnesota,* Vol. 34, 1905.

320 Temin, *Iron and Steel*, 145–52.

321 Chris Lefteri, *Making It: Manufacturing Techniques for Product Design* (Laurence King Publishing, 2019), 42–43, 46–47.

322 Lefteri, *Making It*, 114–21.

323 Boothroyd, *Product Design for Manufacturing and Assembly*, 5–6.

324 Woollard, *Principles of Mass and Flow Production*, 66.

325 Beyond cost minimization, there may also be trade-offs in performance objectives like speed, quality, variety, reliability, and cost. *See* Giovani J.C. da Silveira and Nigel Slack, "Exploring the Trade-Off Concept," *International Journal of Operations & Production Management* 21, no. 7 (July 2001): 949–4, https://www.researchgate.net/publication/235301115_Exploring_the_trade-off_concept.

326 Woollard, *Principles of Mass and Flow Production*, 438.

327 Miles, *Value Analysis and Engineering*, 32–35.

328 Also known as carbon equivalent, a metric consisting of the carbon percentage plus one-third of the silicon percentage plus one-third of the phosphorus percentage.

329 Potter, "The Rise of Steel, Part I," Construction Physics, January 5, 2023, https://www.construction-physics.com/p/the-rise-of-steel-part-i.

330 McDonald, "Evolution of Intel's Copy EXACTLY!"

331 A similar concept is the idea of technology readiness levels, or TRL, in aerospace engineering. The TRL scale identifies nine levels of technological development, starting at level one—observing the basic principles or effects—and ending at level nine—a production version operating in a real environment. John C. Mankins, "Technology Readiness Levels," NASA Office of Space Access and Technology, April 6, 1995, http://www.artemisinnovation.com/images/TRL_White_Paper_2004-Edited.pdf.

332 Potter, "Building Fast and Slow, Part 1: The Empire State Building and the World Trade Center," Construction Physics, November 23, 2022, https://www.construction-physics.com/p/building-fast-and-slow-the-empire.

333 Walter G. Vincenti, *What Engineers Know and How They Know It: Analytical Studies from Aeronautical History* (Johns Hopkins University Press, 1993), 170–99.

334 Vincenti, *What Engineers Know*, 178, 194.

335 This is why machine designs can be patented, just as texts can be copyrighted, even though the design of any machine is, in a sense, latent within the existing library of components.

336 William T. Hogan, *Economic History of the Iron and Steel Industry in the United States* (Heath, 1971), 1520, 1544.

337 Hogan, *Iron and Steel Industry*, 1520, 1527; "Facts About American Steel Sustainability,"

American Iron and Steel Institute, https://www.steel.org/wp-content/uploads/2021/11/AISI_FactSheet_SteelSustainability-11-3-21.pdf.

338 Joseph A. Schumpeter, *Capitalism, Socialism and Democracy* (Routledge, 2013), 83.

339 Ted Cushman, "Southern Pine 2×4 Lumber Downgraded," *The Journal of Light Construction*, June 1, 2012, https://www.jlconline.com/how-to/framing/southern-pine-2x4-lumber-downgraded_o.

340 Potter, "Why Did We Wait So Long for Wind Power?," Construction Physics, September 20, 2022, https://www.construction-physics.com/p/why-did-we-wait-so-long-for-wind.

341 Brandon N. Owens, *The Wind Power Story: A Century of Innovation That Reshaped the Global Energy Landscape* (Wiley, 2019), 191–205.

342 James J. McNerney, Doyne Farmer, and Jessika E. Trancik, "Historical Costs of Coal-Fired Electricity and Implications for the Future," *Energy Policy* 39, no. 6 (June 2011): 3042–54, https://doi.org/10.1016/j.enpol.2011.01.037.

343 Norihiko Shirouzu, "How Toyota Thrives When the Chips Are Down," Reuters, March 19, 2021, https://www.reuters.com/article/us-japan-fukushima-anniversary-toyota-in-idUSKBN2B1005.

344 Sandy Munro and Cory Steuben, "Giga Castings with Sandy: Evolution of Tesla Bodies in White," Munro Live, July 29, 2022, video, 17 min., 53 sec., https://www.youtube.com/watch?v=WNWYk4DdT_E., at 20–2:50.

345 Riz Akhtar, "New Tesla Model 3 Begins Production at Scale with Giant Giga-Casting Tech," *The Driven*, August 14, 2023, https://thedriven.io/2023/08/15/new-tesla-model-3-begins-production-at-scale-with-giant-giga-casting-tech/.

346 Alicia Hartlieb and Martin Hartlieb, "The Impact of Giga-Castings on Car Manufacturing and Aluminum Content," Light Metal Age, July 10, 2023, https://www.lightmetalage.com/news/industry-news/automotive/the-impact-of-giga-castings-on-car-manufacturing-and-aluminum-content/; Dan Carney, "Tesla's Giga Press Die Castings for Model 3 Eliminate 370 Parts," Light Metal Age, July 10, 2023, https://www.lightmetalage.com/news/industry-news/automotive/the-impact-of-giga-castings-on-car-manufacturing-and-aluminum-content/.

347 Sandy Munro, "Why Sandy Is Captivated by Castings: IDRA Conference Recap," Munro Live, November 1, 2023, video, 13 min., 20 sec., https://www.youtube.com/watch?v=q7ExW-ZwuuCl, at 1:39, 4:23, 5:20, 6:00.

348 Hartlieb and Hartlieb, "Impact of Giga-Castings."

349 In HPDC, the molten aluminum is injected rapidly into the mold under high pressure. This is in contrast to other types of casting, such as investment casting, which uses the force of gravity to fill the casting mold.

350 Hartlieb and Hartlieb, "Impact of Giga-Castings."

351 William Andresen, *Die Cast Engineering: A Hydraulic, Thermal, and Mechanical Process* (CRC Press, 2004), 40–45.

352 Sandy Munro, "Tesla Cybertruck Deep Dive with 5 Tesla Executives!" Munro Live, December 11, 2023, video, 57 min., 54 sec., https://www.youtube.com/watch?v=J5zDNaY1fvI.

353 Andresen, *Die Cast Engineering*, 43.

354 Bühler Die Casting, "Mega Casting Considerations for Die Casters," 2022 Die Casting Congress & Tabletop, September 2022, https://www.diecasting.org/archive/transactions/T22-032ppt.pdf; "OL 5500 CS Technical Data," Idra Group, May 28, 2020, https://web.archive.org/web/20200528002904/https://www.idragroup.com/images/pdf/OL5500CSTechDatasheet.pdf; Andresen, *Die Cast Engineering*, 50.

355 Andresen, *Die Cast Engineering*, 50; Idra Group, "OL 550 CS Technical Data."

356 "Giga Press Uncovered: Chapter 3/5—The Technological Challenges, and the Target Audience," Idra, October 18, 2021, video, 4 min., 49 sec., https://www.youtube.com/watch?v=Y-whGS3My9bM; Dan Carney, "Tesla's Giga Press Die Castings for Model 3 Eliminate 370 Parts," DesignNews, August 13, 2021, https://www.designnews.com/automotive-engineering/tesla-s-switch-to-giga-press-die-castings-for-model-3-eliminates-370-parts.

357 Shirouzu, "Tesla Reinvents Carmaking with Quiet Breakthrough," Reuters, September 14, 2023, https://www.reuters.com/technology/gigacasting-20-tesla-reinvents-carmaking-with-quiet-breakthrough-2023-09-14/.

358 Tesla, "Tesla Battery Day."

359 The degree of similarity between the car bodies remains a matter of debate. For example, if a large cast part is damaged in a collision, it's likely uneconomical to repair. While some argue that this represents a clear drawback compared to stamped-steel bodies, others maintain that the amount of force needed to damage a large casting would also damage a stamped-steel body beyond repair. Sandy Munro, "Tesla Bodies in White"; Hartlieb and Hartlieb, "Impact of Giga-Castings."

360 Buhler Die Casting, "Mega-Casting

Considerations."

361 Bühler Die Casting, "Mega-Casting Considerations."

362 Between 1969 and 1974 alone, AT&T spent more than $300 billion in 2023 dollars on building telephone infrastructure. That's more than was spent on NASA's Apollo program. AT&T, American Telephone & Telegraph Annual Reports: 1907–1911, 1913–1977, accessed February 8, 2024, http://archive.org/details/americantelephoneandtelegraphannualreports.

## CHAPTER 4

363 Pig iron can then be used directly by pouring it into molds, becoming cast iron, or it can be further processed into other materials, such as wrought iron or steel.

364 Tylecote, *History of Metallurgy*, 75–99.

365 Michael Arndt, "A Hellish Job: 'I Love It'," *Chicago Tribune*, February 7, 1999, https://www.chicagotribune.com/news/ct-xpm-1999-02-07-9902070360-story.html. A single large blast furnace today makes more than 160 times as much pig iron in a year as all of Britain made in 1720.

366 Chandler Jr., *Visible Hand*, 258–69; Temin, *Iron and Steel*, 284–85.

367 Terence E. Dancy, "The Evolution of Iron Making," *Metallurgical Transactions* B 8, no. 1 (March 1977): 201–13, https://doi.org/10.1007/BF02657648; Temin, *Iron and Steel*, 157–63.

368 Chandler Jr., *Scale and Scope*, 21–27.

369 John Haldi and David Whitcomb, "Economies of Scale in Industrial Plants," *Journal of Political Economy* 75, no. 4 (August 1967): 373–85, https://doi.org/10.1086/259293; Benôit Robidoux and John Lester, "Econometric Estimates of Scale Economies in Canadian Manufacturing," *Applied Economics* 24, no. 1 (January 1992): 113–22, https://doi.org/10.1080/00036849200000109; L. Bruni, "Internal Economies of Scale with a Given Technique," *The Journal of Industrial Economics* 12, no. 3 (1964): 175–90, https://doi.org/10.2307/2097656.

370 Frederick T. Moore, "Economies of Scale: Some Statistical Evidence," *Quarterly Journal of Economics* 73, no. 2 (May 1959): 232–245. https://msuweb.montclair.edu/~lebelp/mooreecsscaleqje1959.pdf.

371 Hogan, *Iron and Steel Industry*, 398.

372 Dudley Jackson, *Profitability, Mechanization and Economies of Scale* (Routledge, 2018), Table 9.2.

373 The cost breakdown for ships has changed over time. Immediately after WWII, many cargo ships in service were surplus ships purchased cheaply from the US government, and the variable costs of operating the vessel made up the majority of the shipping costs. But as container shipping emerged in the 1950s and '60s, shipping companies spent millions to build large ships specially designed to hold containers and the fixed cost of the ship itself began to dominate. These high fixed costs and low variable costs, combined with an oversupply of cargo capacity, incentivized shipowners to reduce their rates to ensure ships sailed as full as possible, resulting in a shipping rate war in the late 1960s. Levinson, *The Box*, 213–214, 222–223.

374 Chandler Jr., *Scale and Scope*, 25–26.

375 Hogan, *Iron and Steel Industry*, 217.

376 "World Container Index," Drewry, accessed February 23, 2023, https://www.drewry.co.uk/supply-chain-advisors/supply-chain-expertise/world-container-index-assessed-by-drewry.

377 Lefteri, *Making It*, 194–96, 244–47.

378 Karen Dawson and Piyush Sabharwall, "A Review of Light Water Reactor Costs and Cost Drivers," USDOE Office of Nuclear Energy, September 1, 2017, https://doi.org/10.2172/1466793.

379 "Electric Generator Dispatch Depends on System Demand and the Relative Cost of Operation," US Energy Information Administration, August 17, 2012, https://www.eia.gov/todayinenergy/detail.php?id=7590.

380 William Murray, "Economies of Scale in Container Ship Costs," United States Merchant Marine Academy, 2016, https://www.usmma.edu/sites/usmma.dot.gov/files/docs/CMA%20Paper%20Murray%201%20%282%29.pdf.

381 M.N. Dastur & Co, "Economies of Scale at Small Integrated Steelworks," United Nations Economic and Social Council, January 1967 https://repositorio.cepal.org/bitstream/handle/11362/29208/S6700235_en.pdf.

382 Nguyen Khoi Tran and Hans-Dietrich Haasis, "An Empirical Study of Fleet Expansion and Growth of Ship Size in Container Liner Shipping," *International Journal of Production Economics* 159 (January 2015): 241–53, https://doi.org/10.1016/j.ijpe.2014.09.016.

383 Jan Owen Jansson and Dan Shneerson, "Economies of Scale of General Cargo Ships," *The Review of Economics and Statistics* 60, no. 2 (1978): 287–93, https://doi.org/10.2307/1924982; Tran and Haasis, "Container Liner Shipping."

384 Jeffrey L. Funk, *Technology Change and the*

*Rise of New Industries* (Stanford University Press, 2013); A. Bejan, A. Almerbati, and S. Lorente, "Economies of Scale: The Physics Basis," *Journal of Applied Physics* 121, no. 4 (January 28, 2017): 044907. https://doi.org/10.1063/1.4974962; Garrett, *Chemical Engineering Economics* (Springer Netherlands, 1989). For purchased goods like steam turbines and chemical processing equipment, part of the cost reduction for higher-capacity equipment likely comes from fixed cost scaling on the supplier's end. The cost of a turbine is a combination of the supplier's fixed costs, such as overhead from making a sale, and variable costs, primarily the costs of constructing the turbine itself. Larger pieces of equipment not only benefit from geometric scaling but also spread the fixed costs of production more thinly, lowering the cost per unit of equipment capacity.

385 Consider a piece of equipment that doubles in size from 10 to 20 units of capacity. Per the scaling exponent equation, the cost ratio is C2/C2 = 20^0.6 / 10^0.6 = 2^0.6, for a cost ratio of ~= 1.51. In other words, doubling the size of the equipment increases costs by roughly 50 percent.

386 Garrett, *Chemical Engineering Economics*, 255–353; author's calculations.

387 For both a sphere and a cube, doubling the volume increases surface area by a factor of roughly 1.59. Using our scaling exponent equation, we get 1.59 ~= 2^x, which yields ~0.67 for the exponent $x$.

388 J. Samuel Walker and Thomas R. Wellock, "A Short History of Nuclear Regulation, 1946–2009," US Nuclear Regulatory Commission, October 2010, https://www.nrc.gov/docs/ML1029/ML102980443.pdf.

389 Devendra Sahal, *Patterns of Technological Innovation* (Addison-Wesley, 1981), 254–255.

390 Walker and Wellock, "History of Nuclear Regulation," 31–37.

391 Robert C. Allen, "Collective Invention," *Journal of Economic Behavior & Innovation*, 4, no. 1 (March 1983): 1–24, https://doi.org/10.1016/0167-2681(83)90023-9.

392 Owens, *Wind Power Story*, 148–154.

393 Klein, *The Power Makers*, 404.

394 Fred H. Colvin, *Sixty Years with Men and Machines: An Autobiography*, collab. D.J. Duffin (McGraw-Hill, 1947), 100–01.

395 T.M. Whitin and M.H. Peston, "Random Variations, Risk, and Returns to Scale," *The Quarterly Journal of Economics* 68, no. 4 (November 1954): 603, https://doi.org/10.2307/1881879.

396 Potter, "Why Are There So Few Economies of Scale in Construction?" Construction Physics, August 24, 2022, https://www.construction-physics.com/p/why-are-there-so-few-economies-of.

397 Toyota, "Supplier Quality Assurance Manual (SQAM)," MinebeaMitsumi Quality Assurance Headquarters, June 26, 2020, https://www.minebeamitsumi.com/english/corp/company/procurements/quality/__icsFiles/afieldfile/2020/12/14/sqam_en.pdf; Apple, "Apple Supplier Code of Conduct," January 1, 2018, Apple-Supplier-Code-of-Conduct-and-Supplier-Responsibility-Standards.pdf

398 Jeffrey M. Loiter and Vicki Norberg-Bohm, "Technology Policy and Renewable Energy: Public Roles in the Development of New Energy Technologies," *Energy Policy* 27, no. 2 (February 1999), 85–97, https://doi.org/10.1016/S0301-4215(99)00013-0; Alan McDonald and Leo Schrattenholzer, "Learning Rates for Energy Technologies," *Energy Policy* 29, no. 4 (March 2001): 255–61, https://doi.org/10.1016/S0301-4215(00)00122-1.

399 David Besanko et al., *Economics of Strategy*, 5th ed. (Wiley, 2010), 48.

400 At one point, funds were so tight that FedEx's founder took the company's last $5,000 to Vegas in the hopes of winning enough money to pay for jet fuel. He succeeded. Roger Frock, *Changing How the World Does Business: FedEx's Incredible Journey to Success: The Inside Story* (Publishers Group West, 2006), 101.

401 Frock, FedEx, 83.

402 Frock, FedEx, 83, 94–95.

403 "AT&T Wants Phase Out of Old Telephone Technologies," Broadband Breakfast, December 31, 2009, https://broadbandbreakfast.com/att-wants-phase-out-of-old-telephone-technologies/; Gregory F. Nemet, *How Solar Energy Became Cheap: A Model for Low-Carbon Innovation* (Taylor & Francis, 2019), 168.

404 J. Shaver and John Mezias, "Diseconomies of Managing in Acquisitions: Evidence from Civil Lawsuits," *Organization Science* 20 (February 2009): 206–22, https://doi.org/10.1287/orsc.1080.0378.

405 Hogan, *Iron and Steel Industry*, 1481–84.

406 "US Statistical Abstract 1957—Mining," US Census Bureau, 1957, 741, https://www2.census.gov/library/publications/1957/compendia/statab/78ed/1957-12.pdf; "US Statistical Abstract 1980—Mining," US Census Bureau, 1980, 764, https://www2.census.gov/library/publications/1980/compendia/statab/101ed/1980-08.pdf.

407  Hogan, *Iron and Steel Industry*, 1481.
408  Candice C. Tuck, "2018 Minerals Yearbook: Iron Ore," US Geological Survey, September 2021, https://pubs.usgs.gov/myb/vol1/2018/myb1-2018-iron-ore.pdf.
409  Paraphrased from Potter, "Why Skyscrapers Are So Short," Works in Progress, January 21, 2022, https://worksinprogress.co/issue/why-skyscrapers-are-so-short/.
410  Potter, "Why Skyscrapers Are So Short."
411  R.A.W. Green and A.A. Leadbeater, "The Reliability of [the] Power Station Plant," in *Steam Plant Operation: A Convention Sponsored by the Verein Deutscher Ingenieure and the Steam Plant Group of the Institution of Mechanical Engineers, 2nd–4th May 1973* (Institution of Mechanical Engineers, 1974), 1–7; G. Jan van Helden and Joan Muysken, "Diseconomies of Scale for Plant Utilisation in Electricity Generation," *Economics Letters* 11, no. 3 (January 1983): 285–89, https://doi.org/10.1016/0165-1765(83)90149-0.
412  Green and Leadbeater, "Power Station Plant," 2.
413  Bela Gold, "Evaluating Scale Economies: The Case of Japanese Blast Furnaces," *The Journal of Industrial Economics* 23, no. 1 (September 1974): 1, https://doi.org/10.2307/2098209.
414  Aubrey Silberston, "Economies of Scale in Theory and Practice." *The Economic Journal* 82, no. 325 (1972): 369–91. https://doi.org/10.2307/2229943.
415  Manfred Fleischer, *The Inefficiency Trap: Strategy Failure in the German Machine Tool Industry* (Ed. Sigma, 1997), 44–50.
416  Silberston, "Economies of Scale"; Robidoux and Lester, "Scale Economies in Canadian Manufacturing"; "Global Aircraft Fleet: Deliveries by Manufacturer 1999-2021," Statista, accessed September 16, 2024. https://www.statista.com/statistics/622779/number-of-jets-delivered-global-aircraft-fleet-by-manufacturer/.
417  Adam Smith, *The Wealth of Nations* (Modern Library, 1937), 4–5. It's worth noting that Smith wasn't necessarily completely accurate in explaining the individual operations of the pin factory and how they were divided. See Jean-Louis Peaucelle and Cameron Guthrie, "How Adam Smith Found Inspiration in French Texts on Pin Making in the Eighteenth Century," *History of Economic Ideas* 19, no. 3 (January 2011): 41–67.
418  Smith, *Wealth of Nations*, 7–8.
419  Charles Babbage, *On the Economy of Machinery and Manufactures* (Charles Knight & Co, 1841), 170–171.
420  Mary Walton, *Car: A Drama of the American Workplace* (W.W. Norton & Company, 1997), xvii–xix.
421  Joseph Wickham Roe and Charles Walter Lytle, *Factory Equipment* (International Textbook Company), 6–9.
422  Sean P. Adams, *Home Fires: How Americans Kept Warm in the 19th Century* (Johns Hopkins University Press, 2014), 23.
423  Howell J. Harris, "Conquering Winter: US Consumers and the Cast-Iron Stove," *Building Research & Information* 36, no. 4 (2008): 337–50, https://doi.org/10.1080/09613210802117411.
424  Adams, *Home Fires*, 27.
425  Ruth Schwartz Cowan, *A Social History of American Technology* (Oxford University Press, 1997), 104.
426  Adams, *Home Fires*, 27.
427  Chauncey Mitchell Depew, *1795–1895: One Hundred Years of American Commerce* (D.O. Haynes & Company, 1895), 361.
428  Temin, *Iron and Steel*, 39. Peter Temin notes that 30 percent of castings were made by manufacturers who primarily made stoves, not necessarily that 30 percent of casting was stoves. So the actual value will be somewhat less depending on what else they were making.
429  Howell J. Harris, "Inventing the U.S. Stove Industry, c. 1815–1875: Making and Selling the First Universal Consumer Durable," *Business History Review* 82, no. 4 (2008): 701–33, https://doi.org/10.1017/S0007680500063170.
430  Adams, *Home Fires*, 35.
431  Harris, "Inventing the U.S. Stove Industry."
432  Harris, "Conquering Winter."
433  Chandler Jr., *Scale and Scope*, 53.
434  Chandler Jr., *Visible Hand*, 240; Chandler Jr., *Scale and Scope*, 62.
435  Levinson, *The Box*, 265.
436  Potter, "How to Build a $20 Billion Semiconductor Fab," Construction Physics, May 3, 2024, https://www.construction-physics.com/p/how-to-build-a-20-billion-semiconductor.
437  Anton Shilov, "AMD Is on Track to Become a Top 3 Fabless Chip Designer," Tom's Hardware, June 10, 2022, https://www.tomshardware.com/news/amd-is-on-track-to-become-a-top-3-fabless-chip-designer; Shilov, "Nvidia Became World's Largest Fabless Chip Designer by Revenue in 2023 Thanks to AI Boom," Tom's Hardware, May 9, 2024, https://www.tomshardware.com/tech-industry/nvidia-became-worlds-largest-fabless-chip-designer-by-revenue-in-2023-thanks-to-ai-boom.
438  Daniel Nenni and Paul Michael McLellan, Fabless: *The Transformation of the*

Semiconductor Industry (SemiWiki.com Project, 2014), 65–69.

439 James P. Womack et al., *The Machine That Changed the World* (Scribner, 1990), 52.

440 Womack et al., *Machine That Changed the World*, 78.

441 Dancy, "Evolution of Iron Making."

442 Gipe, *Wind Energy*, 102.

443 "Wind Turbines: The Bigger, the Better," Office of Energy Efficiency & Renewable Energy, August 24, 2023, https://www.energy.gov/eere/articles/wind-turbines-bigger-better.

444 Walter Musial et al., "Offshore Wind Market Report: 2021 Edition," US Department of Energy, August 2021, https://www.energy.gov/sites/default/files/2021-08/Offshore%20Wind%20Market%20Report%202021%20Edition_Final.pdf., 72.

445 Paul Veers et al., "Grand Challenges in the Science of Wind Energy," *Science* 366, no. 6464 (October 2019): eaau2027, https://doi.org/10.1126/science.aau2027.

446 Veers et al., "Science of Wind Energy."

447 Paraphrased from Potter, "Why Did We Wait so Long for Wind Power?"; "Levelized Cost of Energy by Technology," Our World in Data, accessed January 26, 2024, https://ourworldindata.org/grapher/levelized-cost-of-energy.

448 Different people define the term "bottleneck" in slightly different ways. Some identify the bottleneck as the process in front of which a large amount of work in process accumulates, or the process with an insufficient production rate to keep up with demand. Others define the bottleneck as the step with the highest long-term utilization. Others define it as the process step most likely to block an upstream process and starve a downstream process. Eliyahu M. Goldratt and Jeff Cox, *The Goal: A Process of Ongoing Improvement* (North River Press, 2004), 145–150; Wallace J. Hopp and Marc Spearman, *Factory Physics* (McGraw-Hill, 2008), 231; Jingshan Li, "Continuous Improvement at Toyota Manufacturing Plant: Applications of Production Systems Engineering Methods," *International Journal of Production Research* 51, no. 23–24 (November 2013): 7235–49. https://doi.org/10.1080/00207543.2012.753166.

449 C. Carl Pegels and Craig Watrous, "Application of the Theory of Constraints to a Bottleneck Operation in a Manufacturing Plant," *Journal of Manufacturing Technology Management* 16, no. 3 (April 2005): 302–11, https://doi.org/10.1108/17410380510583617.

450 Kevin Fox, "TOC Stories #2: 'Blue Light' Creating Capacity for Nothing," Theory of Constraints, June 15, 2007, https://theoryofconstraints.blogspot.com/2007/06/toc-stories-2-blue-light-creating.html.

451 Li, "Continuous Improvement."

452 Theory of Constraints Institute. "Theory of Constraints of Eliyahu M. Goldratt." TOC Institute, accessed September 16, 2024. https://www.tocinstitute.org/theory-of-constraints.html.

453 James E. Kelley and Morgan R. Walker, "Critical-Path Planning and Scheduling," *In Papers Presented at the December 1–3, 1959*, Eastern Joint IRE-AIEE-ACM Computer Conference (ACM Press, 1959): 160–73, https://doi.org/10.1145/1460299.1460318.

454 F.K. Levy, G.L. Thompson, and J.D. Wiest, "The ABCs of the Critical Path Method," *Harvard Business Review*, September 1, 1963, https://hbr.org/1963/09/the-abcs-of-the-critical-path-method.

## CHAPTER 5

455 Jeffrey K. Liker, *The Toyota Way: 14 Management Principles from the World's Greatest Manufacturer* (McGraw-Hill, 2004), 27.

456 "Value is not produced at all, and cannot be produced. We never produce anything but forms, shapes of materials, combinations of material, that is to say, things, goods. These goods can of course be goods possessing value, but they do not bring value with them ready made, as something inherent that results from production. They only acquire value from the wants and satisfactions of the economic world." Eugen von Böhm-Bawerk, *Capital and Interest* (Macmillan, 1890), 90.

457 Liker, *Toyota Way*, 88.

458 Liker, *Toyota Way*, 88.

459 Womack and Jones, *Lean Thinking: Banish Waste and Create Wealth in Your Corporation* (Simon & Schuster, 2013), 43.

460 Liker, *Toyota Way*, 30.

461 This distinction between removing value-adding and non-value-adding steps parallels the breakdown of input cost reduction strategies into ones that require product design changes and ones that don't, as discussed in Chapter 3.

462 Maureen K. Phillips, "'Mechanic Geniuses and Duckies,' a Revision of New England's Cut Nail Chronology before 1820," *APT Bulletin: The Journal of Preservation Technology*, 25, no. 3/4 (1993), 4–16, https://doi.org/10.2307/1504461.

463 Merritt Roe Smith, *Harpers Ferry Armory and the New Technology: The Challenge of Change*

(Cornell University Press, 2015), loc 1357.

464 Henry Ford and Samuel Crowther, *My Life and Work* (Doubleday, Page, 1922), 80.

465 Horace B. Drury, *Scientific Management: A History and Criticism* (AMS Press, 1922), 99.

466 Frederick Winslow Taylor, *The Principles of Scientific Management* (Harper & Row, 1911), 25.

467 Hugh Aitken, *Scientific Management in Action: Taylorism at Watertown Arsenal, 1908–1915* (Princeton University Press, 1985), 22.

468 Robert Kanigel, "Taylor-Made," *The Sciences* 32, no. 3 (May–June 1997): 18–23, https://doi.org/10.1002/j.2326-1951.1997.tb03309.x.

469 Frank Bunker Gilbreth, *Primer of Scientific Management* (D. Van Nostrand Company, 1912), 8.

470 Gilbreth, Bricklaying System (M.C. Clark Publishing Company, 1909).

471 Taylor, *Principles of Scientific Management*, 78.

472 Frank Bunker Gilbreth and Lillian Moller Gilbreth, *Applied Motion Study: A Collection of Papers on the Efficient Method to Industrial Preparedness* (Sturgis & Walton, 1917), 42.

473 These symbols would be officially adopted by the American Society of Mechanical Engineers in 1947. See Richard J. Schonberger, "Lean Production—Perspectives from Its Primary Caretaker, Industrial Engineering," in *The Cambridge International Handbook of Lean Production*, ed. Thomas Janoski and Darina Lepadatu (Cambridge University Press, 2021).

474 Lawrence S. Aft, *Productivity Measurement and Improvement* (Reston Publishing Company, 1983), 98–99.

475 Aft, *Work Measurement and Methods Improvement* (Wiley, 2000), 109–12.

476 John Connelly, *Technique of Production Processes* (McGraw-Hill, 1943), 333.

477 Alford, *Production Handbook*, 599.

478 Richard Otto Schmid, "An Analysis of Predetermined Time Systems," (PhD thesis, New Jersey Institute of Technology, 1957).

479 Jefferson Cowie, *Capital Moves: RCA's Seventy-Year Quest for Cheap Labor* (Cornell University Press, 2019), 65-66.

480 Schmid, "Predetermined Time Systems."

481 Gerald Nadler, *Work Simplification* (McGraw-Hill, 1957).

482 James Milo Cox, *Work Simplification: Management's Tool for Economy* (George Washington University, 1963), 7, 26.

483 Nadler, *Work Simplification*, 39–42.

484 David Paul Kirchner, *A Discriminative Study of Navy Work Simplification and Systems Analysis Programs*, (Naval Postgraduate School, 1960).

485 US Department of the Air Force, *Work Simplification: Fundamentals and Techniques for More Effective Use of Manpower, Equipment, Materials, Space* (US Government Printing Office, 2011); US Navy Department, *Work Simplification for Naval Units: Work Distribution, Work Count, Flow Process, Motion Economy, Space Layout* (Personnel Research Division, Bureau of Naval Personnel, 1965).

486 Roe and Lytle, *Factory Equipment*, 467.

487 Miles, *Value Analysis and Engineering*, 33.

488 Gordon M. Ranson, *Group Technology: A Foundation for Better Total Company Observation* (McGraw-Hill, 1972), 60-64.

489 Dan Charnas, *Work Clean: The Life-Changing Power of Mise-en-Place to Organize Your Life, Work, and Mind* (Penguin Books, 2016), 65.

490 More generally, economy of motion is closely tied to the very idea of physical skill: The greater your skill at a task, the fewer motions it takes to accomplish it. A study of arthroscopic surgeons, for example, found that surgeons performed surgery tasks with less movement and fewer motions than nonsurgeons, and experienced surgeons used fewer motions than less experienced surgeons. Nick R. Howells et al., "Motion Analysis: A Validated Method for Showing Skill Levels in Arthroscopy," *The Journal of Arthroscopy and Related Surgery* 24, no. 3 (March 2008): 335–42, https://www.arthroscopyjournal.org/article/S0749-8063(07)00844-4/abstract.

491 "Blickensderfer No. 5 Typewriter," Smithsonian National Museum of American History, https://www.si.edu/object/blickensderfer-no-5-typewriter%3Anmah_849910.

492 Nick Baker, "Why Do We All Use Qwerty Keyboards?" BBC, August 11, 2010, https://www.bbc.co.uk/news/technology-10925456.

493 Yasuhiro Monden, *Toyota Production System: An Integrated Approach to Just-in-Time* (Taylor & Francis, 2012), 3; Liker, *Toyota Way*, 27–29.

494 Liker, *Toyota Way*, 31.

495 Liker, *Becoming Lean*, 195.

496 Lean practitioners are quick to point out that methods like the Toyota production system or lean aren't simply a collection of tools but an entire system of improvement, and the tools must be used together to achieve the greatest gains. Liker, *Becoming Lean*.

497 Shingo, *Toyota Production System*, 81.

498 Jigs and fixtures, for example, which have been used by manufacturers for centuries, can be thought of as poka-yokes. Likewise, go-no-go gauges that prevent mis-sized parts from proceeding through the process were used at

the Springfield Armory in the 1820s. Smith, *Harpers Ferry Armory*, loc 1706.

499 Richard B. Chase and Douglas M. Stewart, *Mistake-Proofing: Designing Errors Out* (Productivity Press, 2010), 37.

500 Chase and Stewart, *Mistake-Proofing*, 30–31.

501 Boothroyd, Dewhurst, and Knight, *Product Design for Manufacture and Assembly*, 10. To minimize the number of parts required, DFMA provides a methodology for finding the theoretical minimum part count—an obvious parallel to breaking down a production process into value-adding and non-value-adding steps and attempting to minimize or remove the non-value-adding ones. A part that moves, must be of a separate material, or must be separate to allow for assembly is required to be a separate part. Parts that don't mezet any of these criteria are candidates for being combined with other parts.

502 Lionel Alexander Bethune Pilkington, "Review Lecture: The Float Glass Process," *Proceedings of the Royal Society* 314, no. 1516 (December 1969), https://doi.org/10.1098/rspa.1969.0212.

503 Pilkington, "Float Glass Process."

504 James M. Utterback, *Mastering the Dynamics of Innovation: How Companies Can Seize Opportunities in the Face of Technological Change* (Harvard Business School Press, 1994), 112–13.

505 David J. Teece, *Managing Intellectual Capital: Organizational, Strategic, and Policy Dimensions* (Oxford University Press, 2000), 237.

506 David A. Hounshell, *From the American System to Mass Production, 1800–1932: The Development of Manufacturing Technology in the United States* (Johns Hopkins University Press, 1984), 49.

507 The removal of these fitting operations was considered a defining feature of mass production methods. "In mass production," wrote Henry Ford, "there are no fitters." Hounshell, *From the American System to Mass Production*, 49.

508 Hounshell, *From the American System to Mass Production*, 49.

509 Joseph Vincent Woodsworth, *American Tool Making and Interchangeable Manufacturing* (N.W. Henley Publishing Company, 1905), 22–23; John Paxton, "Mr. Taylor, Mr. Ford, and the Advent of High-Volume Mass Production: 1900–1912," *Economics & Business Journal: Inquiries and Perspectives* 4, no. 1 (October 2012): 74–90.

510 Roe and Lytle, *Factory Equipment*, 7.

511 Alford, *Production Handbook*, 477.

512 Alford, *Production Handbook*, 477.

513 Tim Dodd, "Starbase Tour with Elon Musk," *Everyday Astronaut*, August 3, 2021, video, 53 min., 16 sec., https://www.youtube.com/watch?v=t705r8ICkRw.

514 Cox, "Work Simplification," 23.

515 Here are Musk's five rules for manufacturing improvement, which should be followed in order: 1) Make your requirements less dumb. 2) Try and delete the part or process. 3) Simplify the design. 4) Reduce cycle time. 5) Automate. Interestingly, the first three rules are about removing unnecessary steps.

516 Utterback, *Dynamics of Innovation*, 118.

517 Shingo, *SMED System*, 315–33.

518 Antoine de Saint-Exupéry, *Wind, Sand and Stars*, trans. Lewis Galantière (Harcourt Brace & Company, 1992), 41.


## CHAPTER 6

517 "Understanding Injection Molding Tolerances," Protolabs, February 7, 2023, https://www.protolabs.com/resources/blog/understanding-injection-molding-tolerances/.

520 "ASTM A6/A6M-21: Standard Specification for General Requirements for Rolled Structural Steel Bars, Plates, Shapes, and Sheet Piling," ASTM International, last updated January 11, 2023, https://www.astm.org/a0006_a0006m-21.html.

521 Cyril Stanley Smith, "The Discovery of Carbon in Steel," *Technology and Culture* 5, no. 2 (1964): 149–75, https://doi.org/10.2307/3101159.

522 Daniel Kedzie, "Skillful Means: Ancient Process Control as Exemplified by the Manufacture of Japanese Sword, Nihonto," 2011, 28–32.

523 Buckingham, *Production Engineering*, 167–68.

524 Ealey, *Quality by Design*, 74–80.

525 Henry R. Neave, *The Deming Dimension* (SPC Press, 1990), 159–160.

526 Ealey, *Quality by Design*, 68.

527 Tylecote, *History of Metallurgy*, 126–29.

528 W.K.V. Gale, "Wrought Iron: A Valediction: Presidential Address," *Transactions of the Newcomen Society* 36, no. 1 (January 1963): 1–11, https://doi.org/10.1179/tns.1963.001; J.R. Harris, "Skills, Coal and British Industry in the Eighteenth Century," *History* 61, no. 202 (June 1976): 167–82, https://doi.org/10.1111/j.1468-229X.1976.tb01337.x.

529 Harris, "Skills, Coal, and British Industry."

530 Joel Mokyr, *The Gifts of Athena: Historical Origins of the Knowledge Economy* (Princeton University Press, 2002), 32.

531  Barbara Whitney Keyser, "Between Science and Craft: The Case of Berthollet Dyeing," *Annals of Science* 47, no. 3 (1990): 213–60, https://philpapers.org/rec/KEYBSA.

532  Harris, "Skills, Coal, and British Industry."

533  Charles Joseph Singer, *A History of Technology*, Vol. 2 (Oxford University Press, 1954), 350.

534  David S. Landes, *The Unbound Prometheus: Technological Change and Industrial Development in Western Europe from 1750 to the Present* (Cambridge University Press, 1969), 251–52; Smil, *Still the Iron Age*, 11.

535  Singer, *History of Technology*, 57; Tylecote, *History of Metallurgy*, 105–6.

536  Brenda J. Buchanan, "Charcoal: 'The Largest Single Variable in the Performance of Black Powder'," *Icon* 14 (2008): 3–29; Thilo Rehren and Marcos Martinón-Torres, *Naturam Ars Imitata: European Brassmaking between Craft and Science* (Routledge, 2008) 170.

537  Mokyr, *Gifts of Athena*, 44–49.

538  Mokyr, *Gifts of Athena*, 38.

539  Taylor, *On the Art of Cutting Metals* (American Society of Mechanical Engineers, 1906), 7.

540  Taylor, *Cutting Metals*, 146; Hugh G.J. Aitken, *Taylorism at Watertown Arsenal: Scientific Management in Action, 1908–1915* (Princeton University Press, 1985), 33.

541  John Joseph Beer, "The Emergence of the German Dye Industry to 1925" (PhD diss., University of Illinois, 1956), 144–45, ProQuest (0019797).

542  Mokyr, *Gifts of Athena*, 56–67.

543  David Cressy, *Saltpeter: The Mother of Gunpowder* (OUP Oxford, 2013), 18.

544  Perazich et al., *Industrial Instruments and Changing Technology* (Works Progress Administration, National Research Project, 1938), xiii.

545  John A. Chaldecott, "Josiah Wedgwood (1730–95)—Scientist," *The British Journal for the History of Science* 8, no. 1 (March 1975): 1–16, https://doi.org/10.1017/S0007087400013674; R.S. Medlock, "The Historical Development of Flow Metering," *Measurement and Control* 19, no. 5 (June 1986): 11–22, https://doi.org/10.1177/002029408601900502.

546  Perazich et al., *Industrial Instruments and Changing Technology*, 2.

547  Mokyr, *Gifts of Athena*, 19.

548  Darrell A. Russel and Gerald G. Williams, "History of Chemical Fertilizer Development," *Soil Science Society of America Journal* 41, no. 2 (March 1977): 260–65, https://doi.org/10.2136/sssaj1977.03615995004100020020x.

549  Harold C. Livesay, *Andrew Carnegie and the Rise of Big Business* (HarperCollins, 1975), 113–14.

550  Camp and Francis, *Steel*.

551  Mokyr, *Gifts of Athena*, 32.

552  Perazich et al., *Industrial Instruments and Changing Technology*, 2.

553  "Heat Treatment of Steel," in *Machinery's Handbook*, 1924, ZiaNet, accessed September 15, 2024, https://www.zianet.com/ebear/metal/heattreat3.html.

554  Perazich et al., *Industrial Instruments and Changing Technology*, 86; *Machinery's Handbook*, 1289.

555  Walter A. Shewhart, *Economic Control of Quality of Manufactured Product* (American Society for Quality Control, 1980), 10–15.

556  This and the following paragraphs draw from my essay "The Science of Production," Construction Physics, March 31, 2022, https://www.construction-physics.com/p/the-science-of-production.

557  *Western Electric Statistical Quality Control Handbook* (Western Electric Co., 1954), 66–72.

558  Neave, *Deming Dimension*, 89–100.

559  David McConnell, *British Smooth-Bore Artillery: A Technological Study to Support Identification, Acquisition, Restoration, Reproduction, and Interpretation of Artillery at National Historic Parks in Canada* (Canadian Government Publishing Centre, 1988), 17.

560  Beer, "German Dye Industry," 7.

561  Perazich et al., *Industrial Instruments and Changing Technology*, 88.

562  Samuel Hollander, *The Sources of Increased Efficiency: A Study of DuPont Rayon Plants* (MIT Press, 1965), 1–150.

563  Peter H. Spitz, *Petrochemicals: The Rise of an Industry* (Wiley, 1988), 96.

564  Gail Cooper, *Air-Conditioning America: Engineers and the Controlled Environment, 1900–1960* (JHU Press, 1998), 29–51.

565  Stuart Bennett, *A History of Control Engineering, 1800–1955* (IET, 1993), 10; Kris De Decker, "Wind Powered Factories: History (and Future) of Industrial Windmills," Low-tech Magazine, October 8, 2009, https://solar.lowtechmagazine.com/2009/10/wind-powered-factories-history-and-future-of-industrial-windmills/.

566  Roger E. Bohn and R. Jaikumar, *From Filing and Fitting to Flexible Manufacturing* (Now, 2005).

567  Automatic control devices had been used in things like water clocks for hundreds of years, but they weren't used for production processes until much later. Otto Mayr, "The Origins of Feedback Control," *Scientific American* 223,

no. 4 (1970): 110–19.

568 Bennett, *History of Control Engineering*, 9–10.

569 Bennett, *History of Control Engineering*, 24.

570 Perazich et al., *Industrial Instruments and Changing Technology*, 74.

571 Paul Truesdell, *National Petroleum News*, February 9, 1927, (McGraw-Hill, 1927), 85–6.

572 Truesdell, *National Petroleum News*, 85–6.

573 Anderson II, "Technological Progress and Industrial Dynamics," 237.

574 Stuart Bennett, "'The Industrial Instrument—Master of Industry, Servant of Management': Automatic Control in the Process Industries, 1900–1940." *Technology and Culture* 32, no. 1 (January 1991): 69.

575 Perazich et al., *Industrial Instruments and Changing Technology*, 5.

576 Briefly, total quality management is based on the idea that quality—meaning products and processes that are free of defects and work reliably—requires the focus of the entire organization to achieve. Six Sigma, broadly speaking, is based on combining the ideas of total quality management with statistical process control.

577 In 1982, when Philip Caldwell, then head of Ford, visited Japan, Toyota president Eiji Toyoda told him, "There is no secret to how we learned to do what we do, Mr. Caldwell. We learned it at the Rouge." David Halberstam, *The Reckoning* (Morrow, 1986), 88; Etsuo Abe, *The Origins of Japanese Industrial Power: Strategy, Institutions, and the Development of Organisational Capability* (F. Cass, 1995), 107–20.

578 Taiichi Ohno, *Toyota Production System—Beyond Large-Scale Production* (CRC Press, 1988), 4.

579 Womack et al., *Machine That Changed the World*, 208.

580 Jon Miller, "I'll Take My Lean with Water, On the Rocks," Gemba Academy, July 28, 2009, https://blog.gembaacademy.com/2009/07/27/ill_take_my_lean_with_water_on_the_rocks/.

581 Ohno, *Toyota Production System*, 123.

582 Ohno, *Toyota Production System*, 17.

583 Joe Clifford, "Toyota Production System—What It All Means," *Toyota UK Magazine*, May 31, 2013, https://mag.toyota.co.uk/toyota-production-system-glossary/.

584 Liker, *Toyota Way*, 33–34.

585 Liker, *Toyota Way*, 223–36.

586 Brad Power, "How Toyota Pulls Improvement from the Front Line," *Harvard Business Review*, June 24, 2011, https://hbr.org/2011/06/how-toyota-pulls-improvement-f.

587 Monden, *Toyota Production System*, 143–59.

588 Womack et al., *Machine That Changed the World*, 144–56.

589 Hopp and Spearman, *Factory Physics*; Katsuki Aoki and Thomas Taro Lennerfors, "The New, Improved Keiretsu," *Harvard Business Review*, September 1, 2013, https://hbr.org/2013/09/the-new-improved-keiretsu.

590 Miles, *Value Analysis and Engineering*, 76.

591 Buckingham, *Principles of Interchangeable Manufacturing*, 35.

592 Bolz, *Production Processes*, 2–12.

593 Potter, "How to Build a $20 Billion Semiconductor Fab."

594 Natalie Wolchover, "Why NASA's James Webb Space Telescope Matters So Much," *Quanta Magazine*, December 3, 2021, https://www.quantamagazine.org/why-nasas-james-webb-space-telescope-matters-so-much-20211203/.

595 Colvin, *Sixty Years*, 198.

596 Misa, *Nation of Steel*, 27.

597 Misa, *Nation of Steel*, 28.

598 Hopp and Spearman, *Factory Physics*, 379–80.

599 Ealey, *Quality by Design*, 124–27.

600 Hopp and Spearman, *Factory Physics*, 363.

601 Hopp and Spearman, *Factory Physics*, 372.

602 Philip B. Crosby, *Quality Is Free: The Art of Making Quality Certain* (New American Library, 1980), 2.

603 Connelly, *Production Processes*, 274–275.

604 Bohn and Jaikumar, *Flexible Manufacturing*, 24.

605 Bohn and Jaikumar, *Flexible Manufacturing*, 52–54.

606 Bohn and Jaikumar, *Flexible Manufacturing*, 74.

607 Samuel K. Moore and David Schneider, "The State of the Transistor in 3 Charts," *IEEE Spectrum*, November 26, 2022, https://spectrum.ieee.org/transistor-density.

608 McDonald, "Evolution of Intel's Copy EXACTLY!"

609 Bernard L. Koff, "Gas Turbine Technology Evolution—A Designer's Perspective," *Journal of Propulsion and Power* 20, no. 4 (July–August 2004), https://doi.org/10.2514/1.4361; Kyra Dempsey, "A Matter of Millimeters: The Story of Qantas Flight 32," *Admiral Cloudberg* (blog), *Medium*, December 9, 2023, https://admiralcloudberg.medium.com/a-matter-of-millimeters-the-story-of-qantas-flight-32-bdaa62dc98e7.

610 Michael Riordan and Lillian Hoddeson, *Crystal Fire: The Birth of the Information Age* (Norton, 1997), 219.

611 Winfred B. Hirschmann, "Profit from the Learning Curve," *Harvard Business Review*, January 1, 1964, https://hbr.org/1964/01/profit-from-the-learning-curve.

612 S. Parekh, V.A. Vinci, and R.J. Strobel, "Improvement of Microbial Strains

and Fermentation Processes," *Applied Microbiology and Biotechnology* 54, no. 3 (September 2000): 287–301, https://doi.org/10.1007/s002530000403.

613 Chin Jian Yang et al., "The Genetic Architecture of Teosinte Catalyzed and Constrained Maize Domestication," *Proceedings of the National Academy of Sciences* 116, no. 12 (March 19, 2019): 5643–52, https://doi.org/10.1073/pnas.1820997116; Michael Blake, *Maize for the Gods: Unearthing the 9,000-Year History of Corn* (University of California Press, 2015); "Genetics of Maize Domestication (1992–2019)," Buckler Lab, accessed September 27, 2022, https://www.maizegenetics.net/copy-of-hybrid-vigor-1.

614 Vinay Ramani, Debabrata Ghosh, and ManMohan S. Sodhi, "Understanding Systemic Disruption from the Covid-19-Induced Semiconductor Shortage for the Auto Industry," *Omega* 113 (December 2022): 102720, https://doi.org/10.1016/j.omega.2022.102720.

615 Shirouzu, "How Toyota Thrives When the Chips Are Down," Reuters, March 9, 2021. https://www.reuters.com/article/us-japan-fukushima-anniversary-toyota-in-idUSKBN2B1005.

616 Shingo, *Toyota Production System*, 90.

617 Atmospheric radiation levels have since declined following the 1963 Test Ban Treaty, reducing the need for low-background steel. Cecil Adams, "Is Steel from Scuttled German Warships Valuable Because It Isn't Contaminated with Radioactivity?" The Straight Dope, December 9, 2010, https://www.straightdope.com/21344067/is-steel-from-scuttled-german-warships-valuable-because-it-isn-t-contaminated-with-radioactivity.

**CHAPTER 7**

618 Mateles, *Penicillin*, 20.

619 Theodore Wright, "Factors Affecting the Cost of Airplanes," *Journal of the Aeronautical Sciences* 3, no. 4 (February 1936): 122–28, https://doi.org/10.2514/8.155.

620 Our World in Data, "Solar (Photovoltaic) Panel Prices." Our World in Data, accessed October 7, 2023, https://ourworldindata.org/grapher/solar-pv-prices.

621 Potter, "How Did Solar Power Get Cheap? Part I," Construction Physics, April 12, 2023, https://www.construction-physics.com/p/how-did-solar-power-get-cheap-part.

622 Abhishek Malhotra and Tobias S. Schmidt, "Accelerating Low-Carbon Innovation," *Joule* 4, no. 11 (November 2020): 2259–67, https://doi.org/10.1016/j.joule.2020.09.004; James McNerney et al., "Role of Design Complexity in Technology Improvement," *PNAS* 108, no. 22 (May 2011): 9008–13, https://doi.org/10.1073/pnas.1017298108.

623 Kazuhiro Mishina, "Learning by New Experiences: Revisiting the Flying Fortress Learning Curve," in *Learning by Doing in Markets, Firms, and Countries*, 145–84 (University of Chicago Press, 1999).

624 Mishina, "Learning by New Experiences."

625 John M. Dutton and Annie Thomas, "Treating Progress Functions as a Managerial Opportunity," *The Academy of Management Review* 9, no. 2 (April 1984): 235, https://doi.org/10.2307/258437; Béla Nagy et al., "Statistical Basis for Predicting Technological Progress," *PLoS ONE* 8, no. 2 (February 2013): e52669, https://doi.org/10.1371/journal.pone.0052669; Boston Consulting Group, *Perspectives on Experience* (The Group, 1968) 70–101.

626 Malhotra and Schmidt, "Accelerating Low-Carbon Innovation"; Brent M. Johnstone, "Improvement Curves: An Early Production Methodology," Lockheed Martin, June 11, 2015, https://www.iceaaonline.com/wp-content/uploads/2015/06/PA03-Presentation-Johnstone-Improvment-Curves.pdf; Hirschmann, "Profit from the Learning Curve"; Sam Korus, "Wright's Law Predicted 109 Years of Autos Gross Margin, and Now Tesla's," ARK Invest, September 4, 2019, https://ark-invest.com/articles/analyst-research/wrights-law-predicts-teslas-gross-margin/.

627 Hirschmann, "Profit from the Learning Curve."

628 Nagy et al., "Predicting Technological Progress," 4–6, 15.

629 James Bessen, *Learning by Doing: The Real Connection Between Innovation, Wages, and Wealth* (Yale University Press, 2015).

630 Gregory Clark, "Why Isn't the Whole World Developed? Lessons from the Cotton Mills," *The Journal of Economic History* 47, no. 1 (March 1987): 141–73, https://www.jstor.org/stable/2121943.

631 Mishina, "Learning by New Experiences."

632 Mishina, "Learning by New Experiences."

633 I.M. Laddon, "Reduction of Man-Hours in Aircraft Production." *Aviation* 42, no. 5 (1943).

634 Boeing's manufacturing practices also illustrate the risks of the single-minded pursuit of cost reduction. The challenges it experienced with the 737 MAX, including two fatal crashes and an in-air accident that occurred when

improperly installed bolts caused a door plug to detach mid-flight, can be traced in part to a culture that pursued low costs and high profits at the expense of the product's safety. Peter Robison, *Flying Blind: The 737 MAX Tragedy and the Fall of Boeing* (Penguin Business, 2021).

635 Tim Worstall, "The Story of Henry Ford's $5 a Day Wages: It's Not What You Think," Forbes, December 10, 2021, https://www.forbes.com/sites/timworstall/2012/03/04/the-story-of-henry-fords-5-a-day-wages-its-not-what-you-think/.

636 William J. Abernathy and Kenneth Wayne, "Limits of the Learning Curve," *Harvard Business Review*, September 1974, https://hbr.org/1974/09/limits-of-the-learning-curve.

637 Linda Argote, Sara L. Beckman, and Dennis Epple, "The Persistence and Transfer of Learning in Industrial Settings," *Management Science* 36, no. 2 (February 1990): 17; Levitt, "Toward an Understanding of Learning by Doing: Evidence from an Automobile Assembly Plant," *Management Science* 36, no. 2 (February 1990): 17; Hollander, Increased Efficiency.

638 Gary P. Pisano, *The Development Factory: Unlocking the Potential of Process Innovation* (Harvard Business School Press, 1997), 12–13.

639 Eric von Hippel and Marcie J. Tyre, "How Learning by Doing Is Done: Problem Identification in Novel Process Equipment," *Research Policy* 24, no. 1 (January 1995): 1–12, https://doi.org/10.1016/0048-7333(93)00747-H.

640 Kim B. Clark, Product Development Performance: Strategy, *Organization, and Management in the World Auto Industry* (Harvard Business School Press, 1991), 192.

641 Christian Terwiesch and Roger E. Bohn, "Learning and Process Improvement during Production Ramp-Up," *International Journal of Production Economics* 70, no. 1 (March 2001): 1–19, https://doi.org/10.1016/S0925-5273(00)00045-1.

642 Thomas, *What Machines Can't Do*, 159.

643 Dorothy Leonard-Barton, *Wellsprings of Knowledge: Building and Sustaining the Sources of Innovation* (Harvard Business School Press, 1995), 11.

644 Hirschmann, "Profit from the Learning Curve."

645 Sahal, *Patterns of Technological Innovation*, 195.

646 Sahal, *Patterns of Technological Innovation*, 237.

647 Sahal, *Patterns of Technological Innovation*, 240.

648 Von Hippel, *The Sources of Innovation* (Oxford University Press, 1988), 174–75.

649 Von Hippel, *Sources of Innovation*, 185.

650 Mishina, "Learning by New Experiences."

651 Gavin Sinclair, Steven Klepper, and Wesley Cohen, "What's Experience Got to Do with It? Sources of Cost Reduction in a Large Specialty Chemicals Producer," *Management Science* 46, no. 1 (January 2000): 28–45, https://doi.org/10.1287/mnsc.46.1.28.15133.

652 Graham and Pruitt, *R&D for Industry*, 75–100.

653 Nagy et al., "Predicting Technological Progress."

654 François Lafond et al., "How Well Do Experience Curves Predict Technological Progress? A Method for Making Distributional Forecasts," *Technological Forecasting and Social Change* 128 (March 2018): 104–17, https://doi.org/10.1016/j.techfore.2017.11.001.

655 Mishina, "Learning by New Experiences"; Peter Thompson, "How Much Did the Liberty Shipbuilders Learn? New Evidence for an Old Case Study," *The Journal of Political Economy* 109, no. 1 (2001): 103–37.

656 François Lafond, Diana Greenwald, and J. Doyne Farmer, "Can Stimulating Demand Drive Costs Down? World War II as a Natural Experiment," *The Journal of Economic History* 84, no. 3 (September 2022): 727–64, https://doi.org/10.1017/S0022050722000249; Matt Clancy, "Learning Curves Are Tough to Use," *New Things Under the Sun*, March 4, 2021, https://www.newthingsunderthesun.com/pub/4xnyepnn/release/9.

657 William Lazonick and Thomas Brush, "The 'Horndal Effect' in Early U.S. Manufacturing," *Explorations in Economic History* 22, no. 1 (January 1985): 53–96, https://doi.org/10.1016/0014-4983(85)90021-X; Bronwyn H. Hall and Nathan Rosenberg, *Handbook of the Economics of Innovation* (Elsevier North Holland, 2010), 430–71; Haim Barkai and David Levhari, "The Impact of Experience on Kibbutz Farming," *The Review of Economics and Statistics* 55, no. 1 (1973): 56–63, https://doi.org/10.2307/1927994.

658 Hollander, *Sources of Increased Efficiency*, 1, 119–20.

659 David Montgomery and George Day, "Diagnosing the Experience Curve," *Journal of Marketing* 47 (April 1983), https://doi.org/10.2307/1251492.

660 Graham and Pruitt, *R&D for Industry*, 75–100.

661 Potter, "The Story of Titanium," Construction Physics, July 7, 2023, https://www.construction-physics.com/p/the-story-of-titanium.

662 Potter, "Story of Titanium."

663 Hirschmann, "Profit from the Learning Curve";

Johnstone, "Improvement Curves"; Louis E. Yelle, "The Learning Curve: Historical Review and Comprehensive Survey," *Decision Sciences* 10, no. 2 (April 1979): 302–28, https://doi.org/10.1111/j.1540-5915.1979.tb00026.x.

664 Johnstone, "Improvement Curves"; Yelle, "Learning Curve"; Boston Consulting Group, *Perspectives on Experience.*

665 This observation is reminiscent of Mar's Law of spacecraft design: Everything is linear if plotted log-log with a fat magic marker.

666 Our World in Data, "Solar (Photovoltaic) Panel Prices."

667 Levitt, "Learning by Doing."

668 Similar models have been studied elsewhere. See McNerney et al., "Design Complexity"; Philip Auerswald et al., "The Production Recipes Approach to Modeling Technological Innovation: An Application to Learning by Doing," *Journal of Economic Dynamics and Control* 24, no. 3 (March 2000): 389–450, https://doi.org/10.1016/S0165-1889(98)00091-8.

669 Malhotra and Schmidt, "Accelerating Low-Carbon Innovation."

670 In this case, the cost of home construction isn't dominated by framing costs per se, likely because many other trades are similarly tightly coupled.

671 Thomas Hughes, *Networks of Power: Electrification in Western Society, 1880–1930* (JHU Press, 1993), 14.

672 G. Nadler and W.D. Smith, "Manufacturing Progress Functions for Types of Processes," *International Journal of Production Research* 2, no. 2 (January 1963): 115–35, https://doi.org/10.1080/00207546308947818.

673 Harold Asher, "Cost-Quantity Relationships in the Airframe Industry," The Rand Corporation, July 1, 1956, https://www.rand.org/content/dam/rand/pubs/reports/2007/R291.pdf. Interestingly, if we consider the learning curve to be the sum of several learning curves for individual operations (machining, assembly, and so on) that each have different slopes, we will not get a straight line on a log-log plot but a slightly curved one. This also suggests that the traditional learning curve formulation is something of a simplification.

674 Lenovo, "Lenovo Manufacturing Sites and Suppliers," Lenovo, 2020, https://static.lenovo.com/ww/docs/sustainability/list-of-lenovo-mfg-sites-and-suppliers-27oct.pdf.

675 Nicholas Baloff, "Startup Management," *IEEE Transactions on Engineering Management* EM-17, no. 4 (1970): 132–41, https://doi.org/10.1109/TEM.1970.6448538.

676 Baloff, "Startup Management."

677 Levitt, "Learning by Doing."

678 American Institute of Industrial Engineers, *The Journal of Industrial Engineering* 12, no. 1 (February 1961), http://archive.org/details/sim_journal-of-industrial-engineering_january-february-1961_12_1.

679 Baloff, "Startup Management."

680 McDonald, "Evolution of Intel's Copy EXACTLY!"

681 Baloff, "Startup Management."

682 Roberta Pellegrino et al., "Construction of Multi-Storey Concrete Structures in Italy: Patterns of Productivity and Learning Curves," *Construction Management and Economics* 30, no. 2 (February 2012): 103–15, https://doi.org/10.1080/01446193.2012.660776.

683 Hirschmann, "Profit from the Learning Curve."

684 C. Lanier Benkard, "Learning and Forgetting: The Dynamics of Aircraft Production," *American Economic Review* 90, no. 4 (September 2000): 1034–5,. https://doi.org/10.1257/aer.90.4.1034.; Argote, Beckman, and Epple, "Learning in Industrial Settings."

685 Potter, "Why Does Nuclear Power Plant Construction Cost So Much?" Institute for Progress, May 1, 2023, https://progress.institute/nuclear-power-plant-construction-costs/.

686 Liker, *Toyota Way*, 142.

687 Potter, "Why Are Nuclear Power Construction Costs So High?" Construction Physics, July 1, 2022, https://www.construction-physics.com/p/why-are-nuclear-power-construction-c3c.

688 John F. Schank, *Learning from Experience Vol III* (RAND National Defense Research Institute, 2011), 39–43.

689 Boston Consulting Group, *Perspectives on Experience.*

690 Miguel Ángel Reguero, "An Economic Study of the Military Airframe Industry" (PhD diss., New York University, 1957), 234–235, https://babel.hathitrust.org/cgi/pt?id=uc1.b3289166&seq=260&view=1up.

691 McNerney, Doyne Farmer, and Trancik, "Costs of Coal-Fired Electricity."

CHAPTER 8

692 Norman Beasley, *Knudsen: A Biography* (McGraw-Hill, 1947), 66.

693 Hounshell, *From the American System to Mass Production*, 248.

694 Brinkley, *Wheels for the World*, 153.

695 Hounshell, *From the American System to Mass Production*, 237.

696 Hounshell, *From the American System to Mass Production*, 250.

697 Misa, *Nation of Steel*, 247; Kennteth Warren, *A Century of American Steel: The Strip Mill and the Transformation of an Industry* (Rowman & Littlefield, 2019), 2.

698 There are, of course, ways that removing a step can increase the probability of failure, such as cutting out an inspection step that frequently finds errors.

699 Cornwall, Diorazio, and Monks, "Route Design."

700 C.H. Stapper et al., "Evolution and Accomplishments of VLSI Yield Management at IBM," *IBM Journal of Research and Development* 26, no. 5 (September 1982): 532–45, https://doi.org/10.1147/rd.265.0532.

701 Anderson II, "Technological Progress and Industrial Dynamics," 237, 239.

702 Leonard-Barton, *Wellsprings of Knowledge*, 230.

703 Luis Deflorez, "Automatic Control and the Chemical Industries," *Industrial & Engineering Chemistry* 29, no. 11 (November 1937): 1210–13, https://doi.org/10.1021/ie50335a003.

704 The need for additional machining steps in metal casting and forging has spurred the development of near-net-shape production methods, which produce parts very close to their final dimensions and require fewer subsequent operations. Daniele Marini, David Cunningham, and Jonathan R. Corney, "Near Net Shape Manufacturing of Metal: A Review of Approaches and Their Evolutions," *Proceedings of the Institution of Mechanical Engineers, Part B: Journal of Engineering Manufacture* 232, no. 4 (March 2018): 650–69, https://doi.org/10.1177/0954405417708220.

705 Liker, *Becoming Lean*, 329–30.

706 As we saw in Chapters 2 and 7, a similar effect can take place early in the development of a technology, where fixing one problem tends to reveal more problems.

707 Boothroyd and Redford, *Mechanized Assembly*, 172.

708 Bennett, *History of Control Engineering*, 13.

709 Bennett, *History of Control Engineering*, 13; Allan G. Bogue, "Changes in Mechanical and Plant Technology: The Corn Belt, 1910–1940," *The Journal of Economic History* 43, no. 1 (March 1983): 1–25, https://doi.org/10.1017/S0022050700028953.

710 Nevins, *Ford*, 337.

711 Brinkley, *Wheels for the World*, 87.

712 Brinkley, *Wheels for the World*, 87; Nevins, *Ford*, 350; Misa, *Nation of Steel*, 224.

713 Nevins, *Ford*, 325–27, 368.

714 Hounshell, *From the American System to Mass Production*, 221. It's worth noting that Ford probably advertised this level of uniformity before actually achieving it.

715 Nevins, *Ford: The Times*, 324–326.

716 Nevins, *Ford*, 335; Hounshell, *From the American System to Mass Production*, 222.

717 Nevins, *Ford*, 391; Bruce W. McCalley, *Model T Ford: The Car That Changed the World* (Krause Publications, 1994), 23.

718 Brinkley, *Wheels for the World*, 107.

719 Beasley, *Knudsen*, 40–41.

720 Hounshell, *From the American System to Mass Production*, 233.

721 Brinkley, *Wheels for the World*, 103.

722 Nevins, *Ford*, 351.

723 Hounshell, *From the American System to Mass Production*, 229.

724 Brinkley, *Wheels for the World*, 121.

725 McCalley, *Model T Ford*, 23.

726 Hounshell, *From the American System to Mass Production*, 225.

727 McCalley, *Model T Ford*, 29; Beasley, *Knudsen*, 28.

728 McCalley, *Model T Ford*, 43.

729 Brinkley, *Wheels for the World*, 110.

730 Nevins, *Ford*, 394–97, 410.

731 Nevins, *Ford*, 501.

732 Nevins, *Ford*, 333, 452.

733 Nevins, *Ford*, 453–454; Hounshell, *From the American System to Mass Production*, 226.

734 Hounshell, *From the American System to Mass Production*, 228; Charles E. Sorensen, *My Forty Years with Ford* (Wayne State University Press, 2006), 127.

735 Nevins, *Ford*, 364, 454.

736 Hounshell, *From the American System to Mass Production*, 227; Nevins, *Ford*, 464.

737 Nevins, *Ford*, 464.

738 Nevins, *Ford*, 463.

739 Hounshell, *From the American System to Mass Production*, 229.

740 Nevins, *Ford*, 466. This is also an early example of SMED-style setup reduction techniques.

741 McCalley, *Model T Ford*, 23, 33, 109, 149–52, 428.

742 Nevins, *Ford*, 475.

743 Hounshell, *From the American System to Mass Production*, 247.

744 Nevins, *Ford*, 472.

745 Hounshell, *From the American System to Mass Production*, 248.

746 Hounshell, *From the American System to Mass Production*, 248–49, 254–56.

747 Nevins, *Ford*, 507. It should be noted that while the assembly line greatly increased productivity, it also created much worse working

conditions. Keith Sward notes that "so great was labor's distaste for the new machine system that toward the close of 1913 every time the company wanted to add 100 men to its factory personnel, it was necessary to hire 963." Ford's famous five-dollar day, which was more than double the previous minimum wage, was implemented in part to combat worker turnover, and was only possible due to the increased productivity and projected enormous production volumes resulting in greater economies of scale. Hounshell, *From the American System to Mass Production*, 257–58; Sorensen, *Forty Years with Ford*, 138–39.

748  Nevins, *Ford*, 505.

749  Nevins, *Ford*, 506; McCalley, *Model T Ford*, 146.

750  Nevins, *Ford*, 507; Ford and Crowther, *My Life and Work*, 87–88.

751  McCalley, *Model T Ford*, 157.

752  Nevins, *Ford*, 509.

753  McCalley, *Model T Ford*, 149.

754  Ford and Crowther, *My Life and Work*, 85.

755  Beasley, *Knudsen*, 73.

756  Dan Hutcheson, "Graphic: Transistor Production Has Reached Astronomical Scales," *IEEE Spectrum*, April 2, 2015, https://spectrum.ieee.org/transistor-production-has-reached-astronomical-scales.

## CHAPTER 9

757  Judith A. McGaw, *Most Wonderful Machine: Mechanization and Social Change in Berkshire Paper Making, 1801–1885* (Princeton University Press, 1987), 41–43.

758  McGaw, *Most Wonderful Machine*, 96.

759  McGaw, *Most Wonderful Machine*, 103–105.

760  McGaw, *Most Wonderful Machine*, 108, 174.

761  McGaw, *Most Wonderful Machine*, 100; Avi J. Cohen, "Technological Change as Historical Process: The Case of the U.S. Pulp and Paper Industry, 1915–1940," *The Journal of Economic History* 44, no. 3 (September 1984): 775–99, https://doi.org/10.1017/S0022050700032368.

762  "Rockies Express Pipeline," Wikipedia, accessed August 13, 2023. https://en.wikipedia.org/w/index.php?title=Rockies_Express_Pipeline&oldid=1170124176.

763  Chevron, "Safely and Efficiently Transporting Natural Gas," Chevron, accessed September 9, 2024, https://www.chevron.com/what-we-do/energy/oil-and-natural-gas/liquefied-natural-gas-lng.

764  Richard J. Schonberger, *Japanese Manufacturing Techniques: Nine Hidden Lessons in Simplicity* (Free Press, 1982), 121.

765  Schonberger, *Japanese Manufacturing Techniques*, 122.

766  Though, as we discussed in Chapter 4, these might be false economies of scale if the costs of extra inventory outweigh the energy savings, if the larger furnace is being used to produce parts that aren't needed yet, or if setup times can be reduced.

767  "Job Shop," Inc, *February* 6, 2020, https://www.inc.com/encyclopedia/job-shop.html.

768  The batch-versus-continuous distinction is only one of many dimensions that determine the overall degree of production flexibility. Consider automation equipment as an example. One method of automation is to use flexible, multipurpose equipment (such as six-axis robot arms or CNC routers) that have been programmed to complete a specific task. This sort of automation allows manufacturers to change what they produce by reprogramming the equipment or changing out the end effectors on the tools. But if production volumes are high enough, this may justify the development of "hard automation," or equipment designed to perform the specific steps of one specific process. Hard automation can produce high volumes at low unit costs but can't easily produce different products.

769  Alan Heaton, ed., *An Introduction to Industrial Chemistry* (Springer Netherlands, 1996), 9.

770  "Coffee Roasting: Roasting Equipment," *Coffee Review*, accessed July 20, 2023. https://www.coffeereview.com/coffee-reference/from-crop-to-cup/professional-coffee-roasting/roasting-equipment/.

771  Christophe Lécuyer and David C. Brock, *Makers of the Microchip: A Documentary History of Fairchild Semiconductor* (MIT Press, 2010), 20.

772  "JBT Fruit and Vegetables Processing—ReadyGo Juice," *JBT*, February 5, 2020, video, 2 min., 12 sec. https://www.youtube.com/watch?v=3B6DEHmsUGQ.

773  Misa, *Nation of Steel*, 28.

774  R.L. Storch, *Ship Production* (Cornell Maritime Press, 1995), 46–47.

775  Potter, "Lessons from Shipbuilding Productivity, Part II," Construction Physics, March 10, 2022, https://www.construction-physics.com/p/lessons-from-shipbuilding-productivity-4b9.

776  John Storck, *Flour for Man's Bread* (The University of Minnesota Press, 1952), 161.

777  Paolo Gaiardelli, "'Faster, Better, Cheaper' in the History of Manufacturing: From the Stone Age to Lean Manufacturing and Beyond,"

*Production Planning & Control* 29, no. 2 (January 2018): 184–184, https://doi.org/10.1080/09537287.2017.1368429.

778 Chandler Jr., *Visible Hand*, 55.

779 John Storck, *Flour for Man's Bread*, 160; Chandler Jr., *Visible Hand*, 55; Gaiardelli, "Faster, Better, Cheaper," 248.

780 Loveday, "Cut Nail Industry," 36–40.

781 Steven Lubar, "Culture and Technological Design in the 19th-Century Pin Industry: John Howe and the Howe Manufacturing Company," *Technology and Culture* 28, no. 2 (April 1987): 253, https://doi.org/10.2307/3105567.

782 Hubert Dunell, *British Wire-Drawing and Wire-Working Machinery* (Constable Limited, 1925), 36; N.K. Laman, "The Development of the Wire-Drawing Industry," *Metallurgist* 3, no. 6 (1959): 267–70, https://doi.org/10.1007/BF00740175; Joseph Mcfadden, "From Invention to Monopoly: The History of the Consolidation of the Barbed Wire Industry" (1968), 1873–1899; Earl W. Hayter, "Barbed Wire Fencing—A Prairie Invention," *Agricultural History* 13, no. 4 (October 1939): 189–207, https://www.jstor.org/stable/3739686. Here we can see how the distinction between "automatic machinery" and "continuous process" can be somewhat fuzzy. A sufficiently advanced automatic machine can blur the boundaries between separate operations by combining previously distinct steps into a single machine.

783 Chandler Jr., *Scale and Scope*, 26.

784 Chandler Jr., *Visible Hand*, 249–50.

785 Chandler Jr., *Visible Hand*, 250; Betsy H. Bradely, *The Works: The Industrial Architecture of the United States* (Oxford University Press, 1999), 28.

786 "Automatic Screw Making Machine," Smithsonian National Museum of American History, accessed July 21, 2023, https://americanhistory.si.edu/collections/search/object/nmah_1203377.

787 Chandler Jr., *Visible Hand*, 253; Gaiardelli, "Faster, Better, Cheaper," 252–53.

788 Chandler Jr., *Visible Hand*, 296–97; Gaiardelli, "Faster, Better, Cheaper," 253.

789 Cook, "LeBlanc Soda Process"; L.E. Scriven, "On the Emergence and Evolution of Chemical Engineering," in *Advances in Chemical Engineering*, 16 (1991): 3–40, https://doi.org/10.1016/S0065-2377(08)60141-6; George Stephanopoulos and Gintaras V. Reklaitis, "Process Systems Engineering: From Solvay to Modern Bio- and Nanotechnology: A History of Development, Successes and Prospects for the Future," *Chemical Engineering Science*, 66, no. 19 (October 2011): 4272–4306, https://doi.org/10.1016/j.ces.2011.05.049.

790 Wikipedia, for instance, gives the following definition of mass production: "Mass production, also known as flow production or continuous production, is the production of substantial amounts of standardized products in a constant flow, including and especially on assembly lines." "Mass Production," Wikipedia, last updated May 18, 2024, https://en.wikipedia.org/wiki/Mass_production.

791 John Herman Lorant, *The Role of Capital-Improving Innovations in American Manufacturing During the 1920s* (Arno Press, 1975), 187.

792 Enos, *Petroleum Progress and Profits*, 52, 61.

793 Lorant, *Capital-Improving Innovations*, 187; Warren, *American Steel*, 68.

794 Graham and Pruitt, *R&D for Industry*, 14; N.D.C. Hodges, "Aluminium and Its Manufacture by the Deville-Castner Process," *Science* 13, no. 322 (1889): 260–65.

795 John Jewkes, *The Sources of Invention* (Springer, 1969), 241–43.

796 Sau L. Lee et al., "Modernizing Pharmaceutical Manufacturing: From Batch to Continuous Production," *Journal of Pharmaceutical Innovation* 10, no. 3 (September 2015): 191–99, https://doi.org/10.1007/s12247-015-9215-8; András Domokos et al., "Integrated Continuous Pharmaceutical Technologies—A Review," *Organic Process Research & Development* 25, no. 4 (April 16, 2021): 721–39, https://doi.org/10.1021/acs.oprd.0c00504; Adam Zamecnik, "Continuous Manufacturing Builds on Hype but Adoption Remains Gradual," *Pharmaceutical Technology*, May 20, 2022, https://www.pharmaceutical-technology.com/features/continuous-manufacturing-builds-on-hype-but-adoption-remains-gradual/.

797 Chandler Jr., *Visible Hand*, 249; ASME, "Corning Ribbon Machine"; Warren, *American Steel*, 71.

798 Scott Horsley, "How the U.S. Got into This Baby Formula Mess," NPR, May 19, 2022. https://www.npr.org/2022/05/19/1099748064/baby-infant-formula-shortages

799 "Slow Steaming in Container Shipping," Port Economics, Management and Policy, February 10, 2020, https://porteconomicsmanagement.org/pemp/contents/part1/ports-and-container-shipping/slow-steaming-container-shipping/.

800 Hounshell, *From the American System to Mass Production*, 11; David E. Nye, *America's Assembly Line* (MIT Press, 2013), 8.

801 Nye, *America's Assembly Line*, 8.

802  Chandler Jr., *Scale and Scope*, 62.

## CHAPTER 10

803  "Rolex Price Evolution," Minus 4 Plus 6, accessed August 16, 2023, https://www.minus4plus6.com/PriceEvolution.php. It's worth noting that producers still have incentives to reduce their production costs even if they maintain or raise their sales prices, since those savings would simply become higher profits rather than being passed on to the consumer.

804  James B. Bailey, "Can Health Spending Be Reined in Through Supply Constraints? An Evaluation of Certificate-of-Need Laws," *SSRN Electronic Journal*, 2018, https://doi.org/10.2139/ssrn.3211647; Nick Thomas, "Georgia Bill to Replace Certificate of Need Advances," Becker's Hospital CFO Report, February 22, 2023, https://www.beckershospitalreview.com/finance/georgia-bill-to-replace-certificate-of-need-advances.html; Matthew D. Mitchell, "Do Certificate-of-Need Laws Limit Spending?" Mercatus Working Paper, Mercatus Center at George Mason University, Arlington, VA, September 2016, https://doi.org/10.2139/ssrn.2871325.

805  Somi Seong et al., *Titanium: Industrial Base, Price Trends, and Technology Initiatives* (RAND Corporation, 2009), 13–14.

806  Alwyn Scott, "Boeing Looks at Pricey Titanium in Bid to Stem 787 Losses," Reuters, July 24, 2015, https://www.reuters.com/article/us-boeing-787-titanium-insight-idUSKCN-0PY1PL20150724.

807  Seong et al., *Titanium*, 9–10.

808  Potter, "The Story of Titanium," Construction Physics, July 7, 2023, https://www.construction-physics.com/p/the-story-of-titanium.

809  "Residential Code, 2015 (IRC 2015)," UpCodes, accessed February 29, 2024, https://up.codes/code/international-residential-code-irc-2015.

810  Of course, even performance-based regulations are more restrictive than no regulations at all and will often represent a large fixed cost that is best overcome by large production volume.

811  Hartlieb and Hartlieb, "Impact of Giga-Castings."

812  Scott Breneman, Matt Timmers, and Dennis Richardson, "Tall Wood Buildings in the 2021 IBC," WoodWorks, 2022, https://www.woodworks.org/wp-content/uploads/wood_solution_paper-tall-wood.pdf.

813  T.W. Leslie, "'Dry and Ready in Half the Time': Gypsum Wallboard's Uneasy History," in *History of Construction Cultures*, João Mascarenhas-Mateus, Ana Paula Pires, Manuel Marques Caiado, and Ivo Veiga, eds., 682–87 (CRC Press, 2021); Potter, "How Building Codes Work in the US," Construction Physics, July 29, 2022, https://www.construction-physics.com/p/how-building-codes-work-in-the-us; J.K. Dineen, "San Francisco, Trade Unions at Odds over Modular Construction—Even for Homeless Projects," *San Francisco Chronicle*, November 27, 2020. https://www.sfchronicle.com/bayarea/article/San-Francisco-trade-unions-at-odds-over-modular-15755264.php.

814  Levinson, *The Box*, 101–26, 203–4; Bob Woods, "A West Coast Port Worker Union Is Fighting Robots. The Stakes for the Supply Chain Are High," CNBC, July 23, 2022, https://www.cnbc.com/2022/07/23/a-west-coast-port-worker-union-is-fighting-robots-the-stakes-are-high.html.

815  "How ASML Builds a $150 Million EUV Machine," Asianometry, May 24, 2021, video, 14 min., 51 sec. https://www.youtube.com/watch?v=jJIO7aRXUCg; Katie Tarasov, "ASML Is the Only Company Making the $200 Million Machines Needed to Print Every Advanced Microchip. Here's an Inside Look," CNBC, March 23, 2022, https://www.cnbc.com/2022/03/23/inside-asml-the-company-advanced-chipmakers-use-for-euv-lithography.html

816  Walton, *Car*, 180.

817  John Perlin, *From Space to Earth: The Story of Solar Electricity* (Aatec Publications, 1999).

818  Potter, "How Did Solar Power Get Cheap? Part I," Construction Physics, April 12, 2023, https://www.construction-physics.com/p/how-did-solar-power-get-cheap-part; Potter, "Why Did We Wait so Long for Wind Power?" Construction Physics, September 20, 2022, https://www.construction-physics.com/p/why-did-we-wait-so-long-for-wind.

819  Kylie Conrad and John D. Graham, "The Benefits and Costs of Automotive Regulations for Low-Income Americans," *Journal of Benefit-Cost Analysis* 12, no. 3 (2021): 518–49, https://doi.org/10.1017/bca.2021.12.

820  McNerney, Doyne Farmer, and Trancik, "Costs of Coal-Fired Electricity."

821  W. Mark Crain and Nicole V. Crain, "The Cost of Federal Regulation to the US Economy, Manufacturing and Small Business," National Association of Manufacturers, September 10, 2014, https://www.nam.org/wp-content/

uploads/2019/05/Federal-Regulation-Full-Study.pdf. Hopefully it's obvious that although regulations can be a burden in the sense that they increase production costs, the greater safety they ensure for consumers is a highly desirable trade-off in many cases.

822  Michael Maiello, "Diagnosing William Baumol's Cost Disease," Chicago Booth Review, May 18, 2017, https://www.chicagobooth.edu/review/diagnosing-william-baumols-cost-disease.

823  "Big Dave Bohannon: Operative Builder by the California Method," Fortune, April 1946.

824  "Characteristics of New Housing: Current Data," US Census Bureau, 2023, https://www.census.gov/construction/chars/current.html.

825  Potter, "On Klein on Construction," Construction Physics, February 18, 2023, https://www.construction-physics.com/p/on-klein-on-construction.

826  "Construction Cost Indices." Statistics Norway, accessed August 23, 2023, https://www.ssb.no/en/priser-og-prisindekser/byg-gekostnadsindekser; Construction Costs 1910–2023," Statistics Sweden, accessed August 23, 2023, https://www.scb.se/en/find-ing-statistics/statistics-by-subject-area/prices-and-consumption/building-price-in-dex-and-construction-cost-index-for-bu/construction-cost-index-for-buildings-cci-input-price-index/pong/tables-and-graphs/construction-costs/; "ABEX Index," ABEX, accessed August 23, 2023, https://www.abex.be/fr/indice-abex/; "Construction Producer Price and Construction Cost Indices Overview," Eurostat, July 2023, https://ec.europa.eu/eurostat/statistics-explained/index.php?ti-tle=Construction_producer_price_and_con-struction_cost_indices_overview; US Census Bureau, "New Housing."

827  Richard Pray, ed., 2022 National Construction Estimator (Craftsman Book Company, 2022), 306.

828  Charlsy Panzino, "GM, Ford, Fiat Chrysler to See Increased Hourly Labor Costs through 2023," S&P Global, January 15, 2020, https://www.spglobal.com/mar-ketintelligence/en/news-insights/trend-ing/Fic7Dwvvxuh14hs9rmjwJw2; "Average New-Vehicle Prices Up 2% Year-Over-Year in July 2020, According to Kelley Blue Book," PR Newswire, August 3, 2020, https://www.prnewswire.com/news-releases/average-new-vehicle-prices-up-2-year-over-year-in-july-2020-according-to-kelley-blue-book-301104310.html.

829  For certain types of construction, such as park-ing garages, prefab can be competitive on price.

830  Danushka Nanayakkara-Skillington, "Modular and Other Non-Site Built Housing in 2021," NAHB Eye on Housing, September 8, 2022. https://eyeonhousing.org/2022/09/modular-and-other-non-site-built-housing-in-2021/; US Census Bureau, "New Housing."

831  Steve Kerch, "Big Builder Getting It Together Again," Chicago Tribune, July 26, 1987, https://www.chicagotribune.com/news/ct-xpm-1987-07-26-8702240616-story.html; Davies, The Prefabricated Home (Reaktion Books, 2005), 53–54; Hounshell, From the American System to Mass Production, 310–15; "Stirling Homex Corporation," Wikipedia, accessed September 17, 2021. https://en.wikipedia.org/w/index.php?title=Stirling_Homex_Corp oration&ol-did=1044836378.; Thomas T. Fetters, The Lustron Home The History of a Postwar Prefabricated Housing Experiment (McFarland, 2015); Potter, "Another Day in Katerradise," Construction Physics, June 29, 2021, https://www.construction-physics.com/p/another-day-in-katerradise.

832  Potter, "Operation Breakthrough: America's Failed Government Program to Industrialize Home Production," Construction Physics, February 17, 2021, https://www.construction-physics.com/p/operation-breakthrough-americas-failed.

833  Kerch, "Big Builder"; Ned Eichler, The Merchant Builders (MIT Press, 1982), 81.

834  Stacey Freed, "The Nordic Track," Offsite Builder, July 15, 2022, https://offsitebuilder.com/the-nordic-track/.

835  Statistics Sweden, "Newly Produced Dwellings"; US Census Bureau, "New Housing." Swedish homes may be better than US homes in other ways, such as being more energy efficient.

836  Ohno, Toyota Production System.

837  "Toyota Housing Corporation," Toyota, accessed October 12, 2023, https://www.toy-ota-global.com/company/history_of_toyota/75years/data/business/housing/toyotahome.html.

838  US Census Bureau, "New Housing"; John Murphy, "Toyota Throws More Weight Behind Its Homes Unit," Wall Street Journal, July 2, 2008, https://www.wsj.com/articles/SB121496449430221935.

839  2022 Manufactured Housing Facts (Manufactured Housing Institute, 2022).

840  Christopher Herbert, Chadwick Reed, and James Shen, "Comparison of the Costs of Manufactured and Site-Built Housing," Joint Center for Housing Studies, Harvard University, 2023, https://www.jchs.harvard.

edu/sites/default/files/research/files/harvard_jchs_pew_report_1_updated_0.pdf. Despite the cost advantage, manufactured homes only make up a small fraction of US housing construction, on the order of 10 percent. This is in part due to political limitations like zoning rules that govern where manufactured homes can be placed and financing rules that govern mortgage lending.

841 Recall that the dollar density of assembled cars is low enough that it was worth it for Ford to build a series of local assembly plants close to the point of sale. The dollar density of buildings is even lower.

842 There are some cases where prefabrication can exceed these transportation limits while still reducing costs. This most often happens when the factory is located in a low-cost labor area and the building is shipped to a high-cost labor area. The modular construction company Autovol touts its cost savings, made possible because it builds its homes using low-cost labor in a factory in Utah and ships them several hundred miles to higher-cost locations in California. Potter, "Autovol, Stack Modular, and Labor Arbitrage," Construction Physics, April 6, 2021, https://www.construction-physics.com/p/autovol-stack-modular-and-labor-arbitrage.

843 US Census Bureau and US Department of Housing and Urban Development, "New Privately-Owned Housing Units Started: Single-Family Units," FRED, Federal Reserve Bank of St. Louis, January 1, 1959, https://fred.stlouisfed.org/series/HOUST1F; Mathilde Carlier, "Light Vehicle Retail Sales in the United States from 1976 to 2023," Statista, last updated February 28, 2024, https://www.statista.com/statistics/199983/us-vehicle-sales-since-1951/; "North America Smartphone Shipment down 6% in Q2 2022 as Demand Dampens," Canalys, August 22, 2022, https://www.canalys.com/newsroom/north-america-smartphone-market-share-Q2-2022.

844 Tom Randall and Demetrios Pogkas, "Tesla Now Runs the Most Productive Auto Factory in America," Bloomberg, January 24, 2022, https://www.bloomberg.com/graphics/2022-tesla-factory-california-texas-car-production/?embedded-checkout=true

845 US Census Bureau, "Building Permits Survey," US Census Bureau, accessed October 12, 2023. https://www.census.gov/construction/bps/about_the_surveys/index.html.

846 Potter, "Comparing Process Improvement in Manufacturing and Construction: Duco vs. Drywall," Construction Physics, November 11, 2022, https://www.construction-physics.com/p/comparing-process-improvement-in.

847 NAHB Research Center, "Steel vs. Wood Cost Comparison: Beaufort Demonstration Homes," US Department of Housing and Urban Development, January 2002, https://www.huduser.gov/portal/Publications/pdf/steel_vs_wood1.pdf.

848 Potter, "The Rise and Fall of the Manufactured Home, Part II," Construction Physics, July 22, 2022, https://www.construction-physics.com/p/the-rise-and-fall-of-the-manufactured; Nanayakkara-Skillington, "Modular and Other Non-Site Built Housing."

849 Frederick H. Abernathy, Bigger Isn't Necessarily Better: Lessons from the Harvard Home Builder Study, (Lexington Books, 2012).

850 Potter, "Where Are My Damn Learning Curves?" Construction Physics, December 1, 2021, https://www.construction-physics.com/p/where-are-my-damn-learning-curves.

851 Potter, "Why It's Hard to Innovate in Construction," Construction Physics, June 30, 2021, https://www.construction-physics.com/p/why-its-hard-to-innovate-in-construction.

852 Potter, "How Valuable Are Building Methods That Use Fewer Materials?" Construction Physics, February 6, 2024, https://www.construction-physics.com/p/how-valuable-are-building-methods; Potter, "An Overview of Concrete Forming Technology," Construction Physics, November 3, 2022, https://www.construction-physics.com/p/an-overview-of-concrete-forming-technology.

853 Potter, "Where Are the Robotic Bricklayers?" Construction Physics, July 29, 2021, https://www.construction-physics.com/p/where-are-the-robotic-bricklayers.

854 Potter, "Does Construction Ever Get Cheaper?" Construction Physics, February 1, 2023, https://www.construction-physics.com/p/does-construction-ever-get-cheaper.

855 There are some cases where this natural limit is beneficial. For example, the proliferation of nuclear weapons has been hindered by the fact that producing enriched uranium or plutonium suitable for bombs is difficult and expensive.

856 Joseph A. DiMasi, Henry G. Grabowski, and Ronald W. Hansen, "Innovation in the Pharmaceutical Industry: New Estimates of R&D Costs," Journal of Health Economics 47 (May 2016): 20–33, https://doi.org/10.1016/j.jhealeco.2016.01.012.

857 Eric Helland and Alex Tabarrok, Why Are the Prices So Damn High? Health, Education, and

the *Baumol Effect* (Mercatus Center, 2019); Tabarrok, "Physician and Nurse Incomes Have Increased Tremendously," Marginal Revolution, May 29, 2019, https://marginalrevolution. com/marginalrevolution/2019/05/physician-and-nurse-incomes-have-increased-tremendously.html.

## CONCLUSION

858  L.D. Chirillo and R.D. Chirillo, "The History of Modern Shipbuilding Methods: The U.S.-Japan Interchange," *Journal of Ship Production* 1, no. 01 (February 1, 1985): 1–6, https://doi. org/10.5957/jsp.1985.1.1.1

859  Moran, *Printing Presses*, 173–75.

860  A printing press, in turn, is much more flexible than something like plastic injection molding. By rearranging the movable type, a rotary press can produce any sort of text. In contrast, changing what an injection molding process produces requires laboriously machining a new mold.

861  Though modern presses typically use a technology known as offset lithography, which forms the image to be printed on an aluminum plate covered with a thin layer of polymer.

862  A. Gasparetto and L. Scalera, "A Brief History of Industrial Robotics in the 20th Century," *Advances in Historical Studies* 8, no. 1 (January 2019): 24–35, https://doi.org/10.4236/ ahs.2019.81002.

863  Robert S. Woodbury, *Studies in the History of Machine Tools* (MIT Press, 1972), 17.

864  "The Haas VF-2—Small VMC Workhorse," Haas Automation, Inc., September 21, 2021, video, 3 min., 6 sec., https://www.youtube.com/ watch?v=LN_LYo84Oko.

865  There are a few robotic welding systems that can use a combination of part designs and computer vision to plan a weld sequence, but as of this writing they are a small niche, and most automated welding is human-programmed.

866  Potter, "Welding and the Automation Frontier," Construction Physics, November 22, 2023, https://www.construction-physics.com/p/ welding-and-the-automation-frontier.

867  Heath Raftery, "Plyprinting, for Fun and Profit," EmpiricalEE, March 4, 2023, https://www. empirical.ee/plyprinting/.

868  Stefan H. Thomke, *Experimentation Matters: Unlocking the Potential of New Technologies for Innovation* (Harvard Business Press, 2003), 30–36.

869  Mujeeb Malik and Dennis Bushnell, eds., "Role of Computational Fluid Dynamics and Wind Tunnels in Aeronautics R&D," NASA, September 2012, https://ntrs.nasa. gov/api/citations/20120016316/downloads/20120016316.pdf.

870  P.R. Spalart and V. Venkatakrishnan, "On the Role and Challenges of CFD in the Aerospace Industry," *The Aeronautical Journal* 120, no. 1223 (January 2016): 209–32, https://doi. org/10.1017/aer.2015.10.

871  Spalart and Venkatakrishnan, "CFD in the Aerospace Industry."

872  Manuel Ansede, "Four Mathematicians Demonstrate It Is Impossible to Predict Where 29,000 Rubber Ducks in the Sea Will Wash Up," *El País*, May 12, 2021, https://english.elpais. com/usa/2021-05-12/four-mathematicians-demonstrate-it-is-impossible-to-predict-where-rubber-ducks-in-the-sea-will-wash-up. html.

873  Clay Mathematics Institute, "Navier-Stokes Equation," accessed March 3, 2024. https://www.claymath.org/millennium/ navier-stokes-equation/.

874  Derek Lowe, "AI and Drug Discovery: Attacking the Right Problems," *Science*, March 19, 2021, https://www.science.org/content/blog-post/ ai-and-drug-discovery-attacking-right-problems.

# B

# Bibliography

# A

Abe, Etsuo. *The Origins of Japanese Industrial Power: Strategy, Institutions, and the Development of Organisational Capability*. F. Cass, 1995.

Abernathy, Frederick H. *Bigger Isn't Necessarily Better: Lessons from the Harvard Home Builder Study*. Lexington Books, 2012.

Abernathy, William J. *The Productivity Dilemma: Roadblock to Innovation in the Automobile Industry*. Johns Hopkins University Press, 1978.

Abernathy, William J., and Kenneth Wayne. "Limits of the Learning Curve." *Harvard Business Review*, September 1974. https://hbr.org/1974/09/limits-of-the-learning-curve.

ABEX. "ABEX Index." Accessed August 23, 2023. https://www.abex.be/fr/indice-abex/.

Adams, Cecil. "Is Steel from Scuttled German Warships Valuable Because It Isn't Contaminated with Radioactivity?" The Straight Dope, December 9, 2010. https://www.straightdope.com/21344067/is-steel-from-scuttled-german-warships-valuable-because-it-isn-t-contaminated-with-radioactivity.

Adams, Sean P. *Home Fires: How Americans Kept Warm in the Nineteenth Century*. Johns Hopkins University Press, 2014.

Adedeji, W.A. "The Treasure Called Antibiotics." *Annals of Ibadan Postgraduate Medicine* 14, no. 2 (December 2016): 56–57. https://pmc.ncbi.nlm.nih.gov/articles/PMC5354621/.

Adner, Ron, and Rahul Kapoor. "Innovation Ecosystems and the Pace of Substitution: Re-Examining Technology S-Curves." *Strategic Management Journal* 37, no. 4 (April 2016): 625–48. https://doi.org/10.1002/smj.2363.

Aft, Lawrence S. *Productivity Measurement and Improvement*. Reston Publishing Company, 1983.

Aft, Lawrence S. *Work Measurement and Methods Improvement*. Wiley, 2000.

AI Impacts. "Historic Trends in Book Production." February 7, 2020. https://aiimpacts.org/historic-trends-in-book-production/.

AI Impacts. "Historic Trends in Flight Airspeed Records." February 7, 2020. https://aiimpacts.org/historic-trends-in-flight-airspeed-records/.

Aitken, Hugh. *Scientific Management in Action: Taylorism at Watertown Arsenal, 1908–1915*. Princeton University Press, 1985.

Akhtar, Riz. "New Tesla Model 3 Begins Production at Scale with Giant Giga-Casting Tech." The Driven, August 14, 2023. https://thedriven.io/2023/08/15/new-tesla-model-3-begins-production-at-scale-with-giant-giga-casting-tech/.

Alfani, Guido, and Cormac Ó Gráda. "The Timing and Causes of Famines in Europe." *Nature Sustainability* 1, no. 6 (June 2018): 283–88. https://doi.org/10.1038/s41893-018-0078-0.

Alford, Leon Pratt. *Production Handbook*. Ronald Press Company, 1944.

Allen, Robert C. "Collective Invention." *Journal of Economic Behavior & Organization* 4, no. 1 (March 1983): 1–24.

Allen, Robert C. "Economic Structure and Agricultural Productivity in Europe, 1300–1800." *European Review of Economic History* 4, no. 1 (April 2000): 1–25. https://doi.org/10.1017/S1361491600000125.

Alumina Limited. "The AWAC Business: Global Business." Archived October 25, 2006. https://web.archive.org/web/20061025041242/http://www.aluminalimited.com/index.php?s=awac_biz&ss=global&p=global_op#Wagerup.

American Institute of Industrial Engineers. *The Journal of Industrial Engineering* 12, no. 1 (February 1961). http://archive.org/details/sim_journal-of-industrial-engineering_january-february-1961_12_1.

American Management Association, R.F. Whistler, and J.L. Cochran. *Production Planning and Control for the Mass Production Plant: Mass Production Series 1–7*. Literary Licensing, LLC, 2013.

American Society of Mechanical Engineers. "The Corning Ribbon Machine for Incandescent Light Bulb Blanks." American Society of Mechanical Engineers. 1983. https://www.asme.org/wwwasmeorg/media/resourcefiles/aboutasme/who%20we%20are/engineering%20history/landmarks/81-corning-ribbon-machine.pdf.

*American Telephone & Telegraph Annual Reports: 1907–1911, 1913–1977*. Accessed February 8, 2024. http://archive.org/details/americantelephoneandtelegraphannualreports.

Anderson, Philip Calvin, II. *On the Nature of Technological Progress and Industrial Dynamics*. PhD diss., Columbia University, 1988. ProQuest (8905998).

Andresen, William. *Die Cast Engineering: A Hydraulic, Thermal, and Mechanical Process*. Routledge, 2004.

Ansede, Manuel. "Four Mathematicians Demonstrate It Is Impossible to Predict Where 29,000 Rubber Ducks in the Sea Will Wash Up." *El País*, May 12, 2021. https://english.elpais.com/usa/2021-05-12/four-mathematicians-demonstrate-it-is-impossible-to-predict-where-rubber-ducks-in-the-sea-will-wash-up.html.

Aoki, Katsuki, and Thomas

Taro Lennerfors. "The New, Improved Keiretsu." *Harvard Business Review*, September 1, 2013. https://hbr.org/2013/09/the-new-improved-keiretsu.

Appelbaum, Binyamin. "American Companies Still Make Aluminum. In Iceland." *The New York Times*, July 1, 2017. https://www.nytimes.com/2017/07/01/us/politics/american-companies-still-make-aluminum-in-iceland.html.

"Apple Supplier Code of Conduct." Apple, 2022. https://www.apple.com/supplier-responsibility/pdf/Apple-Supplier-Code-of-Conduct-and-Supplier-Responsibility-Standards.pdf.

Argote, Linda, Sara L. Beckman, and Dennis Epple. "The Persistence and Transfer of Learning in Industrial Settings." *Management Science* 36, no. 2 (February 1990): 17.

Argyres, Nicholas. "Evidence on the Role of Firm Capabilities in Vertical Integration Decisions." *Strategic Management Journal* 17, no. 2 (February 1996): 129–50.

Arndt, Michael. "A Hellish Job: 'I Love It'." *Chicago Tribune*, February 7, 1999. https://www.chicagotribune.com/news/ct-xpm-1999-02-07-9902070360-story.html.

Asher, Harold. "Cost-Quantity Relationships in the Airframe Industry." The Rand Corporation, July 1, 1956. https://www.rand.org/content/dam/rand/pubs/reports/2007/R291.pdf.

"ASTM A6/A6M-21: Standard Specification for General Requirements for Rolled Structural Steel Bars, Plates, Shapes, and Sheet Piling." ASTM International, last updated January 11, 2023. https://www.astm.org/a0006_a0006m-21.html.

"AT&T Wants Phase Out of Old Telephone Technologies." Broadband Breakfast, December 31, 2009. https://broadbandbreakfast.com/att-wants-phase-out-of-old-telephone-technologies/.

Auerswald, Philip, Stuart Kauffman, José Lobo, and Karl Shell. "The Production Recipes Approach to Modeling Technological Innovation: An Application to Learning by Doing." *Journal of Economic Dynamics and Control* 24, no. 3 (March 2000): 389–450. https://doi.org/10.1016/S0165-1889(98)00091-8.

"Automatic Screw Making Machine." Smithsonian National Museum of American History, accessed July 21, 2023. https://americanhistory.si.edu/collections/search/object/nmah_1203377.

"Average New-Vehicle Prices Up 2% Year-Over-Year in July 2020, According to Kelley Blue Book." PR Newswire, August 3, 2020. https://www.prnewswire.com/news-releases/average-new-vehicle-prices-up-2-year-over-year-in-july-2020-according-to-kelley-blue-book-301104310.html.

Ayres, R.U. "Technological Transformations and Long Waves. Part I." *Technological Forecasting and Social Change* 37, no. 1 (March 1990): 1–37. https://doi.org/10.1016/0040-1625(90)90057-3.

**B**

Babbage, Charles. *On the Economy of Machinery and Manufactures by Charles Babbage.* C. Knight, 1841.

Bailey, James B. "Can Health Spending Be Reined in Through Supply Constraints? An Evaluation of Certificate-of-Need Laws." *SSRN Electronic Journal*, 2018. https://doi.org/10.2139/ssrn.3211647.

Baker, Nick. "Why Do We All Use Qwerty Keyboards?" BBC, August 11, 2010. https://www.bbc.co.uk/news/technology-10925456.

Baloff, Nicholas. "Startup Management." *IEEE Transactions on Engineering Management* EM-17, no. 4 (1970): 132–41. https://doi.org/10.1109/TEM.1970.6448538.

Barbosa, Filipe, Jonathan Woetzel, Jan Mischke, Maria João Ribeirinho, Mukund Sridhar, Matthew Parsons, Nick Bertram, and Stephanie Brown. "Reinventing Construction: A Route to Higher Productivity." McKinsey & Company, February 2017. https://www.mckinsey.com/~/media/mckinsey/business%20functions/operations/our%20insights/reinventing%20construction%20through%20a%20productivity%20revolution/mgi-reinventing-construction-a-route-to-higher-productivity-full-report.pdf.

Barkai, Haim, and David Levhari. "The Impact of Experience on Kibbutz Farming." *The Review of Economics and Statistics* 55, no. 1 (1973): 56–63. https://doi.org/10.2307/1927994.

Beach, Frederick Converse, and George Edwin Rines, eds. *The Encyclopedia Americana: A Universal Reference Library Comprising the Arts and Sciences, Literature, History, Biography, Geography, Commerce, Etc., of the World.* Scientific American Compiling Department, 1905.

Beasley, Norman. *Knudsen: A Biography.* McGraw-Hill, 1947.

Beer, John Joseph. "The Emergence of the German Dye Industry to 1925." PhD diss., University of Illinois, 1956. Proquest (0019797).

Bejan, A., A. Almerbati, and S. Lorente. "Economies of Scale: The Physics Basis." *Journal of Applied Physics* 121, no. 4 (January 28, 2017): 044907. https://doi.

org/10.1063/1.4974962.

Benkard, C. Lanier. "Learning and Forgetting: The Dynamics of Aircraft Production." *American Economic Review* 90, no. 4 (September 2000): 1034–54. https://doi.org/10.1257/aer.90.4.1034.

Bennett, Stuart. *A History of Control Engineering, 1800–1955*. IET, 1993.

Bennett, Stuart. "'The Industrial Instrument—Master of Industry, Servant of Management': Automatic Control in the Process Industries, 1900–1940." *Technology and Culture* 32, no. 1 (January 1991): 69–81. https://doi.org/10.2307/3106009.

Berman, Constance H. "Women's Work in Family, Village, and Town after 1000 CE: Contributions to Economic Growth?" *Journal of Women's History* 19, no. 3 (2007): 10–32. https://doi.org/10.1353/jowh.2007.0052.

Besanko, David, David Dranove, Mark Shanley, and Scott Schaefer. *Economics of Strategy*. 5th ed. Wiley, 2010.

Bessen, James. *Learning by Doing: The Real Connection Between Innovation, Wages, and Wealth*. Yale University Press, 2015.

Bettis, Richard A., Stephen P. Bradley, and Gary Hamel. "Outsourcing and Industrial Decline." *Academy of Management Perspectives* 6, no. 1 (February 1992): 7–22. https://doi.org/10.5465/ame.1992.4274298.

"Big Dave Bohannon: Operative Builder by the California Method." *Fortune*, April 1946.

Billard, Aude, and Danica Kragic. "Trends and Challenges in Robot Manipulation." *Science* 364, no. 6446 (June 21, 2019): eaat8414. https://doi.org/10.1126/science.aat8414.

Blacker, A. John, and Mike T. Williams, eds. *Pharmaceutical Process Development: Current Chemical and Engineering Challenges*. Royal Society of Chemistry, 2011.

Blake, Michael. *Maize for the Gods: Unearthing the 9,000-Year History of Corn*. University of California Press, 2015.

Blake-Coleman, Barrie Charles. *Copper Wire and Electrical Conductors: The Shaping of a Technology*. Harwood Academic Publishers, 1992.

"Blickensderfer No. 5 Typewriter." Smithsonian National Museum of American History. https://www.si.edu/object/blickensderfer-no-5-typewriter%3Anmah_849910.

Blondal, Petur. "Icelandic Aluminium Production Is Nearly 900,000 Tonnes, the Second Biggest in Europe after Norway." Interview by AlCircle, April 17, 2023. https://www.alcircle.com/interview/detail/92549/icelandic-aluminium-production-is-nearly-900-000-tonnes-the-second-biggest-in-europe-after-norway-petur-blondal-managing-director-icelandic-associatio.

Bogue, Allan G. "Changes in Mechanical and Plant Technology: The Corn Belt, 1910–1940." *The Journal of Economic History* 43, no. 1 (March 1983): 1–25. https://doi.org/10.1017/S0022050700028953.

Bohn, Roger E., and R. Jaikumar. *From Filing and Fitting to Flexible Manufacturing*. Now, 2005.

Bolz, Roger William. *Production Processes: The Productivity Handbook*. Industrial Press, 1981.

Boothroyd, Geoffrey, Peter Dewhurst, and Winston A. Knight. *Product Design for Manufacture and Assembly*. CRC Press, 2010.

Boothroyd, Geoffrey, and A.H. Redford. *Mechanized Assembly*. McGraw-Hill, 1968.

Boston Consulting Group. *Perspectives on Experience*. The Group, 1968.

"Boulton and Watt Rotative Steam Engine." American Society of Mechanical Engineers, April 17, 1986. https://www.asme.org/wwwasmeorg/media/resourcefiles/aboutasme/who%20we%20are/engineering%20history/landmarks/111-boulton-watt-rotative-steam-engine.pdf.

Bourdain, Anthony. *Kitchen Confidential: Adventures in the Culinary Underbelly*. Bloomsbury, 2013.

Bradley, Betsy H. *The Works: The Industrial Architecture of the United States*. Oxford University Press, 1999.

Braudel, Fernand. *Civilization and Capitalism, 15th–18th Century, Volume I: The Structures of Everyday Life: The Limits of the Possible*. University of California Press, 1992.

Breneman, Scott, Matt Timmers, and Dennis Richardson. "Tall Wood Buildings in the 2021 IBC." WoodWorks, 2022. https://www.woodworks.org/wp-content/uploads/wood_solution_paper-tall-wood.pdf.

Bricknell, David J. *Float: Pilkingtons' Glass Revolution*. Crucible, 2009.

Bright, Arthur Aaron. *The Electric-Lamp Industry: Technological Change and Economic Development from 1800 to 1947*. Macmillan, 1949.

Bright, James. *Automation and Management*. Plimpton Press, 1958.

Brinkley, Douglas. *Wheels for the World: Henry Ford, His Company, and a Century of Progress*. Penguin Publishing Group, 2004.

Brooks, Elizabeth. "Wool in Ancient Rome." Dulcia's Roman Closet, 2020. https://sites.google.com/view/dulciasromancloset/roman-textiles/wool-in-ancient-rome.

Brown, George J. *The U.S. Automobile Industry: Will It Survive Increasing International Competition?* Defense

Technical Information Center, April 22, 1991. http://archive.org/details/DTIC_ADA237808.

Brown, James. *Value Engineering: A Blueprint*. Industrial Press, 1992.

Brown, John K. "Design Plans, Working Drawings, National Styles: Engineering Practice in Great Britain and the United States, 1775–1945." *Technology and Culture* 41, no. 2 (2000): 195–238.

Brown, Warren. "America's Auto Industry Geared Up a Century Ago." *Washington Post*, July 10, 1996. https://www.washingtonpost.com/archive/1996/07/10/americas-auto-industry-geared-up-a-century-ago/4cbae9ab-136a-4c70-8890-992a50681ac4/.

Bruni, L. "Internal Economies of Scale with a Given Technique." *The Journal of Industrial Economics* 12, no. 3 (1964): 175–90. https://doi.org/10.2307/2097656.

Bruno, Lars, and Stig Tenold. "The Basis for South Korea's Ascent in the Shipbuilding Industry, 1970–1990." *The Mariner's Mirror* 97, no. 3 (January 2011): 201–17. https://doi.org/10.1080/00253359.2011.10708948.

Bryan, Ford R. *Beyond the Model T: The Other Ventures of Henry Ford*. Wayne State University Press, 1997.

Bryan, Ford R. *Rouge: Pictured in Its Prime*. Wayne State University Press, 2003.

Buchanan, Brenda J. "Charcoal: 'The Largest Single Variable in the Performance of Black Powder.'" *Icon* 14 (2008): 3–29.

Buckingham, Earle. *Principles of Interchangeable Manufacturing*. Industrial Press, 1921.

Buckingham, Earle. *Production Engineering*. John Wiley & Sons, 1942.

Bühler Die Casting, "Mega Casting Considerations for Die Casters," 2022 Die Casting Congress & Tabletop, September 2022, https://www.diecasting.org/archive/transactions/T22-032ppt.pdf

Buringh, Eltjo, and Jan Luiten van Zanden. "Charting the 'Rise of the West': Manuscripts and Printed Books in Europe, A Long-Term Perspective from the Sixth through Eighteenth Centuries." *The Journal of Economic History* 69, no. 02 (June 2009): 409. https://doi.org/10.1017/S0022050709000837.

## C

Camp, James McIntyre, and Charles Blaine Francis. *The Making, Shaping and Treating of Steel*. Carnegie Steel Company, 1920.

Carlier, Mathilde. "Light Vehicle Retail Sales in the United States from 1976 to 2023." Statista, last updated February 28, 2024. https://www.statista.com/statistics/199983/us-vehicle-sales-since-1951/.

Carney, Dan. "Tesla's Giga Press Die Castings for Model 3 Eliminate 370 Parts." DesignNews, August 13, 2021. https://www.designnews.com/automotive-engineering/tesla-s-switch-to-giga-press-die-castings-for-model-3-eliminates-370-parts.

Carr, Matthew A. "Thermodynamic Analysis of a Newcomen Steam Engine." *The International Journal for the History of Engineering & Technology* 83, no. 2 (July 2013): 187–208. https://doi.org/10.1179/1758120613Z.00000000024.

Caves, Richard E., and David R. Barton. *Efficiency in U.S. Manufacturing Industries*. MIT Press, 1990.

Chaldecott, John A. "Josiah Wedgwood (1730–95) — Scientist." *The British Journal for the History of Science* 8, no. 1 (March 1975): 1–16.

https://doi.org/10.1017/S0007087400013674.

Chandler, Alfred D., Jr. *Giant Enterprise: Ford, General Motors, and the Automobile Industry: Sources and Readings*. Harcourt, Brace & World, 1964.

Chandler, Alfred D., Jr. *Scale and Scope: The Dynamics of Industrial Capitalism*. With the assistance of Takashi Hikino. Belknap Press, 1990.

Chandler, Alfred D., Jr. *Strategy and Structure: Chapters in the History of the Industrial Enterprise*. MIT Press, 1962.

Chandler, Alfred D., Jr. *The Visible Hand: The Managerial Revolution in American Business*. Belknap Press, 1977.

Chang, Morris. "Can Semiconductor Manufacturing Return to the US?" Brookings Institution, April 19, 2022. Video, 42 min., 20 sec. https://www.youtube.com/watch?v=NwCWYcag5RE.

"Characteristics of New Housing: Current Data." US Census Bureau, 2023. https://www.census.gov/construction/chars/current.html.

Charnas, Dan. *Work Clean: The Life-Changing Power of Mise-en-Place to Organize Your Life, Work, and Mind*. Penguin Books, 2016.

Chase, Herbert. *Handbook of Designing for Quantity Production*. McGraw-Hill, 1944.

Chase, Richard B., and Douglas M. Stewart. *Mistake-Proofing: Designing Errors Out*. Productivity Press, 2010.

Chessa, Bernardo, Filipe Pereira, Frederick Arnaud, et al. "Revealing the History of Sheep Domestication Using Retrovirus Integrations." *Science* 324, no. 5926 (April 2009): 532–36. https://doi.org/10.1126/science.1170587.

Chirillo, L.D., and R.D. Chirillo. "The History of Modern Shipbuilding Methods: The U.S.-Japan Interchange."

*Journal of Ship Production* 1, no. 1 (February 1985): 1–6. https://doi.org/10.5957/jsp.1985.1.1.1.

Christensen, Clayton M. "Exploring the Limits of the Technology S-Curve. Part I: Component Technologies." *Production and Operations Management* 1, no. 4 (1991): 334–57. https://doi.org/10.1111/j.1937-5956.1992.tb00001.x.

Christensen, Clayton M. *The Innovator's Dilemma: When New Technologies Cause Great Firms to Fail*. Harvard Business School Press, 1997.

Clancy, Matt. "Learning Curves Are Tough to Use." New Things Under the Sun, March 4, 2021. https://www.newthingsunderthesun.com/pub/4xnyepnn/release/9.

Clark, Gregory. "The British Industrial Revolution, 1760–1860." UC Davis, Spring 2005. https://faculty.econ.ucdavis.edu/faculty/gclark/ecn110b/readings/ecn110b-chapter2-2005.pdf.

Clark, Gregory. "Lifestyles of the Rich and Famous: Living Costs of the Rich versus the Poor in England, 1209–1869." UC Davis, August 2004. https://www.semanticscholar.org/paper/Lifestyles-of-the-Rich-and-Famous%3A-Living-Costs-of-Clark/15e66c96dd401ba845af3d6e8e3dabf7e8dd9707.

Clark, Gregory. "Why Isn't the Whole World Developed? Lessons from the Cotton Mills." *The Journal of Economic History* 47, no. 1 (March 1987): 141–73. https://www.jstor.org/stable/2121943.

Clark, Kim B. *Product Development Performance: Strategy, Organization, and Management in the World Auto Industry*. Harvard Business School Press, 1991.

Clark, Kim B. and Robert H. Hayes. "Exploring the Sources of Productivity Differences at the Factory Level." In *The Uneasy Alliance: Managing the Productivity-Technology Dilemma*, edited by Kim B. Clark, Robert H. Hayes, and Christopher Lorenz. Harvard Business School Press, 1985.

Clifford, Joe. "Toyota Production System—What It All Means." *Toyota UK Magazine*, May 31, 2013. https://mag.toyota.co.uk/toyota-production-system-glossary/.

Coase, Ronald. "The Nature of the Firm." *Economica* 4, no. 16 (November 1937): 386–405. https://doi.org/10.1111/j.1468-0335.1937.tb00002.x.

Cochrane, Willard Wesley. *The Development of American Agriculture: A Historical Analysis*. University of Minnesota Press, 1993.

"Coffee Roasting: Roasting Equipment." Coffee Review, accessed July 20, 2023. https://www.coffeereview.com/coffee-reference/from-crop-to-cup/professional-coffee-roasting/roasting-equipment/.

Cohen, Avi J. "Technological Change as Historical Process: The Case of the U.S. Pulp and Paper Industry, 1915–1940." *The Journal of Economic History* 44, no. 3 (September 1984): 775–99. https://doi.org/10.1017/S0022050700032368.

Colbert, Thomas Burnell. "Iowa Farmers and Mechanical Corn Pickers, 1900-1952." *Agricultural History* 74, no. 2 (2000): 530–44.

Colvin, Fred H. *Sixty Years with Men and Machines: An Autobiography*. In collaboration with D.J. Duffin. McGraw-Hill, 1947.

Connelly, John. *Technique Of Production Processes*. McGraw-Hill, 1943.

Conrad, Kylie, and John D. Graham. "The Benefits and Costs of Automotive Regulations for Low-Income Americans." *Journal of Benefit-Cost Analysis* 12, no. 3 (2021): 518–49. https://doi.org/10.1017/bca.2021.12.

"Construction Cost Indices." Statistics Norway, accessed August 23, 2023. https://www.ssb.no/en/priser-og-prisindekser/byggekostnadsindekser.

"Construction Costs 1910–2023." Statistics Sweden, accessed August 23, 2023. https://www.scb.se/en/finding-statistics/statistics-by-subject-area/prices-and-consumption/building-price-index-and-construction-cost-index-for-bu/construction-cost-index-for-buildings-cci-input-price-index/pong/tables-and-graphs/construction-costs/.

"Construction Producer Price and Construction Cost Indices Overview." Eurostat, July 2023. https://ec.europa.eu/eurostat/statistics-explained/index.php?title=Construction_producer_price_and_construction_cost_indices_overview.

Cook, Michael. "The LeBlanc Soda Process: A Gothic Tale for Freshman Engineers." *Chemical Engineering Education* 32, no. 2 (Spring 1998).

Cooper, Gail. *Air-Conditioning America: Engineers and the Controlled Environment, 1900-1960*. JHU Press, 1998.

Cornwall, Philip, Louis J. Diorazio, and Natalie Monks. "Route Design, the Foundation of Successful Chemical Development." *Bioorganic & Medicinal Chemistry* 26, no. 14 (August 2018): 4336–47. https://doi.org/10.1016/j.bmc.2018.06.006.

"Cost of Space Launches to Low Earth Orbit." Our World in Data, 2023. https://ourworldindata.org/grapher/cost-space-launches-low-earth-orbit.

"Cost-of-Construction Index (CCI)." Insee. Accessed August 23, 2023. https://

www.insee.fr/en/statistiques/
serie/000008630.

Cowan, Ruth Schwartz. *A Social History of American Technology*. Oxford University Press, 1997.

Cowie, Jefferson. *Capital Moves: RCA's Seventy-Year Quest for Cheap Labor.* Cornell University Press, 2019.

Cox, James Milo. *Work Simplification: Management's Tool for Economy*. George Washington University, 1963.

Crain, W. Mark, and Nicole V. Crain. "The Cost of Federal Regulation to the US Economy, Manufacturing and Small Business." National Association of Manufacturers, September 10, 2014. https://www.nam.org/wp-content/uploads/2019/05/Federal-Regulation-Full-Study.pdf.

Cressy, David. *Saltpeter: The Mother of Gunpowder*. OUP Oxford, 2013.

Crosby, Philip B. *Quality Is Free: The Art of Making Quality Certain*. New American Library, 1980.

Curry, Kenneth C. "Iron and Steel Slag." US Geological Survey, January 2020. https://pubs.usgs.gov/periodicals/mcs2020/mcs2020-iron-steel-slag.pdf.

Cushman, Ted. "Southern Pine 2x4 Lumber Downgraded." The Journal of Light Construction, June 1, 2012. https://www.jlconline.com/how-to/framing/southern-pine-2x4-lumber-downgraded_o.


D

da Silveira, Giovani J.C., and Nigel Slack. "Exploring the Trade-Off Concept." *International Journal of Operations & Production Management* 21, no. 7 (July 2001): 949–4. https://www.researchgate.net/publication/235301115_Exploring_the_trade-off_

concept.

Dancy, Terence E. "The Evolution of Iron Making." *Metallurgical Transactions* 8, no. 1 (March 1977): 201–13. https://doi.org/10.1007/BF02657648.

David, Paul. "The Dynamo and the Computer: A Historical Perspective on the Modern Productivity Paradox." *The American Economic Review* 80, no. 2 (May 1990): 355–61. http://digamo.free.fr/david90.pdf.

Davies, Colin. *The Prefabricated Home*. Reaktion Books, 2005.

Dawson, Karen, and Piyush Sabharwall. "A Review of Light Water Reactor Costs and Cost Drivers." USDOE Office of Natural Energy, September 1, 2017. https://doi.org/10.2172/1466793.

De Decker, Kris. "Wind Powered Factories: History (and Future) of Industrial Windmills." Low-tech Magazine, October 8, 2009. https://solar.lowtechmagazine.com/2009/10/wind-powered-factories-history-and-future-of-industrial-windmills/.

de Saint-Exupéry, Antoine. *Wind, Sand and Stars*. Translated by Lewis Galantière. Harcourt Brace & Company, 1992.

de Waal, Alex. "Ending Mass Atrocity and Ending Famine." *The Lancet* 386, no. 10003 (October 2015): 1528–29. https://doi.org/10.1016/S0140-6736(15)00480-8.

Deavers, Kenneth L. "Outsourcing: A Corporate Competitiveness Strategy, not a Search for Low Wages." *Journal of Labor Research* 18, no. 4 (December 1997): 503–19. https://doi.org/10.1007/s12122-997-1019-2.

Defense Technical Information Center. *DTIC AD0890839: Engineering Design Handbook. Design Guidance for Producibility*. US Army Materiel Command, August 31, 1971. http://archive.org/details/DTIC_AD0890839.

Deflorez, Luis. "Automatic Control and the Chemical Industries." *Industrial & Engineering Chemistry* 29, no. 11 (November 1937): 1210–13. https://doi.org/10.1021/ie50335a003.

Dempsey, Kyra. "A Matter of Millimeters: The Story of Qantas Flight 32." *Admiral Cloudberg* (blog), *Medium*, December 9, 2023. https://admiralcloudberg.medium.com/a-matter-of-millimeters-the-story-of-qantas-flight-32-bdaa62dc98e7.

Depew, Chauncey Mitchell. *1795-1895: One Hundred Years of American Commerce: A History of American Commerce by One Hundred Americans*. D.O. Haynes, 1895.

Desrochers, Pierre. "Victorian Pioneers of Corporate Sustainability." *Business History Review* 83, no. 4 (2009): 703–29. https://doi.org/10.1017/S000768050000088X.

Devereaux, Bret. "Collections: Iron, How Did They Make It? Part II, Trees for Blooms." A Collection of Unmitigated Pedantry. September 25, 2020. https://acoup.blog/2020/09/25/collections-iron-how-did-they-make-it-part-ii-trees-for-blooms/.

Diamond, Jared. "Ten Thousand Years of Solitude." *Discover*, March 1, 1993. https://www.discovermagazine.com/the-sciences/ten-thousand-years-of-solitude.

DiMasi, Joseph A., Henry G. Grabowski, and Ronald W. Hansen. "Innovation in the Pharmaceutical Industry: New Estimates of R&D Costs." *Journal of Health Economics* 47 (May 2016): 20–33. https://doi.org/10.1016/j.jhealeco.2016.01.012.

Dineen, J.K. "San Francisco, Trade Unions at Odds over Modular Construction—Even for Homeless Projects." *San Francisco Chronicle*, November 27, 2020. https://

www.sfchronicle.com/bayarea/article/San-Francisco-trade-unions-at-odds-over-modular-15755264.php.

Dodd, Tim. "Starbase Tour with Elon Musk." Everyday Astronaut, August 3, 2021. Video, 53 min., 16 sec. https://www.youtube.com/watch?v=t705r8ICkRw.

Domokos, András, Brigitta Nagy, Botond Szilágyi, György Marosi, and Zsombor Kristóf Nagy. "Integrated Continuous Pharmaceutical Technologies—A Review." Organic Process Research & Development 25, no. 4 (April 2021): 721–39. https://doi.org/10.1021/acs.oprd.0c00504.

Drury, Horace B. Scientific Management: A History and Criticism. AMS Press, 1922.

Dunell, Hubert. British Wire-Drawing and Wire-Working Machinery. Constable Limited, 1925.

Dutton, John M., and Annie Thomas. "Treating Progress Functions as a Managerial Opportunity." The Academy of Management Review 9, no. 2 (April 1984): 235. https://doi.org/10.2307/258437.

Dyer, Davis, and Daniel Gross. The Generations of Corning: The Life and Times of a Global Corporation. Oxford University Press, 2001.

## E

Ealey, Lance A. Quality by Design: Taguchi Methods and U.S. Industry. ASI Press, 1988.

"Economics of Nuclear Power." World Nuclear Association. Last updated September 29, 2023. https://world-nuclear.org/information-library/economic-aspects/economics-of-nuclear-power.aspx.

"The Edison Electric Light." Journal of the Franklin Institute 109, no. 3 (1880): 145–56.

Eichler, Ned. The Merchant Builders. MIT Press, 1982.

"Electric Generator Dispatch Depends on System Demand and the Relative Cost of Operation." US Energy Information Administration, August 17, 2012. https://www.eia.gov/todayinenergy/detail.php?id=7590.

Engelhardt, Andreas W. "Textiles 2025." International Fiber Journal, April 2, 2020. https://www.fiberjournal.com/textiles-2025/.

Enos, John Lawrence. Petroleum Progress and Profits: A History of Process Innovation. MIT Press, 1962.

"Ernst Heinkel." San Diego Air & Space Museum. Accessed September 6, 2024. https://sandiegoairandspace.org/hall-of-fame/honoree/ernst-heinkel.

Eyre, Manning. "The Incandescent Lamp." Electrical World XXV, no. 1 (1895): 9–13.

## F

"Facts About American Steel Sustainability." American Iron and Steel Institute, November 2021. https://www.steel.org/wp-content/uploads/2021/11/AISI_FactSheet_SteelSustainability-11-3-21.pdf.

Fawcett, J.R. Designing for Mass Production: An Introduction. Sir Isaac Pitman & Sons, 1939.

Feix, Thomas, Javier Romero, Heinz-Bodo Schmiedmayer, Aaron M. Dollar, and Danica Kragic. "The GRASP Taxonomy of Human Grasp Types." IEEE Transactions on Human-Machine Systems 46, no. 1 (February 2016): 66–77. https://doi.org/10.1109/THMS.2015.2470657.

Felix, Richard. "Making Penicillin in Milk Bottles: A Tale of Wartime Antibiotic Manufacture." The Pharmaceutical Journal (May 25, 2018). https://pharmaceutical-journal.com/article/opinion/making-penicillin-in-milk-bottles-a-tale-of-wartime-antibiotic-manufacture.

Fetters, Thomas T. The Lustron Home: The History of a Postwar Prefabricated Housing Experiment. McFarland, 2015.

Fleischer, Manfred. The Inefficiency Trap: Strategy Failure in the German Machine Tool Industry. Ed. Sigma, 1997.

"The FMC Whole Citrus Juice Extractor." American Society of Mechanical Engineers, March 24, 1983. https://www.asme.org/wwwasmeorg/media/resourcefiles/aboutasme/who%20we%20are/engineering%20history/landmarks/82-fmc-citrus-juice-extractor.pdf.

Ford Dealer and Service Field. Trade Press Publishing Company, 1927.

"Ford Model T." Wikipedia, last updated October 10, 2024. https://en.wikipedia.org/wiki/Ford_Model_T.

"Ford Model T Original Prices." Mitch Taylor's Ford Model T.net, accessed September 13, 2024. https://www.fordmodelt.net/model-t-ford-prices.htm.

Ford, Henry, and Samuel Crowther. My Life and Work. Doubleday, Page & Company, 1922.

Forgeng, Jeffrey L. Daily Life in Medieval Europe. Greenwood Press, 1999.

"Fortune 500: 1960 Full List." CNN Money, accessed March 3, 2023. https://money.cnn.com/magazines/fortune/fortune500_archive/full/1960/.

Forward, E.A. "The Early History of the Cylinder Boring Machine." Transactions of the Newcomen Society 5, no. 1 (January 1924): 24–38. https://doi.org/10.1179/tns.1924.004.

Fox, Kenneth I., and Jose H Flores. "Changing Technology in Citrus Processing." Proceedings of the Florida State Horticultural Society 105 (1992): 139–44.

Fox, Kevin. "TOC Stories #2: 'Blue Light' Creating Capacity for Nothing." Theory of Constraints, June 15, 2007. http://theoryofconstraints.blogspot.com/2007/06/toc-stories-2-blue-light-creating.html.

Freed, Stacey. "The Nordic Track." Offsite Builder, July 15, 2022. https://offsitebuilder.com/the-nordic-track/.

Friedel, Robert D., Paul Israel, and Bernard S. Finn. *Edison's Electric Light: The Art of Invention*. Johns Hopkins University Press, 2010.

Frock, Roger. *Changing How the World Does Business: FedEx's Incredible Journey to Success: The Inside Story*. Publishers Group West, 2006.

Funk, Jeffrey L. *Technology Change and the Rise of New Industries*. Stanford University Press, 2013.

Füssel, Stephan. *Gutenberg*. Haus Publishing, 2020.


**G**

Gaiardelli, Paolo. "'Faster, Better, Cheaper' in the History of Manufacturing: From the Stone Age to Lean Manufacturing and Beyond." *Production Planning & Control* 29, no. 2 (January 2018): 184–184. https://doi.org/10.1080/09537287.2017.1368429.

Galbraith, Craig S. "Transferring Core Manufacturing Technologies in High-Technology Firms." *California Management Review* 32, no. 4 (July 1990): 56–70. https://doi.org/10.2307/41166628.

Gale, W.K.V. "Wrought Iron: A Valediction: Presidential Address." *Transactions of the Newcomen Society* 36, no. 1 (January 1963): 1–11. https://doi.org/10.1179/tns.1963.001.

Gardiner, James H., ed. *The Chemical News and Journal of Industrial Science*. Chemical News Office, 1921.

Garrett, Donald E. *Chemical Engineering Economics*. Netherlands, 1989.

Gasparetto, A., and L. Scalera. "A Brief History of Industrial Robotics in the 20th Century." *Advances in Historical Studies* 8, no. 1 (January 2019): 24–35. https://doi.org/10.4236/ahs.2019.81002.

Gass, Saul I. *An Illustrated Guide to Linear Programming*. McGraw-Hill, 1970.

Gaynes, Robert. "The Discovery of Penicillin—New Insights After More Than 75 Years of Clinical Use." *Emerging Infectious Diseases* 23, no. 5 (May 2017): 849–53. https://doi.org/10.3201/eid2305.161556.

Gendre, Maxime F. "Two Centuries of Electric Light Source Innovations." Eindhoven University of Technology, January 2003. https://www.einlightred.tue.nl/lightsources/history/light_history.pdf.

"Genetics of Maize Domestication (1992–2019)." Buckler Lab, accessed September 27, 2022. https://www.maizegenetics.net/copy-of-hybrid-vigor-1.

Gerke, B., A. Ngo, A. Alstone, and K. Fisseha. "The Evolving Price of Household LED Lamps: Recent Trends and Historical Comparisons for the US Market." Lawrence Berkeley National Laboratory Report LBNL-6854E, November 14, 2014. https://escholarship.org/uc/item/5mh3f9x1.

Germain, Scott. "World's Fastest Piston-Power Airplane." *Smithsonian Magazine*, 2018. https://www.smithsonianmag.com/air-space-magazine/worlds-fastest-piston-airplane-180969509/.

"Germany Price Index Prefab. One-Fam. Houses without Cellar." DBnomics, accessed October 12, 2023. https://db.nomics.world/DESTATIS/61261BJ007/DG.PRE012.

Giattino, Charlie, Esteban Ortiz-Ospina, and Max Roser. "Working Hours." Our World in Data, January 15, 2024. https://ourworldindata.org/working-hours.

Gies, Frances, and Joseph Gies. *Life in a Medieval Village*. Harper & Row, 1990.

"Giga Press Uncovered Chapter 3/5: The Technological Challenges, and the Target Audience." Idra, October 18, 2021. Video, 4 min., 48 sec. https://www.youtube.com/watch?v=YwhGS3My9bM.

Gilbreth, Frank Bunker. *Bricklaying System*. M.C. Clark Publishing Company, 1909.

Gilbreth, Frank Bunker. *Primer of Scientific Management*. D. Van Nostrand Company, 1912).

Gilbreth, Frank Bunker, and Lillian Moller Gilbreth. *Applied Motion Study: A Collection of Papers on the Efficient Method to Industrial Preparedness*. Sturgis & Walton, 1917.

Gipe, Paul. *Wind Energy Comes of Age*. Wiley, 1995.

"Global Aircraft Fleet: Deliveries by Manufacturer 1999–2021." Statista, accessed September 16, 2024. https://www.statista.com/statistics/622779/number-of-jets-delivered-global-aircraft-fleet-by-manufacturer/

Gold, Bela. "Evaluating Scale Economies: The Case of Japanese Blast Furnaces." *The Journal of Industrial Economics* 23, no. 1 (September 1974): 1. https://doi.org/10.2307/2098209.

Goldratt, Eliyahu M., and Jeff Cox. *The Goal: A Process of Ongoing Improvement*. North River Press, 2004.

Goolsbee, Austan, and Chad Syverson. "The Strange and Awful Path of Productivity in the US Construction Sector," NBER Working Paper 30845, January 2023. http://www.nber.org/papers/w30845.

Gordon, Robert B. "Who Turned the Mechanical Ideal

into Mechanical Reality?" *Technology and Culture* 29, no. 4 (October 1988): 744. https://doi.org/10.2307/3105044.

Gottfried, Joseph. "History Repeating? Avoiding a Return to the Pre-Antibiotic Age." Third year paper, Harvard Law School, 2005. https://dash.harvard.edu/bitstream/handle/1/8889467/Gottfried05.html.

Graham, Margaret B.W., and Bettye H. Pruitt. *R&D for Industry: A Century of Technical Innovation at Alcoa.* Cambridge University Press, 1990.

Green, R.A.W., and A.A. Leadbeater. "The Reliability of [the] Power Station Plant." *Steam Plant Operation: A Convention Sponsored by the Verein Deutscher Ingenieure and the Steam Plant Group of the Institution of Mechanical Engineers, 2nd–4th May 1973.* The Institution of Mechanical Engineers, 1974.

Greenwood, Jeremy, and Gokce Uysal. "New Goods and the Transition to a New Economy." *SSRN Electronic Journal,* October 7, 2003. https://doi.org/10.2139/ssrn.439381.

Grieve, Robert. *The Cotton Centennial, 1790–1890: Cotton and Its Uses, the Inception and Development of the Cotton Industries of America, and a Full Account of the Pawtucket Cotton Centenary Celebration.* J.A. & R.A. Reid, 1891.

Groves, General Leslie R. *Now It Can Be Told: The Story of The Manhattan Project.* Hachette Books, 2009.

"Grumman F8F-2 Bearcat 'Conquest I.'" Smithsonian National Air and Space Museum. Accessed September 6, 2024. https://airandspace.si.edu/collection-objects/grumman-f8f-2-bearcat-conquest-i/nasm_A19770989000.

## H

Haas Automation, Inc. "The Haas VF-2—Small VMC Workhorse." September 21, 2021. Video, 3 min., 6 sec. https://www.youtube.com/watch?v=LN_LYo84Oko.

Halberstam, David. *The Reckoning.* Morrow, 1986.

Haldi, John, and David Whitcomb. "Economies of Scale in Industrial Plants." *Journal of Political Economy* 75, no. 4, Part 1 (August 1967): 373–85. https://doi.org/10.1086/259293.

Hall, Bronwyn H., and Nathan Rosenberg. *Handbook of the Economics of Innovation.* Elsevier North Holland, 2010.

Hall, G.R., and R.E. Johnson. "Transfers of United States Aerospace Technology to Japan." In *The Technology Factor in International Trade.* NBER, 1970. https://www.nber.org/books-and-chapters/technology-factor-international-trade/transfers-united-states-aerospace-technology-japan.

Hallion, Richard P., ed. *The Hypersonic Revolution: Case Studies in the History of Hypersonic Technology*, Vol. 2., *From Scramjet to the National Aero-Space Plane.* Air Force History and Museums Program, 1998.

Haralambides, Hercules E. "Gigantism in Container Shipping, Ports and Global Logistics: A Time-Lapse into the Future." *Maritime Economics & Logistics* 21, no. 1 (March 1, 2019): 1–60. https://doi.org/10.1057/s41278-018-00116-0.

Harrigan, Kathryn Rudie. "Formulating Vertical Integration Strategies." *The Academy of Management Review* 9, no. 4 (October 1984): 638–52. https://doi.org/10.2307/258487.

Harris, Howell J. "Conquering Winter: US Consumers and the Cast-Iron Stove." *Building Research & Information* 36, no. 4 (August 2008): 337–50. https://doi.org/10.1080/09613210 802117411.

Harris, Howell J. "Inventing the U.S. Stove Industry, c. 1815–1875: Making and Selling the First Universal Consumer Durable." *Business History Review* 82, no. 4 (2008): 701–33. https://doi.org/10.1017/S0007680500063170.

Harris, J.R. "Skills, Coal and British Industry in the Eighteenth Century." *History* 61, no. 202 (June 1976): 167–82. https://doi.org/10.1111/j.1468-229X.1976.tb01337.x.

Harris, Susanna. "Flax Fibre: Innovation and Change in the Early Neolithic: A Technological and Material Perspective." In *Textile Society of America 2014 Biennial Symposium Proceedings: New Directions: Examining the Past, Creating the Future,* Los Angeles, California, September 10–14, 2014. https://digitalcommons.unl.edu/tsaconf/913/.

Hartlieb, Alicia, and Martin Hartlieb. "The Impact of Giga-Castings on Car Manufacturing and Aluminum Content." *Light Metal Age,* July 10, 2023. https://www.lightmetalage.com/news/industry-news/automotive/the-impact-of-giga-castings-on-car-manufacturing-and-aluminum-content/.

Hart-Smith, L.J. "Out-Sourced Profits—The Cornerstone of Successful Subcontracting." Presented at the Boeing Third Annual Technical Excellence (TATE) Symposium St. Louis, Missouri, February 14–15, 2001. https://techrights.org/wp-content/uploads/2022/06/2014130646.pdf.

Hayter, Earl W. "Barbed Wire Fencing—A Prairie Invention." *Agricultural History* 13, no. 4 (October 1939): 189–207. https://www.jstor.org/stable/3739686.

"Heat Treatment of Steel." In

*Machinery's Handbook*, 1924. ZiaNet, accessed September 15, 2024. https://www.zianet.com/ebear/metal/heattreat3.html.

Heaton, Alan, ed. *An Introduction to Industrial Chemistry*. Springer Netherlands, 1996. https://doi.org/10.1007/978-94-011-0613-9.

Helland, Eric, and Alex Tabarrok. *Why Are the Prices So Damn High?* Mercatus Center, 2019.

Helper, Susan. "Strategy and Irreversibility in Supplier Relations: The Case of the U.S. Automobile Industry." *Business History Review* 65, no. 4 (1991): 781–824. https://doi.org/10.2307/3117265.

Henderson, Rebecca. "Of Life Cycles Real and Imaginary: The Unexpectedly Long Old Age of Optical Lithography." *Research Policy* 24, no. 4 (July 1995): 631–43. https://doi.org/10.1016/S0048-7333(94)00790-X.

Henrich, Joseph. "Demography and Cultural Evolution: How Adaptive Cultural Processes Can Produce Maladaptive Losses: The Tasmanian Case." *American Antiquity* 69, no. 2 (2004): 197–214. https://doi.org/10.2307/4128416.

Henshaw, John. *Liberty's Provenance: The Evolution of the Liberty Ship from Its Sunderland Origins*. Pen and Sword, 2019.

Herbert, Christopher, Chadwick Reed, and James Shen. "Comparison of the Costs of Manufactured and Site-Built Housing." Joint Center for Housing Studies, Harvard University, 2023. https://www.jchs.harvard.edu/sites/default/files/research/files/harvard_jchs_pew_report_1_updated_0.pdf.

Hills, Richard L. *Power from Steam: A History of the Stationary Steam Engine*. Cambridge University Press, 1993.

Hindle, Brooke, and Stephen Lubar. *Engines of Change: The American Industrial Revolution, 1790–1860*. Smithsonian, 1986.

Hirschmann, Winfred B. "Profit from the Learning Curve." *Harvard Business Review*, January 1, 1964. https://hbr.org/1964/01/profit-from-the-learning-curve.

Hodges, N.D.C. "Aluminium and Its Manufacture by the Deville-Castner Process." *Science* 13, no. 322 (1889): 260–65.

Hodkin, Frederick William, and Arnold Cousen. *A Textbook of Glass Technology*. D. Van Nostrand, 1925.

Hoe, Robert. *A Short History of the Printing Press and of the Improvements in Printing Machinery from the Time of Gutenberg Up to the Present Day*. Robert Hoe, 1902.

Hogan, William T. *Economic History of the Iron and Steel Industry in the United States*. Heath, 1971.

Hollander, Samuel. *The Sources of Increased Efficiency: A Study of DuPont Rayon Plants*. MIT Press, 1965.

Holley, Irving Brinton, Jr. *Buying Aircraft: Matériel Procurement for the Army Air Forces*. United States Army in World War II: Special Studies. Center of Military History, United States Army, 1964.

Holmes, Harry N. "The Story of Aluminum." *Journal of Chemical Education* 7, no. 2 (February 1930): 232. https://doi.org/10.1021/ed007p232.

Holton, Gerald James. *Thematic Origins of Scientific Thought: Kepler to Einstein*. Harvard University Press, 1975.

Hooker, A.H. "Niagara Falls Power and American Industries." *Transactions of the American Electrochemical Society, Twenty-Ninth General Meeting*, Washington, DC, April 27–29, 1916.

Hopp, Wallace J. *Factory Physics*. McGraw-Hill, 2008.

Horsley, Scott. "How the U.S. Got into This Baby Formula Mess." NPR, May 19, 2022. https://www.npr.org/2022/05/19/1099748064/baby-infant-formula-shortages.

Hounshell, David A. *From the American System to Mass Production, 1800–1932: The Development of Manufacturing Technology in the United States*. JHU Press, 1984.

Hounshell, David A., and John Kenly Smith Jr. *Science and Corporate Strategy: Du Pont R&D, 1902–1980*. Cambridge University Press, 1988.

Houson, Ian, ed. *Process Understanding: For Scale-Up and Manufacture of Active Ingredients*. Wiley-VCH, 2011.

"How ASML Builds a $150 Million EUV Machine." Asianometry, May 24, 2021. Video, 14 min., 51 sec. https://www.youtube.com/watch?v=jJIO7aRXUCg.

"How Is Uranium Made into Nuclear Fuel?" World Nuclear Association. 2023. https://world-nuclear.org/nuclear-essentials/how-is-uranium-made-into-nuclear-fuel.aspx.

Howells, Nick R., Mark D. Brinsden, Richie S. Kill, Andrew J. Carr, and Jonathan L. Rees. "Motion Analysis: A Validated Method for Showing Skill Levels in Arthroscopy." *The Journal of Arthroscopy and Related Surgery* 24, no. 3 (March 2008): 335–42. https://www.arthroscopyjournal.org/article/S07498063(07)00844-4/abstract.

Hughes, Thomas. *Networks of Power: Electrification in Western Society, 1880–1930*. JHU Press, 1993.

Humphrey, Arthur E., and S. Edward Lee. "Industrial Fermentation: Principles, Processes, and Products." In *Riegel's Handbook of Industrial Chemistry*, edited by James A. Kent. Springer, 2003.

Hutcheson, Dan. "Graphic: Transistor Production Has Reached Astronomical Scales." *IEEE Spectrum,* April 2, 2015.

https://spectrum.ieee.org/
transistor-production-has-
reached-astronomical-scales.

Hutchings, Matthew I., Andrew W.
Truman, and Barrie Wilkinson.
"Antibiotics: Past, Present and
Future." *Current Opinion in
Microbiology* 51 (October 2019):
72–80. https://doi.org/10.1016/j.
mib.2019.10.008.

I

Innovation Resource Center.
"Evolution of the Clothing
Industry in the United
States." Innovation Resource
Center, March 1, 2023.
https://web.archive.org/
web/20230301133836/https:/
irc4hr.org/resource/evolution-
of-the-clothing-industry-in-the-
united-states/.

Irwin, Douglas A., and Peter J.
Klenow. "Learning-by-Doing
Spillovers in the Semiconductor
Industry." *Journal of Political
Economy* 102, no. 6 (1994):
1200–27.

J

Jackson, Dudley. *Profitability,
Mechanization and Economies
of Scale.* Routledge, 2018.

Jansson, Jan Owen, and Dan
Shneerson. "Economies of
Scale of General Cargo Ships."
*The Review of Economics
and Statistics* 60, no. 2 (1978):
287–93. https://doi.org/10.2307/
1924982.

Jasper, James M. *Nuclear Politics:
Energy and the State in the
United States, Sweden, and
France.* Princeton University
Press, 1990.

"JBT Fruit and Vegetables
Processing—ReadyGo
Juice." JBT, February 5,
2020. Video, 2 min., 12 sec.
https://www.youtube.com/
watch?v=3B6DEHmsUGQ.

Jenkins, Rhys. "Ironfounding
in England, 1490–1603."
*Transactions of the Newcomen
Society* 19, no. 1 (January 1938):
35–49. https://doi.org/10.1179/
tns.1938.002.

Jeremy, David J. *Transatlantic
Industrial Revolution:
The Diffusion of Textile
Technologies between Britain
and America, 1790–1830s.* MIT
Press, 1981.

Jewkes, John. *The Sources of
Invention.* Springer, 1969.

"Job Shop." *Inc*, February 6,
2020. https://www.inc.com/
encyclopedia/job-shop.html.

Johnstone, Brent M. "Improvement
Curves: An Early Production
Methodology." Lockheed
Martin, June 11, 2015. https://
www.iceaaonline.com/wp-
content/uploads/2015/06/
PA03-Presentation-Johnstone-
Improvment-Curves.pdf.

Jones, Bence. *The Royal
Institution: Its Founder and Its
First Professors.* Longmans,
Green, and Co., 1871.

Jones, Daniel T., and James P.
Womack. *Lean Thinking: Banish
Waste and Create Wealth in
Your Corporation.* Simon &
Schuster, 2013.

Josephson, G.W. *Iron Blast-
Furnace Slag Production,
Processing, Properties, and
Uses.* U.S. Government Printing
Office, 1949.

*Journal of the House of
Representatives, During the
Twenty-Fourth Session of
the Legislature of the State
of Minnesota*, Volume 34.
Minnesota Legislature House of
Representatives, 1905.

Jovanovic, Boyan, and Yaw
Nyarko. "A Bayesian Learning
Model Fitted to a Variety of
Empirical Learning Curves."
*Brookings Papers on Economic
Activity. Microeconomics*
1995 (1995): 247. https://doi.
org/10.2307/2534775.

K

Kang, J.Y., Song Kim, Hugh
Murphy, and Stig Tenold.
"Old Methods Versus New:
A Comparison of Very Large
Crude Carrier Construction
at Scott Lithgow and Hyundai
Heavy Industries, 1970–1977."
*The Mariner's Mirror* 101, no.
4 (October 2, 2015): 426–57.
https://doi.org/10.1080/002533
59.2015.1085707.

Kanigel, Robert. "Taylor-Made,"
*The Sciences* 32, no. 3 (May–
June 1997): 18–23, https://doi.
org/10.1002/j.2326-1951.1997.
tb03309.x.

Karpova, Elena, Grace I. Kunz,
and Myrna B. Garner. *Going
Global: The Textile and
Apparel Industry.* Bloomsbury
Publishing USA, 2021.

Katz, Harold. "The Decline of
Competition in the Automobile
Industry, 1920–1940." PhD diss.,
Columbia University, 1970.
ProQuest (7023446).

Keats, Jonathon. "How Alan Turing
Found Machine Thinking in the
Human Mind." New Scientist,
June 29, 2016. https://www.
newscientist.com/article/
mg23130803-200-how-alan-
turing-found-machine-thinking-
in-the-human-mind/.

Kedzie, Daniel. "Skillful Means:
Ancient Process Control as
Exemplified by the Manufacture
of Japanese Sword, Nihonto."
(Master's thesis, California
State University, 2011).
ProQuest (1507397).

Keller, Jim. "Moore's Law
Is Not Dead." Berkeley
EECS, September 18, 2019.
Video, 64 min., 52 sec.
https://www.youtube.com/
watch?v=oIG9ztQw2Gc.

Kelley, James E., and Morgan R.
Walker. "Critical-Path Planning
and Scheduling." In *Papers
Presented at the December
1-3, 1959, Eastern Joint IRE-
AIEE-ACM Computer*, 160–73.
ACM Press, 1959. https://doi.
org/10.1145/1 460299.1460318.

Kelly, Thomas. *Early Public Libraries: A History of Public Libraries in Great Britain before 1850*. Library Association, 1966.

Kerch, Steve. "Big Builder Getting It Together Again." *Chicago Tribune*, July 26, 1987. https://www.chicagotribune.com/news/ct-xpm-1987-07-26-8702240616-story.html.

Keyser, Barbara Whitney. "Between Science and Craft: The Case of Berthollet and Dyeing." *Annals of Science* 47, no. 3 (May 1990): 213–60. https://doi.org/10.1080/00033799000200211.

Kim, Hyunchul, Levi Makaio Miller, Aimen Al-Refai, Moshe Brand, and Jacob Rosen. "Redundancy Resolution of a Human Arm for Controlling a Seven DOF Wearable Robotic System." *Annual International Conference of the IEEE Engineering in Medicine and Biology Society. IEEE Engineering in Medicine and Biology Society* (2011): 3471–74. https://doi.org/10.1109/IEMBS.2011.6090938.

Kirchner, David Paul. *A Discriminative Study of Navy Work Simplification and Systems Analysis Programs*. Naval Postgraduate School, 1960.

Kirwan, Richard, and Sophie Mullins, eds. *Specialist Markets in the Early Modern Book World*. Brill, 2015.

Klein, Maury. *The Power Makers*. Bloomsbury Press, 2008.

Klier, Thomas H., and James M. Rubenstein. *Who Really Made Your Car? Restructuring and Geographic Change in the Auto Industry*. W.E. Upjohn Institute, 2008.

Klipstein, Don. "The Great Internet Light Bulb Book, Part I." 1996. https://donklipstein.com/bulb1.html#eff.

Knight, Philip H. *Shoe Dog: A Memoir by the Creator of Nike*. Scribner, 2016.

Koff, Bernard L. "Gas Turbine Technology Evolution—A Designer's Perspective." *Journal of Propulsion and Power* 20, no. 4 (July–August 2004). https://doi.org/10.2514/1.4361.

Kornmayer, Pall. "Tesla Model 3: Center Underbody." Pall Kornmayer. Accessed January 19, 2024. https://www.pallkornmayer.com/work/tesla-model-3-biw.

Korus, Sam. "Wright's Law Predicted 109 Years of Autos Gross Margin, and Now Tesla's." ARK Invest, September 4, 2019. https://ark-invest.com/articles/analyst-research/wrights-law-predicts-teslas-gross-margin/.

Kuo, T.C. and Hong-Chao Zhang. "Design for Manufacturability and Design for 'X': Concepts, Applications, and Perspectives." In *Seventeenth IEEE/CPMT International Electronics Manufacturing Technology Symposium, Manufacturing Technologies—Present and Future*, 446–59. IEEE, 1995. https://doi.org/10.1109/IEMT.1995.526203.

**L**

Laddon, I.M. "Reduction of Man-Hours in Aircraft Production." *Aviation* 42, no. 5 (1943).

Lafond, François, Aimee Gotway Bailey, Jan David Bakker, Dylan Rebois, Rubina Zadourian, Patrick McSharry, and J. Doyne Farmer. "How Well Do Experience Curves Predict Technological Progress? A Method for Making Distributional Forecasts." *Technological Forecasting and Social Change* 128 (March 2018): 104–17. https://doi.org/10.1016/j.techfore.2017.11.001.

Lafond, François, Diana Greenwald, and J Doyne Farmer. "Can Stimulating Demand Drive Costs Down? World War II as a Natural Experiment." *The Journal of Economic History* 84, no. 3 (September 2022): 727–64. https://doi.org/10.1017/S0022050722000249.

Laman, N.K. "The Development of the Wire-Drawing Industry." *Metallurgist* 3, no. 6 (1959): 267–70. https://doi.org/10.1007/BF00740175.

Landes, David S. *The Unbound Prometheus: Technological Change and Industrial Development in Western Europe from 1750 to the Present*. Cambridge University Press, 1969.

Lane, Kevin, Clara Camarasa, Chiara Delmastro, Noah Sloots, and Fabian Voswinkel. "Lighting." International Energy Agency. Last updated July 11, 2023. https://www.iea.org/reports/lighting.

Langlois, Richard N. "The Vanishing Hand: The Changing Dynamics of Industrial Capitalism." *Industrial and Corporate Change* 12, no. 2 (April 2003): 351–85. https://doi.org/10.1093/icc/12.2.351.

Langlois, Richard N., and Paul L. Robertson. "Explaining Vertical Integration: Lessons from the American Automobile Industry." *The Journal of Economic History* 49, no. 2 (June 1989): 361–75. https://doi.org/10.1017/S0022050700007993.

"Lazard's Levelized Cost of Energy Analysis—Version 15.0." Lazard. October 2021. https://www.lazard.com/media/sptlfats/lazards-levelized-cost-of-energy-version-150-vf.pdf

Lazonick, William, and Thomas Brush. "The 'Horndal Effect' in Early U.S. Manufacturing." *Explorations in Economic History* 22, no. 1 (January 1985): 53–96. https://doi.org/10.1016/0014-4983(85)90021-X.

Leaney, P.G., and Gunter Wittenberg. "Design For Assembling: The Evaluation Methods of Hitachi, Boothroyd and Lucas." *Assembly*

*Automation* 12, no. 2 (February 1992): 8–17. https://doi.org/10.1108/eb004359.

Lebergott, Stanley. "Labor Force and Employment, 1800–1960." In *Output, Employment, and Productivity in the United States after 1800*, edited by Dorothy S. Brady. National Bureau of Economic Research, 1966.

Lécuyer, Christophe, and David C. Brock. *Makers of the Microchip: A Documentary History of Fairchild Semiconductor*. MIT Press, 2010.

Lee, Sau L., Thomas F. O'Connor, Xiaochuan Yang, Celia N. Cruz, Sharmista Chatterjee, Rapti D. Madurawe, Christine M.V. Moore, Lawrence X. Yu, and Janet Woodcock. "Modernizing Pharmaceutical Manufacturing: From Batch to Continuous Production." *Journal of Pharmaceutical Innovation* 10, no. 3 (September 2015): 191–99. https://doi.org/10.1007/s12247-015-9215-8.

Lefteri, Chris. *Making It: Manufacturing Techniques for Product Design*. Laurence King Publishing, 2019.

"Lenovo Manufacturing Sites and Suppliers." Lenovo, 2020. https://static.lenovo.com/ww/docs/sustainability/list-of-lenovo-mfg-sites-and-suppliers-27oct.pdf.

Leonard-Barton, Dorothy. *Wellsprings of Knowledge: Building and Sustaining the Sources of Innovation*. Harvard Business School Press, 1995.

Leslie, T.W. "'Dry and Ready in Half the Time': Gypsum Wallboard's Uneasy History." In *History of Construction Cultures*, edited by João Mascarenhas-Mateus, Ana Paula Pires, Manuel Marques Caiado, and Ivo Veiga, 682–87. CRC Press, 2021.

"Levelized Cost of Energy by Technology." Accessed January 26, 2024. https://ourworldindata.org/grapher/levelized-cost-of-energy.

Levinson, Marc. *The Box: How the Shipping Container Made the World Smaller and the World Economy Bigger*. Princeton University Press, 2016.

Levitt, Steven D., John A. List, and Chad Syverson. "Toward an Understanding of Learning by Doing: Evidence from an Automobile Assembly Plant." *Journal of Political Economy* 121, no. 4 (August 2013). https://doi.org/10.1086/671137.

Levy, F.K., G.L. Thompson, and J.D. Wiest. "The ABCs of the Critical Path Method." *Harvard Business Review*, September 1, 1963. https://hbr.org/1963/09/the-abcs-of-the-critical-path-method.

Li, Jingshan. "Continuous Improvement at Toyota Manufacturing Plant: Applications of Production Systems Engineering Methods." *International Journal of Production Research* 51, no. 23–24 (November 2013): 7235–49. https://doi.org/10.1080/00207543.2012.753166.

Li, Jingshan, and Semyon M. Meerkov. *Production Systems Engineering*. Springer, 2009.

Li, Johnnie Liew Zhong, Mohd Rizal Alkahari, Nor Ana Binti Rosli, Rafidah Hasan, Mohd Nizam Sudin, and Faiz Redza Ramli. "Review of Wire Arc Additive Manufacturing for 3D Metal Printing." *International Journal of Automation Technology* 13, no. 3 (May 2019): 346–53. https://doi.org/10.20965/ijat.2019.p0346.

"Light Bulb Mass Production Process: Last Incandescent Lamp Factory in Korea." All Process of World, April 25, 2022. Video, 12 min., 14 sec. https://www.youtube.com/watch?v=v9Psw6_5lHg.

Liker, Jeffrey K. *Becoming Lean: Inside Stories of U.S. Manufacturers*. CRC Press, 1997.

Liker, Jeffrey K. *The Toyota Way: 14 Management Principles from the World's Greatest Manufacturer*. McGraw-Hill, 2004.

Livesay, Harold C. *Andrew Carnegie and the Rise of Big Business*. HarperCollins, 1975.

Lloyd, Jennifer. "Instructional Bulletin No. 16-01: Value Engineering Policies and Procedures." State of Tennessee Department of Transportation, February 25, 2016. https://www.tn.gov/content/dam/tn/tdot/documents/Instructional BulletinNo_16_01.pdf.

Loiter, Jeffrey M., and Vicki Norberg-Bohm. "Technology Policy and Renewable Energy: Public Roles in the Development of New Energy Technologies." *Energy Policy* 27, no. 2 (February 1999): 85–97. https://doi.org/10.1016/S0301-4215(99)00013-0.

Lorant, John Herman. *The Role of Capital-Improving Innovations in American Manufacturing During the 1920s*. Arno Press, 1975.

Loveday, Amos John. "The Cut Nail Industry 1776–1890: Technology, Cost Accounting and the Upper Ohio Valley." PhD diss., Ohio State University, 1979.

Lovering, Jessica R., Arthur Yip, and Ted Nordhaus. "Historical Construction Costs of Global Nuclear Power Reactors." *Energy Policy* 91 (April 2016): 371–82. https://doi.org/10.1016/j.enpol.2016.01.011.

Lowe, Derek. "AI and Drug Discovery: Attacking the Right Problems." *Science*, March 19, 2021. https://www.science.org/content/blog-post/ai-and-drug-discovery-attacking-right-problems.

Lubar, Steven. "Culture and Technological Design in the 19th-Century Pin Industry: John Howe and the Howe Manufacturing Company."

Technology and Culture 28, no. 2 (April 1987): 253. https://doi.org/10.2307/3105567.

## M

M.N. Dastur & Co. "Economies of Scale at Small Integrated Steelworks." United Nations Economic and Social Council, January 1967. https://repositorio.cepal.org/bitstream/handle/11362/29208/S6700235_en.pdf.

Machinery's Handbook. The Industrial Press, 1924.

MacKenzie, Donald, and Graham Spinardi. "Tacit Knowledge, Weapons Design, and the Uninvention of Nuclear Weapons." American Journal of Sociology 101, no. 1 (1995): 44–99.

Maiello, Michael. "Diagnosing William Baumol's Cost Disease." Chicago Booth Review, May 18, 2017. https://www.chicagobooth.edu/review/diagnosing-william-baumols-cost-disease.

Malhotra, Abhishek, and Tobias S. Schmidt. "Accelerating Low-Carbon Innovation." Joule 4, no. 11 (November 2020): 2259–67. https://doi.org/10.1016/j.joule.2020.09.004.

Malik, Mujeeb, and Dennis Bushnell, eds. "Role of Computational Fluid Dynamics and Wind Tunnels in Aeronautics R&D." NASA, September 2012. https://ntrs.nasa.gov/api/citations/20120016316/downloads/20120016316.pdf.

Malone, Michael S. The Big Score: The Billion Dollar Story of Silicon Valley. Stripe Press, 2021.

Mankins, John C. "Technology Readiness Levels." NASA Office of Space Access and Technology, April 6, 1995. http://www.artemisinnovation.com/images/TRL_White_Paper_2004-Edited.pdf.

Mapelli, Carlo, Walter Nicodemi, Riccardo F. Riva, and Maurizio Vedani. "Analysis of the Nails from the Roman Legionary at Inchtuthil." Steel Research International 79, no. 7 (2008): 569–76. https://doi.org/10.1002/srin.200806168.

Marini, Daniele, David Cunningham, and Jonathan R Corney. "Near Net Shape Manufacturing of Metal: A Review of Approaches and Their Evolutions." Proceedings of the Institution of Mechanical Engineers, Part B: Journal of Engineering Manufacture 232, no. 4 (March 2018): 650–69. https://doi.org/10.1177/0954405417708220.

Mateles, Richard I. Penicillin: A Paradigm for Biotechnology. Candida Corporation, 1998.

Mayr, Otto. "The Origins of Feedback Control." Scientific American 223, no. 4 (1970): 110–19.

Mayr, Otto, and Robert C. Post, eds. Yankee Enterprise: The Rise of the American System of Manufactures. Smithsonian Institution Press, 1981.

McCalley, Bruce W. Model T Ford: The Car That Changed the World. Krause Publications, 1994.

McConnell, David. British Smooth-Bore Artillery: A Technological Study to Support Identification, Acquisition, Restoration, Reproduction, and Interpretation of Artillery at National Historic Parks in Canada. Canadian Government Publishing Centre, 1988.

McDonald, Alan, and Leo Schrattenholzer. "Learning Rates for Energy Technologies." Energy Policy 29, no. 4 (March 2001): 255–61. https://doi.org/10.1016/S0301-4215(00)00122-1.

McDonald, Chris J. "The Evolution of Intel's Copy EXACTLY! Technology Transfer Method." Intel Technology Journal Q4'98, 1998. https://www.intel.com/content/dam/www/public/us/en/documents/research/1998-vol02-iss-4-intel-technology-journal.pdf

McFadden, Joseph. "From Invention to Monopoly: The History of the Consolidation of the Barbed Wire Industry, 1873–1899." PhD diss., Northern Illinois University, 1968. ProQuest (6903104).

McGaw, Judith A. Most Wonderful Machine: Mechanization and Social Change in Berkshire Paper Making, 1801–1885. Princeton University Press, 1987.

McGeer, J.P. "Hall-Héroult: 100 Years of Processes Evolution." JOM Journal of the Minerals, Metals and Materials Society 38 (1986): 27–33.

McMaster-Carr. Accessed October 12, 2023. https://www.mcmaster.com/.

McNerney, James, J. Doyne Farmer, Sidney Redner, and Jessika E. Trancik. "Role of Design Complexity in Technology Improvement." PNAS 108, no. 22 (May 2011): 9008–13. https://doi.org/10.1073/pnas.1017298108.

McNerney, James, J. Doyne Farmer, and Jessika E. Trancik. "Historical Costs of Coal-Fired Electricity and Implications for the Future." Energy Policy 39, no. 6 (June 2011): 3042–54. https://doi.org/10.1016/j.enpol.2011.01.037.

Medlock, R.S. "The Historical Development of Flow Metering." Measurement and Control 19, no. 5 (June 1986): 11–22. https://doi.org/10.1177/002029408601900502.

Meursinge, John H. "Practical Experience in the Transfer of Technology." Technology and Culture 12, no. 3 (1971): 469–70. https://doi.org/10.2307/3103001.

Meyers, Robert A., ed. Encyclopedia of Physical Science and Technology. Academic Press, 2001.

Miles, Lawrence D. *Techniques of Value Analysis and Engineering*. McGraw-Hill, 1961.

Miller, Jon. "I'll Take My Lean with Water, On the Rocks." Gemba Academy, July 28, 2009. https://blog.gembaacademy.com/2009/07/27/ill_take_my_lean_with_water_on_the_rocks/.

Misa, Thomas J. *A Nation of Steel*. Johns Hopkins University Press, 1995.

Mishina, Kazuhiro. "Learning by New Experiences: Revisiting the Flying Fortress Learning Curve." In *Learning by Doing in Markets, Firms, and Countries*, 145–84. University of Chicago Press, 1999.

Mitchell, Matthew D. "Do Certificate-of-Need Laws Limit Spending?" Mercatus Working Paper, Mercatus Center at George Mason University, Arlington, VA, September 2016. https://doi.org/10.2139/ssrn.2871325.

Mokyr, Joel, ed. *The British Industrial Revolution: An Economic Perspective*. Westview Press, 1999.

Mokyr, Joel. *The Gifts of Athena: Historical Origins of the Knowledge Economy*. Princeton University Press, 2002.

Mokyr, Joel. *The Lever of Riches: Technological Creativity and Economic Progress*. Oxford University Press, 1992.

Monden, Yasuhiro. *Toyota Production System: An Integrated Approach to Just-in-Time*. Springer, 2012.

Monks, Sarah. *Toy Town: How a Hong Kong Industry Played a Global Game*. PPP Company, 2010.

Montgomery, David, and George Day. "Diagnosing the Experience Curve." *Journal of Marketing* 47 (April 1983). https://doi.org/10.2307/1251492.

Moore, Frederick T. "Economies of Scale: Some Statistical Evidence." *Quarterly Journal of Economics* 73, no. 2 (May 1959): 232–245. https://msuweb.montclair.edu/~lebelp/mooreecsscaleqje1959.pdf.

Moore, Samuel K., and David Schneider. "The State of the Transistor in 3 Charts." *IEEE Spectrum*, November 26, 2022. https://spectrum.ieee.org/transistor-density.

Moran, James. *Printing Presses: History and Development from the Fifteenth Century to Modern Times*. Faber, 1973.

*Motor West and California Motor* 47, 1927.

Muldrew, Craig. "'Th'ancient Distaff' and 'Whirling Spindle': Measuring the Contribution of Spinning to Household Earnings and the National Economy in England, 1550–1770." *The Economic History Review* 65, no. 2 (May 2012): 498–526. https://doi.org/10.1111/j.1468-0289.2010.00588.x.

Munro, Sandy. "Tesla Cybertruck Deep Dive with 5 Tesla Executives!" Munro Live, December 11, 2023. Video, 57 min., 54 sec. https://www.youtube.com/watch?v=J5zDNaY1fvI.

Munro, Sandy. "Why Sandy Is Captivated By Castings: IDRA Conference Recap." Munro Live, November 1, 2023. Video, 13 mins., 20 sec. https://www.youtube.com/watch?v=q7ExWZwuuCI.

Munro, Sandy, and Cory Steuben. "Giga Castings with Sandy: Evolution of Tesla Bodies in White." Munro Live, July 29, 2022. Video, 17 min., 53 sec. https://www.youtube.com/watch?v=WNWYk4DdT_E.

Murphy, John. "Toyota Throws More Weight Behind Its Homes Unit." *Wall Street Journal*, July 2, 2008. https://www.wsj.com/articles/SB121496449430221935.

Murphy, John Paul. "Energy, Mining, and the Commercial Success of the Newcomen Steam Engine." PhD diss., Northeastern University, 2012. https://doi.org/10.17760/d20002721.

Murray, William. "Economies of Scale in Container Ship Costs." US Merchant Marine Academy, 2016. https://www.usmma.edu/sites/usmma.dot.gov/files/docs/CMA%20Paper%20Murray%201%20%282%29.pdf.

Musial, Walter, Paul Spitsen, Philipp Beiter, Patrick Duffy, Melinda Marquis, Aubryn Cooperman, Rob Hammond, and Matt Shields. "Offshore Wind Market Report: 2021 Edition." US Department of Energy, August 2021. https://www.energy.gov/sites/default/files/2021-08/Offshore%20Wind%20Market%20Report%202021%20Edition_Final.pdf

Musson, A.E. "Joseph Whitworth and the Growth of Mass-Production Engineering." *Business History* 17, no. 2 (July 1975): 109–49. https://doi.org/10.1080/00076750000000018.

N

Nadler, Gerald. *Work Simplification*. McGraw-Hill, 1957.

Nadler, Gerald, and W.D. Smith. "Manufacturing Progress Functions for Types of Processes." *International Journal of Production Research* 2, no. 2 (1963): 115–35. https://doi.org/10.1080/00207546308947818.

Nagy, Béla, J. Doyne Farmer, Quan M. Bui, and Jessika E. Trancik. "Statistical Basis for Predicting Technological Progress." *PLoS ONE* 8, no. 2 (February 2013): e52669. https://doi.org/10.1371/journal.pone.0052669.

NAHB Research Center. "Steel vs. Wood Cost Comparison: Beaufort Demonstration Homes." U.S. Department of Housing and Urban

Development, January 2002. https://www.huduser.gov/portal/Publications/pdf/steel_vs_wood1.pdf.

Nanayakkara-Skillington, Danushka. "Modular and Other Non-Site Built Housing In 2021." NAHB Eye on Housing, September 8, 2022. https://eyeonhousing.org/2022/09/modular-and-other-non-site-built-housing-in-2021/.

"Navier–Stokes Equation." Clay Mathematics Institute, accessed March 3, 2024. https://www.claymath.org/millennium/navier-stokes-equation/.

Neave, Henry R. *The Deming Dimension*. SPC Press, 1990.

Nelson, Lee H. "Nail Chronology as an Aid to Dating Old Buildings." History News Technical Leaflet, 24-1. American Association for State and Local History, 1968.

Nemet, Gregory F. *How Solar Energy Became Cheap: A Model for Low-Carbon Innovation*. Routledge, 2019.

Nenni, Daniel, and Paul Michael McLellan. *Fabless: The Transformation of the Semiconductor Industry*. SemiWiki.com Project, 2014.

Neushul, Peter. "Science, Government and the Mass Production of Penicillin." *Journal of the History of Medicine and Allied Sciences* 48, no. 4 (1993): 371–95. https://doi.org/10.1093/jhmas/48.4.371.

Nevins, Allan. *Ford: Expansion and Challenge, 1915–1933*. Scribner, 1957.

Nevins, Allan. *Ford: The Times, the Man, the Company*. Scribner, 1954.

Newhouse, John. *The Sporty Game: The High-Risk Competitive Business of Making and Selling Commercial Airliners*. Alfred A. Knopf, 1982.

"Nike Manufacturing Map." Nike, accessed January 6, 2023. https://manufacturingmap.nikeinc.com/#.

Noble, David F. *Forces of Production: A Social History of Industrial Automation*. Alfred A. Knopf, 1984.

Norman, Jeremy M. "200–250 Sheets per Hour Remains the Standard for Hand-Press Output." Jeremy Norman's History of Information. Accessed September 12, 2024. https://www.historyofinformation.com/detail.php?id=39.

"North America Smartphone Shipment down 6% in Q2 2022 as Demand Dampens." Canalys, August 22, 2022. https://www.canalys.com/newsroom/north-america-smartphone-market-share-Q2-2022.

Nye, David E. *America's Assembly Line*. Cambridge, Mass: The MIT Press, 2013.

O

Ohno, Taiichi. *Toyota Production System: Beyond Large-Scale Production*. CRC Press, 1988.

"OL 5500 CS Technical Data." Idra, May 28, 2020. https://web.archive.org/web/20200528002904/https://www.idragroup.com/images/pdf/OL5500CSTechDatasheet.pdf.

Oliphant, George W. *An Application of Some Industrial Engineering Principles to an Electrical Maintenance Organization*. Union Carbide Nuclear Company, 1957.

"Operation Warp Speed: Accelerated COVID-19 Vaccine Development Status and Efforts to Address Manufacturing Challenges." US Government Accountability Office, February 2021. https://www.gao.gov/assets/gao-21-319.pdf.

O'Reagan, Douglas Michael. "Science, Technology, and Know-How: Exploitation of German Science and the Challenges of Technology Transfer in the Postwar World." PhD diss., UC Berkeley, 2014. https://escholarship.org/content/qt3w65f1hm/qt3w65f1hm.pdf.

Owens, Brandon N. *The Wind Power Story: A Century of Innovation That Reshaped the Global Energy Landscape*. John Wiley & Sons, 2019.

P

Padilla, Michelle. "Running the Ribbon Machine: Stories from the Team." Corning Museum of Glass, January 9, 2018. https://blog.cmog.org/2018/01/09/running-the-ribbon-machine-stories-from-the-team/.

Panzino, Charlsy. "GM, Ford, Fiat Chrysler to See Increased Hourly Labor Costs through 2023." S&P Global, January 15, 2020. https://www.spglobal.com/marketintelligence/en/news-insights/trending/Fic7Dwvvxuh14hs9rmjwJw2.

Parekh, S., V.A. Vinci, and R.J. Strobel. "Improvement of Microbial Strains and Fermentation Processes." *Applied Microbiology and Biotechnology* 54, no. 3 (September 2000): 287–301. https://doi.org/10.1007/s002530000403.

Parsaei, Hamid R., and William G. Sullivan, eds. *Concurrent Engineering: Contemporary Issues and Modern Design Tools*. Springer US, 1993. https://doi.org/10.1007/978-1-4615-3062-6.

Paxton, John. "Mr. Taylor, Mr. Ford, and the Advent of High-Volume Mass Production: 1900–1912," *Economics & Business Journal: Inquiries and Perspectives* 4, no. 1 (October 2012): 74–90.

Peaucelle, Jean-Louis, and Cameron Guthrie. "How Adam Smith Found Inspiration in French Texts on Pin Making in the Eighteenth Century." HAL,

November 27, 2016. https://hal.
univ-reunion.fr/hal-01403681.

Pegels, C. Carl, and Craig
Watrous. "Application of
the Theory of Constraints
to a Bottleneck Operation
in a Manufacturing Plant."
*Journal of Manufacturing
Technology Management* 16,
no. 3 (April 2005): 302–11.
https://doi.org/10.1108/174103
80510583617.

Pellegrino, Roberta, Nicola
Costantino, Roberto
Pietroforte, and Silvio Sancilio.
"Construction of Multi-Storey
Concrete Structures in Italy:
Patterns of Productivity and
Learning Curves." *Construction
Management and Economics*
30, no. 2 (February 2012):
103–15. https://doi.org/10.1080/
01446193.2012.660776.

Perazich, George, Herbert
Schimmel, and Benjamin
Rosenberg. *Industrial
Instruments and Changing
Technology*. Works Progress
Administration, National
Research Project, 1938.

"Performance Curve Database."
Santa Fe Institute. https://
pcdb.santafe.edu/graph.
php?curve=23.

Perlin, John. *From Space to Earth:
The Story of Solar Electricity*.
Aatec Publications, 1999.

Phillips, Maureen K. "'Mechanic
Geniuses and Duckies,' a
Revision of New England's Cut
Nail Chronology before 1820,"
*APT Bulletin: The Journal of
Preservation Technology*, 25,
no. 3/4 (1993), 4–16, https://doi.
org/10.2307/1504461.

Picard, Liza. *Dr. Johnson's
London: Coffee-Houses and
Climbing Boys, Medicine,
Toothpaste and Gin, Poverty
and Press-Gangs, Freakshows
and Female Education*. St.
Martin's Press, 2014.

Pilkington, Lionel Alexander
Bethune. "Review Lecture:
The Float Glass Process."
*Proceedings of the Royal
Society* 314, no. 1516 (December

1969). https://doi.org/10.1098/
rspa.1969.0212.

Pingali, Prabhu L. "Green
Revolution: Impacts, Limits,
and the Path Ahead." *PNAS*
109, no. 31 (July 2012): 12302–8.
https://doi.org/10.1073/
pnas.0912953109.

Pisano, Gary P. *The Development
Factory: Unlocking the
Potential of Process Innovation*.
Harvard Business School Press,
1997.

Pisano, Gary P. "Knowledge,
Integration, and the Locus of
Learning: An Empirical Analysis
of Process Development."
*Strategic Management Journal*
15, no. S1 (1994): 85–100.
https://doi.org/10.1002/
smj.4250150907.

Pisano, Gary P. "Learning-before-
Doing in the Development of
New Process Technology."
*Research Policy* 25, no. 7
(October 1996): 1097–1119.
https://doi.org/10.1016/S0048-
7333(96)00896-7.

Postrel, Virginia. *The Fabric of
Civilization: How Textiles Made
the World*. Basic Books, 2020.

Potter, Brian. "Another Day in
Katerradise." Construction
Physics, June 9, 2021. https://
www.construction-physics.
com/p/another-day-in-
katerradise.

Potter, Brian. "Autovol, Stack
Modular, and Labor Arbitrage,"
Construction Physics,
April 6, 2021. https://www.
construction-physics.com/p/
autovol-stack-modular-and-
labor-arbitrage.

Potter, Brian. "Building Fast and
Slow, Part 1: The Empire State
Building and the World Trade
Center." Construction Physics,
November 23, 2022. https://
www.construction-physics.
com/p/building-fast-and-slow-
the-empire.

Potter, Brian. "Comparing Process
Improvement in Manufacturing
and Construction: Duco vs.
Drywall." Construction Physics,
November 11, 2022. https://

www.construction-physics.
com/p/comparing-process-
improvement-in.

Potter, Brian. "Does Construction
Ever Get Cheaper?"
Construction Physics,
February 1, 2023. https://www.
construction-physics.com/p/
does-construction-ever-get-
cheaper.

Potter, Brian. "How Building
Codes Work in the US."
Construction Physics, July
29, 2022. https://www.
construction-physics.com/p/
how-building-codes-work-in-
the-us.

Potter, Brian. "How Did Solar
Power Get Cheap? Part
I." Construction Physics,
April 12, 2023. https://www.
construction-physics.com/p/
how-did-solar-power-get-
cheap-part.

Potter, Brian. "How to Build a
$20 Billion Semiconductor
Fab." Construction Physics,
May 3, 2024. https://www.
construction-physics.com/p/
how-to-build-a-20-billion-
semiconductor.

Potter, Brian. "How Valuable Are
Building Methods That Use
Fewer Materials?" Construction
Physics, February 6, 2024.
https://www.construction-
physics.com/p/how-valuable-
are-building-methods.

Potter, Brian. "Lessons from
Shipbuilding Productivity,
Part II," Construction Physics,
March 10, 2022. https://www.
construction-physics.com/p/
lessons-from-shipbuilding-
productivity-4b9.

Potter, Brian. "On Klein on
Construction." Construction
Physics, February 18, 2023.
https://www.construction-
physics.com/p/on-klein-on-
construction.

Potter, Brian. "Operation
Breakthrough: America's
Failed Government Program to
Industrialize Home Production."
Construction Physics,
February 17, 2021. https://www.

construction-physics.com/p/ operation-breakthrough-americas-failed.

Potter, Brian. "An Overview of Concrete Forming Technology." Construction Physics, November 3, 2022. https:// www.construction-physics. com/p/an-overview-of-concrete-forming-technology.

Potter, Brian. "The Rise and Fall of the Manufactured Home, Part II." Construction Physics, July 22, 2022. https://www. construction-physics.com/p/ the-rise-and-fall-of-the-manufactured.

Potter, Brian. "The Rise of Steel, Part I." Construction Physics, January 5, 2023. https://www. construction-physics.com/p/ the-rise-of-steel-part-i.

Potter, Brian. "The Science of Production." Construction Physics, March 31, 2022. https://www.construction-physics.com/p/the-science-of-production.

Potter, Brian. "The Story of Titanium," Construction Physics, July 7, 2023. https:// www.construction-physics. com/p/the-story-of-titanium.

Potter, Brian. "Welding and the Automation Frontier." Construction Physics, November 22, 2023. https:// www.construction-physics. com/p/welding-and-the-automation-frontier.

Potter, Brian. "Where Are My Damn Learning Curves?" Construction Physics, December 1, 2021. https://www. construction-physics.com/p/ where-are-my-damn-learning-curves.

Potter, Brian. "Where Are the Robotic Bricklayers?" Construction Physics, July 29, 2021. https://www. construction-physics.com/p/ where-are-the-robotic-bricklayers.

Potter, Brian. "Why Are Nuclear Power Construction Costs So High? Part III: The Nuclear Navy." Construction Physics, July 1, 2022. https://www. construction-physics.com/p/ why-are-nuclear-power-construction-c3c.

Potter, Brian. "Why Are There So Few Economies of Scale in Construction?" Construction Physics, August 24, 2022. https://www.construction-physics.com/p/why-are-there-so-few-economies-of.

Potter, Brian. "Why Did Agriculture Mechanize and Not Construction?" Construction Physics, June 16, 2021. https:// www.construction-physics. com/p/why-did-agriculture-mechanize-and.

Potter, Brian. "Why Did We Wait So Long for Wind Power? Part I." Construction Physics, September 20, 2022. https:// www.construction-physics. com/p/why-did-we-wait-so-long-for-wind.

Potter, Brian. "Why Did We Wait So Long for Wind Power? Part II." Construction Physics, October 1, 2022. https://www. construction-physics.com/p/ why-did-we-wait-so-long-for-wind-498.

Potter, Brian. "Why Does Nuclear Power Plant Construction Cost So Much?" Institute for Progress, May 1, 2023. https:// progress.institute/nuclear-power-plant-construction-costs/.

Potter, Brian. "Why It's Hard to Innovate in Construction." Construction Physics, June 30, 2021. https://www. construction-physics.com/p/ why-its-hard-to-innovate-in-construction.

Potter, Brian. "Why Skyscrapers Are So Short." Works in Progress, January 21, 2022. https://worksinprogress.co/ issue/why-skyscrapers-are-so-short.

Powell's. "Powell's City of Books." 2023. https://www.powells. com/locations/powells-city-of-books.

Power, Brad. "How Toyota Pulls Improvement from the Front Line." Harvard Business Review, June 24, 2011. https:// hbr.org/2011/06/how-toyota-pulls-improvement-f.

"Powering Up." Wright Brothers Aeroplane Company. Accessed January 1, 2024. https://www. wright-brothers.org/History_ Wing/History_of_the_Airplane/ Century_Before/Powering_up/ Powering_up.htm.

Pray, Richard, ed. 2022 National Construction Estimator. Craftsman Book Company, 2022.

Preston, Richard. American Steel: Hot Metal Men and the Resurrection of the Rust Belt. Prentice Hall Press, 1991.

"Prices for Newly Produced Dwellings." Statistics Sweden, accessed March 27, 2024. https://www.scb.se/en/ finding-statistics/statistics-by-subject-area/housing-construction-and-building/ construction-costs/prices-for-newly-produced-dwellings/.

Priess, Peter. "Wire Nails in North America." Bulletin of the Association for Preservation Technology 5, no. 4 (1973): 87. https://doi. org/10.2307/1493446.

## Q

Quirk, Laura, and Matthew Beaudoin. "If The Shoe Fits... : Analysis of Nineteenth-Century Shoes." KEWA 11, no. 4–6 (2012): 27–32.

## R

Rae, Douglas W. City: Urbanism and Its End. Yale University Press, 2003.

Raftery, Heath. "Plyprinting, for Fun and Profit." EmpiricalEE, March 4, 2023. https://www.

empirical.ee/plyprinting/.

Ram, Gilbert Scott. *The Incandescent Lamp and Its Manufacture.* Electrician Printing and Publishing Company, Limited, 1893.

Ramani, Vinay, Debabrata Ghosh, and ManMohan S. Sodhi. "Understanding Systemic Disruption from the Covid-19-Induced Semiconductor Shortage for the Auto Industry." *Omega* 113 (December 2022): 102720. https://doi.org/10.1016/j.omega.2022.102720.

Randall, Tom, and Demetrios Pogkas. "Tesla Now Runs the Most Productive Auto Factory in America." Bloomberg, January 24, 2022. https://www.bloomberg.com/graphics/2022-tesla-factory-california-texas-car-production/.

Ranson, Gordon M. *Group Technology: A Foundation for Better Total Company Observation.* McGraw-Hill, 1972.

Read, Leonard E. "I, Pencil: My Story as Told to Leonard E. Read." *The Freeman*, December 1958.

"Recent Trends in GE Adoption." USDA Economic Research Service, last updated July 26, 2024. https://www.ers.usda.gov/data-products/adoption-of-genetically-engineered-crops-in-the-u-s/recent-trends-in-ge-adoption/.

Redhead, P.A. "History of Vaccum Devices." National Research Council, May 1999. https://cds.cern.ch/record/455984/files/p281.pdf.

Reed, Roger C. *The Superalloys: Fundamentals and Applications.* Cambridge University Press, 2009.

Reguero, Miguel Ángel. "An Economic Study of the Military Airframe Industry." PhD diss., New York University, 1957. https://babel.hathitrust.org/cgi/pt?id=uc1.b3289166&seq=260&view=1up.

Rehren, Thilo, and Marcos Martinón-Torres. *Naturam Ars Imitata: European Brassmaking between Craft and Science.* Routledge, 2008.

"Residential Code, 2015 (IRC 2015)." UpCodes, accessed February 29, 2024. https://up.codes/code/international-residential-code-irc-2015.

"Responsible Material Sourcing." Ford, accessed September 13, 2024. https://corporate.ford.com/social-impact/sustainability/responsible-material-sourcing.html.

Riordan, Michael, and Lillian Hoddeson. *Crystal Fire: The Birth of the Information Age.* Norton, 1997.

Robidoux, Benoît, and John Lester. "Econometric Estimates of Scale Economies in Canadian Manufacturing." *Applied Economics* 24, no. 1 (January 1992): 113–22. https://doi.org/10.1080/00036849200000109.

Robison, Peter. *Flying Blind: The 737 MAX Tragedy and the Fall of Boeing.* Penguin Business, 2021.

"Rockies Express Pipeline." Wikipedia, accessed August 13, 2023. https://en.wikipedia.org/w/index.php?title=Rockies_Express_Pipeline&oldid=1170124176.

Roe, Joseph Wickham, and Charles Walter Lytle. *Factory Equipment.* International Textbook Company, 1935.

"Rolex Price Evolution." Minus 4 Plus 6, accessed August 16, 2023. https://www.minus4plus6.com/PriceEvolution.php.

Rönnqvist, Mikael, David Martell, and Andres Weintraub. "Fifty Years of Operational Research in Forestry." *International Transactions in Operational Research* 30, no. 6 (2023): 3296–3328. https://doi.org/10.1111/itor.13316.

Rosenberg, Nathan. *Studies on Science and the Innovation Process: Selected Works.* World Scientific, 2010.

Rothschild, Michael, and Gregory J. Werden. "Returns to Scale from Random Factor Services: Existence and Scope." *The Bell Journal of Economics* 10, no. 1 (1979): 329–35. https://doi.org/10.2307/3003334.

Rouwenhorst, Kevin H.R., Yannick Engelmann, Kevin van 't Veer, Rolf S. Postma, Annemie Bogaerts, and Leon Lefferts. "Plasma-Driven Catalysis: Green Ammonia Synthesis with Intermittent Electricity." *Green Chemistry* 22, no. 19 (2020): 6258–87. https://doi.org/10.1039/D0GC02058C.

Russel, Darrell A., and Gerald G. Williams. "History of Chemical Fertilizer Development." *Soil Science Society of America Journal* 41, no. 2 (March-April 1977): 260–65. https://doi.org/10.2136/sssaj1977.03615995004100020020x.

Ruttan, Vernon W. *Technology, Growth, and Development: An Induced Innovation Perspective.* Oxford University Press, 2001.

Ryzewski, Krysta, and Robert Gordon. "Historical Nail-Making Techniques Revealed in Metal Structure." *Historical Metallurgy* 42, no. 1 (2008): 50–64.

S

"Safely and Efficiently Transporting Natural Gas." Chevron, accessed September 9, 2024. https://www.chevron.com/what-we-do/energy/oil-and-natural-gas/liquefied-natural-gas-lng.

Sahal, Devendra. *Patterns of Technological Innovation.* Addison-Wesley Publishing Company, 1981.

Sartin, Jeffrey S. "Infectious Diseases During the Civil War:

The Triumph of the 'Third Army.'" *Clinical Infectious Diseases* 16, no. 4 (April 1993): 580–84. https://doi.org/10.1093/clind/16.4.580.

Schank, John F. *Learning from Experience Vol III*. RAND National Defense Research Institute, 2011.

Schmid, Richard Otto. "An Analysis of Predetermined Time Systems." PhD diss., Newark College of Engineering, 1957. https://digitalcommons.njit.edu/theses/1528/.

Scholand, Michael. "Max Tech and Beyond: Fluorescent Lamps." US Department of Energy, April 1, 2012. https://doi.org/10.2172/1047754.

Schonberger, Richard J. *Japanese Manufacturing Techniques: Nine Hidden Lessons in Simplicity*. Free Press, 1982.

Schonberger, Richard J. "Lean Production—Perspectives from Its Primary Caretaker, Industrial Engineering." In *The Cambridge International Handbook of Lean Production*, edited by Thomas Janoski and Darina Lepadatu. Cambridge University Press, 2021.

Schroeder, Henry. *History of Electric Light*. The Smithsonian Institution, 1923.

Schumpeter, Joseph A. *Capitalism, Socialism and Democracy*. Routledge, 2013.

Scott, Alwyn. "Boeing Looks at Pricey Titanium in Bid to Stem 787 Losses." Reuters, July 24, 2015. https://www.reuters.com/article/us-boeing-787-titanium-insight-idUSKCN0PY1PL20150724.

Scriven, L.E. "On the Emergence and Evolution of Chemical Engineering." In *Advances in Chemical Engineering*, 16 (1991): 3–40. https://doi.org/10.1016/S0065-2377(08)60141-6.

Seong, Somi, Obaid Younossi, Benjamin Goldsmith, Thomas Lang, and Michael Neumann. *Titanium: Industrial Base, Price Trends, and Technology Initiatives*. RAND Corporation, 2009. https://doi.org/10.7249/MG789.

"Setting Records with the SR-71 Blackbird." Smithsonian Air and Space Museum. July 28, 2016. https://airandspace.si.edu/stories/editorial/setting-records-sr-71-blackbird.

Shaver, J., and John Mezias. "Diseconomies of Managing in Acquisitions: Evidence from Civil Lawsuits." *Organization Science* 20 (February 2009): 206–22. https://doi.org/10.1287/orsc.1080.0378.

Shewhart, Walter A. *Economic Control of Quality of Manufactured Product*. American Society for Quality Control, 1980.

Shilov, Anton. "AMD Is on Track to Become a Top 3 Fabless Chip Designer." Tom's Hardware, June 10, 2022. https://www.tomshardware.com/news/amd-is-on-track-to-become-a-top-3-fabless-chip-designer.

Shilov, Anton. "Nvidia Became World's Largest Fabless Chip Designer by Revenue in 2023 Thanks to AI Boom." Tom's Hardware, May 9, 2024. https://www.tomshardware.com/tech-industry/nvidia-became-worlds-largest-fabless-chip-designer-by-revenue-in-2023-thanks-to-ai-boom.

Shingo, Shigeo. *A Revolution in Manufacturing: The SMED System*. Translated by Andrew P. Dillon. Productivity Press, 1985.

Shingo, Shigeo. *A Study of the Toyota Production System: From an Industrial Engineering Viewpoint*. Translated by Andrew P. Dillon. Productivity Press, 1989.

Shirouzu, Norihiko. "How Toyota Thrives When the Chips Are Down." Reuters, March 9, 2021. https://www.reuters.com/article/us-japan-fukushima-anniversary-toyota-in-idUSKBN2B1005.

Shirouzu, Norihiko. "Tesla Reinvents Carmaking with Quiet Breakthrough." Reuters, September 9, 2023. https://www.reuters.com/technology/gigacasting-20-tesla-reinvents-carmaking-with-quiet-breakthrough-2023-09-14/.

Shuler, Michael L., and Fikret Kargi. *Bioprocess Engineering: Basic Concepts*. Prentice Hall, 2002.

Sichel, Daniel E. "The Price of Nails Since 1695: A Window into Economic Change." *Journal of Economic Perspectives* 36, no. 1 (February 2022): 125–50. https://doi.org/10.1257/jep.36.1.125.

Silberston, Aubrey. "Economies of Scale in Theory and Practice." *The Economic Journal* 82, no. 325 (1972): 369–91. https://doi.org/10.2307/2229943.

Simon, Julian Lincoln. *The Ultimate Resource*. Princeton University Press, 1981.

Sinclair, Gavin, Steven Klepper, and Wesley Cohen. "What's Experience Got to Do With It? Sources of Cost Reduction in a Large Specialty Chemicals Producer." *Management Science* 46, no. 1 (January 2000): 28–45. https://doi.org/10.1287/mnsc.46.1.28.15133.

Singer, Charles Joseph. *A History of Technology, Vol. 2*. Oxford University Press, 1954.

Singh, Jagjit. *Great Ideas of Operations Research*. Dover Publications, 1972.

Sloan, Alfred. *My Years with General Motors*. Crown, 1990.

"Slow Steaming in Container Shipping." Port Economics, Management and Policy, February 10, 2020. https://porteconomicsmanagement.org/pemp/contents/part1/ports-and-container-shipping/slow-steaming-container-shipping/.

Smil, Vaclav. *Energy and Civilization: A History*. MIT Press, 2018.

Smil, Vaclav. *Still the Iron Age: Iron and Steel in the*

*Modern World*. Butterworth-Heinemann, 2016.

Smil, Vaclav. "The Vacuum Tube's Power Law." *IEEE Spectrum*, February 2019. https://vaclavsmil.com/wp-content/uploads/2019/03/February2019.pdf.

Smith, Adam. *The Wealth of Nations*. Modern Library, 1937.

Smith, Cyril Stanley. "The Discovery of Carbon in Steel." *Technology and Culture* 5, no. 2 (1964): 149–75. https://doi.org/10.2307/3101159.

Smith, J. Bucknall. *A Treatise Upon Wire, Its Manufacture and Uses, Embracing Comprehensive Descriptions of the Constructions and Applications of Wire Ropes*. Offices of Engineering, 1891.

Smith, Merritt Roe. *Harpers Ferry Armory and the New Technology: The Challenge of Change*. Cornell University Press, 2015.

"Solar (Photovoltaic) Panel Prices." Accessed October 7, 2023. https://ourworldindata.org/grapher/solar-pv-prices.

Sorensen, Charles E. *My Forty Years with Ford*. Wayne State University Press, 2006.

Soyer, Daniel, ed. *A Coat of Many Colors: Immigration, Globalism, and Reform in the New York City Garment Industry*. Fordham University Press, 2005.

Spalart, P.R., and V. Venkatakrishnan. "On the Role and Challenges of CFD in the Aerospace Industry." *The Aeronautical Journal* 120, no. 1223 (January 2016): 209–32. https://doi.org/10.1017/aer.2015.10.

Spitz, Peter H. *Petrochemicals: The Rise of an Industry*. Wiley, 1988.

Stapper, C.H., P.P. Castrucci, R.A. Maeder, W.E. Rowe, and R.A. Verhelst. "Evolution and Accomplishments of VLSI Yield Management at IBM." *IBM Journal of Research*

*and Development* 26, no. 5 (September 1982): 532–45. https://doi.org/10.1147/rd.265.0532.

Stephanopoulos, George, and Gintaras V. Reklaitis. "Process Systems Engineering: From Solvay to Modern Bio- and Nanotechnology: A History of Development, Successes and Prospects for the Future." *Chemical Engineering Science*, Multiscale Simulation, 66, no. 19 (October 2011): 4272–4306. https://doi.org/10.1016/j.ces.2011.05.049.

"Stirling Homex Corporation." Wikipedia, accessed September 17, 2021. https://en.wikipedia.org/w/index.php?title=Stirling_Homex_Corporation&oldid=1044836378.

Storch, R.L. *Ship Production*. Cornell Maritime Press, 1995.

Storck, John. *Flour for Man's Bread*. The University of Minnesota Press, Minneapolis, 1952.

"Supplier Quality Assurance Manual (SQAM)." MinebeaMitsumi Quality Assurance Headquarters, June 26, 2006. https://www.minebeamitsumi.com/english/corp/company/procurements/quality/__icsFiles/afieldfile/2020/12/14/sqam_en.pdf.

Sybridge Technologies. "4 Best Practices for Optimizing Injection Molding Tolerances." December 14, 2020. https://sybridge.com/injection-molding-tolerances-best-practices/.

Szulanski, Gabriel. *Sticky Knowledge: Barriers to Knowing in the Firm*. Sage, 2003.

**T**

Tabarrok, Alex. "Physician and Nurse Incomes Have Increased Tremendously." Marginal Rrevolution, May 29, 2019.

https://marginalrevolution.com/marginalrevolution/2019/05/physician-and-nurse-incomes-have-increased-tremendously.html.

Tarasov, Katie. "ASML Is the Only Company Making the $200 Million Machines Needed to Print Every Advanced Microchip. Here's an Inside Look." CNBC, March 23, 2022. https://www.cnbc.com/2022/03/23/inside-asml-the-company-advanced-chipmakers-use-for-euv-lithography.html.

Taylor, Frederick Winslow. *On the Art of Cutting Metals*. American Society of Mechanical Engineers, 1906.

Taylor, Frederick Winslow. *The Principles of Scientific Management*. Harper & Row, 1911.

Taylor, Jim, and Abijit Biswas. "Profile of Deep Space Communications Capability." NASA Jet Propulsion Laboratory, August 2015. https://descanso.jpl.nasa.gov/performmetrics/profileDSCC.html.

Teece, David J. *Managing Intellectual Capital: Organizational, Strategic, and Policy Dimensions*. Oxford University Press, 2000.

Teece, David J. "Technology Transfer by Multinational Firms: The Resource Cost of Transferring Technological Know-How." *The Economic Journal* 87 no 356 (June 1977): 242–61. https://doi.org/10.2307/2232084.

Teicholz, Paul. "Labor-Productivity Declines in the Construction Industry: Causes and Remedies (Another Look)." AECbytes, March 14, 2013. https://www.aecbytes.com/viewpoint/2013/issue_67.html.

Temin, Peter. *Iron and Steel in Nineteenth-Century America: An Economic Inquiry*. MIT Press, 1964.

Tenold, Stig. "The Declining Role of Western Europe in Shipping and Shipbuilding, 1900–2000." In *Shipping and Globalization in the Post-War Era: Contexts, Companies, Connections*, edited by Niels P. Petersson, Stig Tenold, and Nicholas J. White, 9–36. Springer International Publishing, 2019. https://doi.org/10.1007/978-3-030-26002-6_2.

Terwiesch, Christian, and Roger E. Bohn. "Learning and Process Improvement during Production Ramp-Up." *International Journal of Production Economics* 70, no. 1 (March 2001): 1–19. https://doi.org/10.1016/S0925-5273(00)00045-1.

Tesla. "Tesla Battery Day." Tesla, September 22, 2020. Video, 131 min., 29 sec. https://www.youtube.com/watch?v=l6T9xIeZTds.

"Theory of Constraints of Eliyahu M. Goldratt." Theory of Constraints Institute. Accessed September 16, 2024. https://www.tocinstitute.org/theory-of-constraints.html.

Thomas, Nick. "Georgia Bill to Replace Certificate of Need Advances." Becker's Hospital CFO Report, February 22, 2023. https://www.beckershospitalreview.com/finance/georgia-bill-to-replace-certificate-of-need-advances.html.

Thomas, Robert J. *What Machines Can't Do: Politics and Technology in the Industrial Enterprise*. University of California Press, 1994.

Thomke, Stefan H. *Experimentation Matters: Unlocking the Potential of New Technologies for Innovation*. Harvard Business Press, 2003.

Thompson, Ben. "Intel Unleashed, Gelsinger on Intel, IDM 2.0." Stratechery, March 24, 2021. https://stratechery.com/2021/intel-unleashed-gelsinger-on-intel-idm-2-0/.

Thompson, Peter. "How Much Did the Liberty Shipbuilders Learn?

New Evidence for an Old Case Study." *The Journal of Political Economy* 109, no. 1 (2001): 103–37.

Thwaites, Thomas. *The Toaster Project: Or A Heroic Attempt to Build a Simple Electric Appliance from Scratch*. Chronicle Books, 2012.

Todd, Robert H., Dell K. Allen, and Leo Alting. *Fundamental Principles of Manufacturing Processes*. Industrial Press Inc., 1994.

"Too Cheap to Meter: A History of the Phrase." US Nuclear Regulatory Commission. Last updated September 24, 2021. https://www.nrc.gov/reading-rm/basic-ref/students/history-101/too-cheap-to-meter.html.

"Toyota Housing Corporation." Toyota, accessed October 12, 2023. https://www.toyota-global.com/company/history_of_toyota/75years/data/business/housing/toyotahome.html.

Tran, Nguyen Khoi, and Hans-Dietrich Haasis. "An Empirical Study of Fleet Expansion and Growth of Ship Size in Container Liner Shipping." *International Journal of Production Economics* 159 (January 2015): 241–53. https://doi.org/10.1016/j.ijpe.2014.09.016.

Truesdell, Paul. "Texas Co. Increases Yield from Cracking Stills by Automatic Control." *National Petroleum News*, February 9, 1927.

Tuck, Candice T. "Iron Ore 2018." US Geological Survey, September 2021. https://pubs.usgs.gov/myb/vol1/2018/myb1-2018-iron-ore.pdf.

Tupy, Marian L., and Gale L. Pooley. *Superabundance: The Story of Population Growth, Innovation, and Human Flourishing on an Infinitely Bountiful Planet*. Cato Institute, 2023.

Tylecote, R.F. *A History of*

*Metallurgy*. Institute of Materials, 1992.

Tyre, Marcie J., and Eric von Hippel. "The Situated Nature of Adaptive Learning in Organizations." *Organization Science* 8, no. 1 (February 1997): 71–83. https://doi.org/10.1287/orsc.8.1.71.


U

"Understanding Injection Molding Tolerances." Protolabs, February 7, 2023. https://www.protolabs.com/resources/blog/understanding-injection-molding-tolerances/.

Unland, John. *Warm, Dry, and Comfortable: The Story of Developing the Gore-Tex Brand*. John Unland, 2019.

Uri, John. "120 Years Ago: The First Powered Flight at Kitty Hawk." NASA. December 14, 2023. https://www.nasa.gov/history/120-years-ago-the-first-powered-flight-at-kitty-hawk/.

"US Aluminum Manufacturing: Industry Trends and Sustainability." Congressional Research Service, October 26, 2022. https://sgp.fas.org/crs/misc/R47294.pdf.

US Army Materiel Command. *Inspections and Investigations: Value Analysis*. The Command, 1965.

US Census Bureau. "Building Permits Survey." US Census Bureau. Accessed October 12, 2023. https://www.census.gov/construction/bps/about_the_surveys/index.html.

US Census Bureau. "Total Shipments of New Manufactured Homes: Total Homes in the United States." FRED, Federal Reserve Bank of St. Louis. Accessed October 12, 2023. https://fred.stlouisfed.org/series/SHTSAUS.

US Census Bureau and US Department of Housing and Urban Development. "New

Privately-Owned Housing Units Started: Single-Family Units." FRED, Federal Reserve Bank of St. Louis. Accessed October 12, 2023. https://fred.stlouisfed.org/series/HOUST1F.

US Department of Agriculture. "Ag and Food Sectors and the Economy." USDA Economic Research Service. Last updated April 19, 2024. https://www.ers.usda.gov/data-products/ag-and-food-statistics-charting-the-essentials/ag-and-food-sectors-and-the-economy/.

US Department of Agriculture. "Agricultural Productivity in the United States." USDA Economic Research Service. Last updated January 12, 2024. https://www.ers.usda.gov/data-products/agricultural-productivity-in-the-u-s/.

US Department of Agriculture. *Crop Production Historical Track Records*. National Agricultural Statistics Service, April 2024. https://downloads.usda.library.cornell.edu/usda-esmis/files/c534fn92g/6t055511z/9593wh83b/croptr24.pdf.

US Department of the Air Force. *Work Simplification: Fundamentals and Techniques for More Effective Use of Manpower, Equipment, Materials, Space.* US Government Printing Office, 2011.

US Department of the Interior. "Aluminum—Historical Statistics (Data Series 140)." USGS, February 26, 2024. https://www.usgs.gov/media/files/aluminum-historical-statistics-data-series-140.

US Navy Department. *Work Simplification for Naval Units: Work Distribution, Work Count, Flow Process, Motion Economy, Space Layout.* Personnel Research Division, Bureau of Naval Personnel, 1965.

"US Statistical Abstract 1957—Mining." US Census Bureau, 1957. https://www2.census.gov/library/publications/1957/compendia/statab/78ed/1957-12.pdf.

"US Statistical Abstract 1980—Mining." US Census Bureau, 1980. https://www2.census.gov/library/publications/1980/compendia/statab/101ed/1980-08.pdf.

*US Steel Fiscal 2019 Annual Report*. United States Steel Corporation Form 10-K. Accessed January 12, 2024. https://www.ussteel.com/documents/40705/43725/2019+Annual+Report+on+Form+10-K.pdf/19a3251b-9ba8-02a5-3f82-389a090edf60?t=1682019025404.

US Tariff Commission. *Incandescent Electric Lamps: A Survey of the Production of Incandescent Electric Lamps and Glass Bulbs in the Principal Producting Countries, Processes of Manufacture, Marketing, and Industrial Agreements, with Special Reference to International Trade and Tariff Considerations.* US Government Printing Office, 1939.

Utterback, James M. *Mastering the Dynamics of Innovation: How Companies Can Seize Opportunities in the Face of Technological Change.* Harvard Business School Press, 1994.

Utterback, James M., and Linsu Kim. "Invasion of a Stable Business by Radical Innovation." In *The Management of Productivity and Technology in Manufacturing*, edited by Paul R. Kleindorfer, 113–51. Springer US, 1985.

V

"Value Engineering FY 2022." Georgia Department of Transportation, 2022. https://www.dot.ga.gov/InvestSmart/SSTP/AttachmentD-ValueEngineering-SB200.pdf.

van der Beek, C.P., and J.A. Roels. "Penicillin Production: Biotechnology at Its Best." *Antonie van Leeuwenhoek* 50, no. 5–6 (November 1984): 625–39. https://doi.org/10.1007/BF02386230.

van Helden, G. Jan, and Joan Muysken. "Diseconomies of Scale for Plant Utilisation in Electricity Generation." *Economics Letters* 11, no. 3 (1983): 285–89. https://doi.org/10.1016/0165-1765(83)90149-0.

van Wyk, Rias J., Georges Haour, and Stephen Japp. "Permanent Magnets: A Technological Analysis." *R&D Management* 21, no. 4 (1991): 301–08.

Vaško, Tibor, ed. *The Long-Wave Debate: Selected Papers from an IIASA (International Institute for Applied Systems Analysis) International Meeting on Long-Term Fluctuations in Economic Growth: Their Causes and Consequences, Held in Weimar, GDR, June 10–14, 1985.* Springer–Verlag, 1987.

Veers, Paul, Katherine Dykes, Eric Lantz, Stephan Barth, Carlo L. Bottasso, Ola Carlson, Andrew Clifton, et al. "Grand Challenges in the Science of Wind Energy." *Science* 366, no. 6464 (October 25, 2019): eaau2027. https://doi.org/10.1126/science.aau2027.

Vincenti, Walter G. *What Engineers Know and How They Know It: Analytical Studies from Aeronautical History.* Johns Hopkins University Press, 1993.

Vivian, E. Charles. *A History of Aeronautics.* Harcourt, Brace and Company, 1921. https://www.gutenberg.org/files/874/874-h/874-h.htm#link2H_4_0005.

von Böhm-Bawerk, Eugen. *Capital and Interest: A Critical Theory* of Economic History. Translated by William A. Smart. Macmillan, 1890.

von Hippel, Eric. The Sources of Innovation. Oxford University Press, 1988.

von Hippel, Eric, and Marcie J. Tyre. "How Learning by Doing Is Done: Problem Identification in Novel Process Equipment." Research Policy 24, no. 1 (January 1995): 1–12. https://doi.org/10.1016/0048-7333(93)00747-H.

von Tunzelmann, G.N. Steam Power and British Industrialization to 1860. Clarendon Press, 1978.


**W**

Walker, J. Samuel, and Thomas R. Wellock. "A Short History of Nuclear Regulation, 1946–2009." US Nuclear Regulatory Commission, October 2010. https://www.nrc.gov/docs/ML1029/ML102980443.pdf.

Walker, Jack M., ed. Handbook of Manufacturing Engineering. Manufacturing Engineering and Materials Processing 48. Marcel Dekker, 1996.

Walton, Mary. Car: A Drama of the American Workplace. W.W. Norton & Company, 1997.

Wang, Dan. "How Technology Grows (a Restatement of Definite Optimism)." July 24, 2018. https://danwang.co/how-technology-grows/.

Warren, Kenneth. A Century of American Steel: The Strip Mill and the Transformation of an Industry. Rowman & Littlefield, 2019.

Webb, R.E., and W.M. Bruce. "Redesigning the Tomato for Mechanized Production." Yearbook of Agriculture, 1968: Science for Better Living. US Department of Agriculture, 1968. https://www.cabidigitallibrary.org/doi/full/10.5555/19691601886.

Western Electric Statistical Quality Control Handbook. Western Electric Co., 1954.

"Where Are Clayton Homes Built?" Clayton Homes, accessed January 12, 2024. https://web.archive.org/web/20220707164045/https://www.claytonhomes.com/studio/where-clayton-manufactured-homes-are-built/.

Whitin, T.M., and M.H. Peston. "Random Variations, Risk, and Returns to Scale." The Quarterly Journal of Economics 68, no. 4 (November 1954): 603. https://doi.org/10.2307/1881879.

Wilson, David. In Search of Penicillin. Knopf, 1976.

"Wind Turbines: The Bigger, the Better." Office of Energy Efficiency & Renewable Energy, August 21, 2024. https://www.energy.gov/eere/articles/wind-turbines-bigger-better.

Wittmann, Christoph, and James C. Liao. Industrial Biotechnology: Products and Processes. John Wiley & Sons, 2017.

Wolchover, Natalie. "Why NASA's James Webb Space Telescope Matters So Much." Quanta Magazine, December 3, 2021. https://www.quantamagazine.org/why-nasas-james-webb-space-telescope-matters-so-much-20211203/.

Womack, James P., Daniel T. Jones, Daniel Roos, and Massachusetts Institute of Technology. Machine That Changed the World. Simon and Schuster, 1990.

Woodbury, Robert S. Studies in the History of Machine Tools. MIT Press, 1972.

Woods, Bob. "A West Coast Port Worker Union Is Fighting Robots. The Stakes for the Supply Chain Are High." CNBC, July 23, 2022. https://www.cnbc.com/2022/07/23/a-west-coast-port-worker-union-is-fighting-robots-the-stakes-are-

high.html.

Woodsworth, Joseph Vincent. American Tool Making and Interchangeable Manufacturing. N.W. Henley Publishing Company, 1905.

Woollard, Frank George. Principles of Mass and Flow Production. Mechanical Handling, 1954.

"World Container Index." Drewry, accessed February 23, 2023. https://www.drewry.co.uk/supply-chain-advisors/supply-chain-expertise/world-container-index-assessed-by-drewry.

Worstall, Tim. "The Story of Henry Ford's $5 a Day Wages: It's Not What You Think." Forbes, December 10, 2021. https://www.forbes.com/sites/timworstall/2012/03/04/the-story-of-henry-fords-5-a-day-wages-its-not-what-you-think/.

Wright, T.P. "Factors Affecting the Cost of Airplanes." Journal of the Aeronautical Sciences 3, no. 4 (February 1936): 122–28. https://doi.org/10.2514/8.155.


**Y**

Yang, Chin Jian, Luis Fernando Samayoa, Peter J. Bradbury, Bode A. Olukolu, Wei Xue, Alessandra M. York, Michael R. Tuholski, Weidong Wang, Lora L. Daskalska, et al. "The Genetic Architecture of Teosinte Catalyzed and Constrained Maize Domestication." Proceedings of the National Academy of Sciences 116, no. 12 (March 2019): 5643–52. https://doi.org/10.1073/pnas.1820997116.

Yelle, Louis E. "The Learning Curve: Historical Review and Comprehensive Survey." Decision Sciences 10, no. 2 (April 1979): 302–28. https://doi.org/10.1111/j.1540-5915.1979.tb00026.x.

# Z

Zamecnik, Adam. "Continuous Manufacturing Builds on Hype but Adoption Remains Gradual." Pharmaceutical Technology, May 20, 2022. https://www.pharmaceutical-technology.com/features/continuous-manufacturing-builds-on-hype-but-adoption-remains-gradual/.

Zyga, Lisa. "White LEDs with Super-High Luminous Efficacy Could Satisfy All General Lighting Needs." Phys.org, 2010. https://phys.org/news/2010-08-white-super-high-luminous-efficacy.html.

# C

# Index

227, 231, 233, 236–39, 241–44, 257, 282
    selective assembly and, 180
    Taurus, 124, 150, 268, 269
    variability and, 155, 189
    vertical integration and, 100
Forgeng, Jeffrey, 17
Fukushima nuclear disaster, 101, 189

## G

Gale, W.K.V., 160
Garrett, Donald, 112
gasoline production, 48, 93, 100, 172–73, 258
Gaynes, Robert, 11–12
General Electric (GE), 33, 78, 79, 80, 124
General Houses, 275, 276
General Motors (GM), 52, 55, 88, 89, 189
geometric diseconomies, 120–21
geometric scaling, 111–14
Giga Presses, 103, 104
Gilbert, J.H., 164
Gilbreth, Frank and Lillian, 139–40, 141, 142
glass bottle production, 94–95
GlobalFoundries, 127
Great Famine of 1315–17, 13
Green Revolution, 13
group technology, 144, 255–56, 260
Grove, William, 41–42
gun manufacturing, 54–55, 77–78, 185–86
Gunnison Homes, 275, 276
Gutenberg, Johannes, 16, 18

## H

Haber–Bosch process, 64, 258, 259
Hall-Héroult process, 63, 66, 218, 258, 259
Hayes, Robert, 62
Heatley, Norman, 9, 10
heijunka, 175
Helper, Susan, 87
Herschel, Clemens, 163
Hewlett-Packard (HP), 182, 230
Hippel, Eric von, 200–201
Hirschmann, Winfred, 187
Hitachi, 80
home construction
    codes and regulations, 266–67, 280, 285, 302
    costs, 264, 265, 271–78, 281–86, 302
    in the future, 297–302
    prefabricated, 19, 274–82
Honeywell, 144
Hounshell, David, 227
Howe, John, 256
HPDC (high-pressure die casting), 102–4

## I

IBM, 78, 80
Illinois Steel, 89
improvements
    added production burdens and, 270–86
    blocked paths to, 263–64, 266–69, 286–87
    bundles of, 228–30, 233
    chains of, 228, 230–35, 241–42, 244
    process stability and, 268–69
    repetition and, 290
    small, 216, 218
Industrial Revolution, 15, 58, 61, 126, 256, 289.
    *See also* Second Industrial Revolution
influence scaling, 115–16
information-processing technology, effects of, 291–97
injection molding, 95, 110–11, 153
input cost reduction
    concept of, 27, 75
    coupling and, 96–97
    location change and, 90–92
    organizational boundaries and, 85–90
    output value and, 92–93
    production process redesign and, 81–85
    product redesign and, 75–81
    technological development and, 97–105
    trade-offs and, 93–96
Intel, 60, 63, 127, 186, 219
interchangeable parts, 51, 124, 148–49, 184–86,
    227–28, 231, 233, 236
iron production, 91–93, 96, 107–10, 160, 206

## J

Jerome, Chauncey, 91
job shop process, 252
Jones, Daniel, 128
Juran, Joseph, 174
just-in-time production, 101, 175, 189, 200, 232

## K

kanban production control system, 175–76, 183, 232
Katerra, 275, 276
keiretsu, 178
Keyser, Barbara, 161
Kinnersley, Ebenezer, 41
knowledge
    dissemination of, 163
    about production process, 160–65
    scientific, 163–65
    transferring, 58–59
Kroll process, 208, 266

technological development and, 63, 203, 253, 267

tight tolerances for, 179, 186, 203

Shewhart, Walter, 165, 182

Shingo, Shigeo, 82–83, 145–46, 150

shipbuilding

continuous process methods and, 260

location change and, 56

production facilities for, 58

transfer of knowledge and, 59

shojinka, 178

SIMO (simultaneous-motion) cycle charts, 140, 142

Six Sigma, 174, 201

Slater, Samuel, 55

Smeaton, John, 46

SMED (single-minute exchange of die) methodology, 82–83, 128, 145–46, 150, 189, 232, 255

Smith, Adam, 65, 123

Society of American Value Engineers (SAVE), 78

solar energy, 19, 194, 195, 207–8, 212, 222, 268

Solvay, Ernest, 257

space travel, 19

SpaceX, 19

Spaichl, Hans, 61

standardization, 229

Starr, John, 42

statistical diseconomies, 121

statistical process control (statistical quality control), 166–67, 174, 177, 178, 186

statistical scaling, 114–15

steam engines, 46–47, 49, 58

steel production

Bessemer process, 50, 60, 91, 92, 94, 100, 180, 218, 254

demand effects and, 119–20

electric arc, 20, 100, 129, 132

heat-treating, 165

history of, 161

knowledge and, 163, 164

learning curves and, 219

location and, 91

nuclear weapons testing and, 191

open-hearth process, 88, 91, 94, 100, 168

slag from, 92–93

variation and, 154

vertical integration and, 89

steps

removing non-value-adding, 137–47, 150–51

removing value-adding, 137, 147–49, 151

value-adding vs. non-value-adding, 135–37, 140, 149–50

stereolithography, 111

Stirling Homex, 275, 276

stove production, 125–26

Stringfellow, John, 44

successor technologies, 63–64

Taguchi, Genichi, 155, 182

Taguchi methods, 182–83, 187

Taiho Industry, 145

Taylor, Frederick, 138–39, 151, 162

technology. *See also* S-curve model

change and, 39–40

complexity and, 72–73

defense funding and, 51

efficiency and, 18–20

explore-exploit dynamic of, 97–99

importance of, 190

input cost reduction and, 97–105

mechanization and, 64–72

performance ceilings and, 62–63

resistance to new, 60–62

successor, 63–64

telephones, 104, 116

Tesla, 101–4, 149

textile manufacturing

labor and, 70, 71, 90

learning curves and, 196–97, 219, 221

location change and, 55, 90

production efficiency and, 13–15

resistance to new technologies in, 61

therbligs, 140, 141

Thomas, Robert, 56

3D printing, 51, 95, 103, 104, 111, 132, 293–94

Thwaites, Thomas, 24

titanium production, 207–8, 264, 266, 286

tolerances

extremely tight, 179–80

relaxing, 178–79

stacking, 179, 181

total factor productivity, 18

Toyota

Covid-19 pandemic and, 101, 189

homebuilding division, 276–77

NUMMI joint venture and, 55

production system, 128, 129, 145–47, 151, 174–78, 231, 232, 234–35, 243, 251, 255, 260, 276

transformation method. *See also* S-curve model

changing, 31, 37–39

concept of, 26

differences in, 51–54

TSMC, 55, 127

typewriter keyboard layouts, 144–45

## U

US Air Force, 144
US Army, 79
US Department of Agriculture (USDA), 10
US Navy, 144, 221
US Steel, 89, 119

## V

value engineering (value analysis), 78–79, 144, 147
variability
  compensating for, with control systems, 169–74
  cultivation of, 187–88
  effects of, 154–56
  eliminating source of, 165–68
  good, 187–88
  historical tendency for, 185–87
  making product or process more robust to,
    178–83
  misalignment between demand and production,
    155–56
  pooling, 180, 182
  reducing, 160–65, 174–78, 184–85
  shielding from, 168–69
  sources of, 28, 153–54
  trade-offs, 184–85
  yield and, 28
Volkswagen, 189

## W

water-jet cutting, 94
Watt, James, 49, 172, 173
Watt steam engine, 47, 49, 64, 167
Wedgwood, Josiah, 163
Westinghouse, 78, 80
Whitworth, Joseph, 76
Wilkinson, John, 167
wind energy, 62–63, 100, 114, 129–30, 172, 268
Womack, James, 128, 137
work in process
  as buffer, 156, 158–59
  definition of, 27–28
  reducing, 31–32
work simplification, 84–85, 143–44
World War I, 9
World War II, 12, 45, 57, 77–78, 180, 194, 205
Wright, Theodore, 193–94
Wright brothers, 44–45
Wright's law, 116, 193, 205. *See also* learning curves

## X

Xerox, 80

## Y

yield
  definition of, 28
  increasing, 31

## Acknowledgments

This book was largely written in a flurry of activity between April 2022 and September 2023, and I owe its completion to a great many people. First and foremost is my wife, Katrina, who was a crucial sounding board and took on more than her share of caring for our daughter, Mercy, allowing me to write seven days a week. My publisher, Stripe Press, especially Tamara Winter and Rebecca Hiscott, for giving me the opportunity to write this book and shepherding me and the book through the process. Matt Clancy, Dan Wang, Dwarkesh Patel, Austin Vernon, Tyler Cowen, and others who read early drafts of the book and provided feedback. Alec Stapp and Caleb Watney of the Institute for Progress, who generously offered to support my newsletter, Construction Physics, which is where the seed of this book germinated, and which gave me the writing practice that made it possible to write a book. I'm immensely proud of this book, and I hope you enjoy reading it.

**About *The Origins of Efficiency***

Efficiency is the engine that powers human civilization. It's the reason rates of famine have fallen precipitously, literacy has risen, and humans are living longer, healthier lives compared to preindustrial times. But where do improvements in production efficiency come from? In *The Origins of Efficiency*, Brian Potter argues that improving production efficiency—finding ways to produce goods and services in less time, with less labor, using fewer resources—is the force behind some of the biggest and most consequential changes in human history. The book is punctuated with examples of production efficiency in practice, including how high-yield manufacturing methods made penicillin the miracle drug that reduced battlefield infection deaths by 80 percent during World War II; the 100-year history of process improvements in incandescent light bulb production; and how automakers like Ford, Toyota, and Tesla developed innovative production methods that transformed not just the automotive industry but manufacturing as a whole. *The Origins of Efficiency* is a comprehensive companion for anyone seeking to understand how we arrived at this age of relative abundance—and how we can push efficiency improvements further into domains like housing, medicine, and education, where much work is left to be done.

**About the Author**

Brian Potter is the author of the Construction Physics newsletter and a senior infrastructure fellow at the Institute for Progress. He writes about the technology and economics of building construction with a focus on improving productivity and reducing costs. He previously managed an engineering team at Katerra, a SoftBank-backed construction startup, and has 15 years of experience as a structural engineer. He has a bachelor's in civil engineering from Georgia Tech and a master's in systems engineering from University of Central Florida.